Crisis of the Wasteful Nation

Crisis of the Wasteful Nation

Empire and Conservation in Theodore Roosevelt's America

IAN TYRRELL

The University of Chicago Press
Chicago and London

Ian Tyrrell was the Scientia Professor of History at the University of New South Wales, Sydney, until his retirement in 2012. He is the author of nine books, including *True Gardens of the Gods: Californian-Australian Environmental Reform, 1860–1930* and *Historians in Public*, also published by the University of Chicago Press.

The University of Chicago Press, Chicago 60637
The University of Chicago Press, Ltd., London

Printed in the United States of America

24 23 22 21 20 19 18 17 16 15 1 2 3 4 5

ISBN-13: 978-0-226-19776-0 (cloth)
ISBN-13: 978-0-226-19793-7 (e-book)
DOI: 10.7208/chicago/9780226197937.001.0001

Library of Congress Cataloging-in-Publication Data

Tyrrell, Ian R., author.
Crisis of the wasteful nation : empire and conservation in Theodore Roosevelt's America / Ian Tyrrell.
pages ; cm
Includes bibliographical references and index.
ISBN 978-0-226-19776-0 (cloth : alk. paper)—ISBN 0-226-19776-X (cloth : alk. paper)—ISBN 978-0-226-19793-7 (e-book) 1. Conservation of natural resources—United States—History. 2. Conservation of natural resources—Government policy—United States. 3. Roosevelt, Theodore, 1858–1919—Knowledge—Conservation of natural resources. I. Title.
S930.T97 2015
333.72—dc23

2014023240

♾ This paper meets the requirements of ANSI/NISO Z39.48-1992 (Permanence of Paper).

To my conservation-minded daughters, Jessica and Ellen.
May there still be a livable future world for you and your generation's descendants.

CONTENTS

PREFACE

Crisis of the Wasteful Nation brings to a close a series of five works that I have written on the transnational relationships of the United States and on aspects of American empire. This final installment is not a biography of Theodore Roosevelt, though he is necessarily the central figure in the story. Nor is it political history, "straight" environmental history, or even the history of environmental diplomacy, though all these are part of the analysis. I focus on the hopes and fears for the future of the United States and the world that Roosevelt and other key conservationists around him expressed, and I trace how their ideas fared politically and economically. I am interested in the vicissitudes of conservation reform in action and in traditions established and paths not taken. The actors in this drama did not operate alone, and I do not wish to slight the role of the many, many thousands of Americans and other nationals who worked for conservation in that era. The "movement" they created did not represent a cohesive economic class in the classic sense of political economy. It encompassed cross-class coalitions over particular issues and exhibited intraclass conflict. It is useful to think of its leadership as part of overlapping professional, political, intellectual, and economic elites, for want of a better word. There were divisions among these, and fractious conflicts with business, even as sections of business supported conservationists' attempts to regulate the American economy's resource base.

Readers should also note that I use both "transnational" and "international" as terms for aspects of the relations between the United States and the wider world. Though there is no completely adequate single term to cover the whole gamut of American cross-national relations, these two are the best for modern America, where the nation-state is inescapably a central fact. I prefer "international" for official, governmental interactions of nation-states, and "transnational" to refer to the broader currents and contexts crossing national boundaries, where non-state actors are involved. A great deal has now been written on the theory and practice of transnational methodology (see the notes) and on the related but distinct idea of comparative history. All these concepts are important and may be used in a complementary fashion.

In the course of writing I have had opportunities to present my ideas to audiences at the University of Houston, Texas, where I gave a Tenneco Lecture (2009); Oxford University (2010); Cambridge University (2011); the University of Nottingham (2011); the European University Institute in Florence (2011); the University of Leipzig (2012); and McMaster University (2012), as part of the Hooker Distinguished Visiting Professor program. Other related presentations were made at the Australian Historical Association Conference in Perth, Western Australia (July 2010); the School of Humanities Seminar, University of New South Wales (2010 and 2012); Australian and New Zealand American Studies Association biennial conferences in Adelaide (July 2010) and Brisbane (July 2012); and the Australian Studies Group in Sydney (2013).

I have learned from conversations with Martin Melosi and Kathleen Brosnan (University of Houston); Andrew Preston and John Thompson (Cambridge University); Mary Dudziak (then at the University of Southern California); Bevan Sewell (University of Nottingham); Mario Del Pero (University of Bologna); Kiran Klaus Patel (then at the European University Institute); Matthias Middell and Katja Naumann (University of Leipzig); Marilyn Blatt Young (New York University); H. V. Nelles and John Weaver (McMaster University); and Thomas Adam (University of Texas, Arlington). Students at the University of New South Wales, Oxford University (2010–11), and UCLA (2009) have listened to my ideas on transnational history, as did students and faculty I spoke to at the University of Texas, Arlington, in October 2012.

I am deeply indebted to the anonymous readers at the University of Chicago Press and especially to David Wrobel of the University of Oklahoma for an insightful reading of the manuscript. Others who have helped include Emily Wakild (Boise State University); James Beattie (University of Waikato, New Zealand); Michael Ondaatje (University of Newcastle, Australia); Edith Ziegler (Sydney); and Shane White, Iain McCalman, and Richard White (University of Sydney).

Jay Sexton, Stephen Tuck, and Gareth Davies, my colleagues at Oxford, provided much support in the year that I occupied the Harold Vyvyan Harmsworth Chair of American History, as did the Rothermere American

Institute and its director, Nigel Bowles. The fellows at the Queens College lived up to the reputation of Oxford for brilliance and collegiality. Thanks especially to the provost, Professor Paul Madden. William Beinart of St. Antony's College offered suggestions and delivered a perceptive formal response to my Harmsworth Inaugural Lecture, as did Gareth Davies of St. Anne's. At UCLA, while occupying the Joyce Appleby Chair as a visiting professor, I was assisted in various ways by Caroline Ford, Joyce Appleby, and Ellen DuBois.

My friends Keith and Linda Sipe generously supplied short-term accommodation and camaraderie in Durham, North Carolina; and I received great hospitality from friends John and Joan Weaver of Hamilton, Ontario, and the entire history faculty at McMaster University during my all-too-short Canadian stay in 2012.

Archivists and librarians at the following institutions were most helpful: Library of Congress Manuscripts Division; National Archives and Records Administration II (College Park, Maryland); Charles E. Young Research Library, University of California, Los Angeles; Fisher Library, University of Sydney; University of New South Wales Library; Australian Archives, Canberra; Australian National Library; Library and Archives Canada, Ottawa; Pennsylvania Historical and Museum Commission, Pennsylvania State Archives, Harrisburg; David M. Rubenstein Rare Book and Manuscript Library, Duke University; Southern Historical Collection, University of North Carolina, Chapel Hill; Water Resources Library and Bancroft Library, University of California, Berkeley; Bodleian Library, Oxford University; Plunkett Foundation library, Oxford; New York Public Library; Sterling Memorial Library, Yale University; and Huntington Library, San Marino, California. The digital resources of the Dickinson State University's Theodore Roosevelt Center were also very helpful, as were those of the Theodore Roosevelt Collection, Harvard University.

Tim Mennel and Doug Mitchell of the University of Chicago Press have been exemplary in helping this manuscript along, and especially I owe a debt of gratitude to Tim for his astute reading and helpful advice. Pamela Bruton has once again proven herself to be an outstanding copy editor. From the Australian Research Council I received the Discovery Grant (DP0558136) that enabled me to travel overseas and to obtain sufficient research help. Marie McKenzie served as an able research assistant in 2009 and 2010. Nadine Kavanagh kindly translated German documents.

To Diane Collins I can only express admiration at the superior quality of her prose and thank her for reading drafts of several chapters of the manuscript. To Ellen Tyrrell, Jessica Tyrrell, and Diane, many thanks for love and support over the years, without which I could never have completed this work.

Parts of chapter 8 appeared in modified form in "America's National Parks: The Transnational Creation of National Space in the Progressive Era," *Journal of American Studies* 46 (January 2012): 1–21; and an earlier version of chapter 10 appeared in the *Australasian Journal of Eco-criticism* 2 (2013): 5–16.

ABBREVIATIONS

ACA	American Civic Association
ASHPS	American Scenic and Historic Preservation Society
BP	James Bryce Papers, Bodleian Library, Oxford University
BPI	Bureau of Plant Industry
IIA	International Institute of Agriculture
LC	Manuscripts Room, Library of Congress, Washington, DC
LCPP	Prints and Photographs Division, Library of Congress, Washington, DC
NARA	National Archives and Records Administration, College Park, Maryland
NCA	National Conservation Association
NCC	National Conservation Commission
NYPL	New York Public Library
PSA	Pennsylvania State Archives, Pennsylvania Historical and Museum Commission, Harrisburg
SHC	Southern Historical Collection, University of North Carolina, Chapel Hill
SPWFE	Society for the Preservation of the Wild Fauna of the Empire
USDA	United States Department of Agriculture
USGS	United States Geological Survey

PART ONE

The Origins of Alarm

ONE

Alarmism and the Wasteful Nation

German American journalist and artist Rudolf Cronau loved the land of his adoption. Traveling far and wide as a foreign correspondent for the *Cologne Gazette,* he had observed over many years the grandeur of American scenery and the richness of the nation's wildlife. Yet he was often pained by what he witnessed in the European settlement of the continent, and his anger burst forth clearly in his writings. A key complaint was the thoughtless destruction of forests, which would, he argued, produce a timber famine in the not-too-distant future and steadily rising prices. His indictment was stark on this score: "As man made himself master over everything on the earth, so he won his battle against the forest. The settlers felled it, smashed it, burned it, till they got all the room they wanted. Their children followed this example and destroyed the forest with the same recklessness they would have used against their worst enemy." Surely, he concluded in 1908, "it is a reminiscence of those hard pioneer days, that so many Americans neither love nor respect trees, but have only one thought about them, and that is to cut them down."[1]

Not unique in his observations, Cronau played a bit part within a larger drama of lamentations over natural resource waste. He could easily cite others who anticipated his jeremiad, including Emerson Hough. A journalist for *Forest and Stream* and an "outdoorsman," Hough wrote "The Slaughter of the Trees" for *Everybody's Magazine* in 1908. There, for more than half a million readers, was a clear message of what would follow if the warnings went unheeded. "In fifty years we will have the whole states as bare as China. . . . The Canadian forests north of the great lakes will be swept away," and the alluvial plain of the Yazoo Delta of the Mississippi "ripped apart" by floods. "We shall shiver in a cold and burn in a heat never before felt," he warned. "Like Chinamen our children will rake the soil for fuel or forage or food."[2]

For Hough, and for Cronau, the collapse of American civilization was at hand.

Cronau entitled his 1908 book on the subject of conservation *Our Wasteful Nation* and thus aptly captured the changing mood. Critics noted that Cronau's indictment covered the whole range of resources, not just forests, which were "only one form of the nation's profligacy."[3] The German American fed off a growing sense of alarm at this apparently "wasteful" republic in the years after 1900, particularly from 1906 to 1910. A *New York Times* headline proclaimed "America's Profligacy with Her Heritage."[4] "A Nation's Prodigal Waste," replied the *Washington Post*.[5] The *Chicago Tribune* called Cronau's work an "appallingly truthful statement of facts."[6] "A Continent Despoiled" was the headline for Cronau's work in *McClure's Magazine*. Volumes such as Mary H. Gregory's *Checking the Waste* added further publicity.[7]

This was not merely the rhetoric of scribes. It was a movement with genuine grassroots—if middle-class—support articulated through some of the key social institutions of the day. Schools, Chautauqua assemblies, churches, the Daughters of the American Revolution, women's clubs, and debating groups joined in.[8] Businessmen voted with their wallets and, on the contagious assumption of an impending timber shortage, invested in tree planting. The years 1907–10 witnessed an Australian eucalyptus boom in California. The fast-growing trees would, speculators hoped, fill gaps in the timber supply after the rape of the land. At the same time, American companies began buying up forests in Mexico, fearful that domestic supplies would vanish anyway. Clearly, the alarm over wasted resources was not limited to journalists or a political elite. It could not simply be orchestrated from the top, even as federal government officials worked tirelessly and cleverly to do so by priming prominent figures with facts and, on occasion, even speeches to deliver.[9] Whatever the true state of forest shortage, businessmen and many others made calculations that factored in a jeremiad about resource destruction.

Encouraging Cronau was the work of Theodore Roosevelt, the twenty-sixth president (1901–9), and Gifford Pinchot, chief of the Division of Forestry (later designated as chief of the Forest Service; 1898–1910). The precocious son of a New York philanthropist and merchant, Roosevelt is well known as an advocate of the masculine and strenuous life, a nationalist, an instigator of American empire, a lover of nature and the American West, a hunter, and yet a conservationist without par among American presidents.[10] Deeply controversial and intellectually complex, many adored him, others loathed him, but his "high-octane personality,"[11] ample ego, and far-reaching agenda for conservation were impossible to ignore. By inheritance a wealthy man with a fortune derived in part from his father's wallpaper business, Pinchot is scarcely less famous in conservation circles than the wellborn Roosevelt. He served the president as de facto second-in-command for domestic affairs. Although receiving only one reference in Cronau's book, his work was vital to understanding the alarm over wasted resources. Roosevelt and Pinchot constituted, with Secretary of the Interior James Garfield and others documented in the pages that follow, a band of brothers (the term is appropriate) who exuded a "peculiar intimacy" and noblesse oblige in the service of the nation. Roosevelt himself has been authoritatively described as a reforming member of the old New York social elite. He was a card-carrying "Knickerbocker aristocrat" fashioning a modern public policy response to the obscenely wealthy "parvenus" of the new Industrial Revolution.[12] In May 1908, the president called a widely reported Conference of Governors to dramatize the problem of resource waste and destruction and to push the existing anxiety in the direction of structural reform and national planning. From the governors' conference came the creation of the National Conservation Commission (NCC), to carry out an intellectual stocktaking with a broad interdisciplinary sweep across the Washington bureaucracy. Its research produced a hefty three-volume government report covering all aspects of conservation, the first such inventory in American history, and further meetings ensued. A Joint Conservation Conference in December 1908 considered the draft report of the NCC, established a committee with federal and state representatives to propose reforms, and endorsed conservationist objectives to continue the work beyond Roosevelt's term. These objectives were expressed in the Conference of Governors Declaration of Principles, a far-reaching statement intended to guide the committee. Then a North American Conservation Conference called by Roosevelt met in February 1909 and initiated continental cooperation on resource allocation. As a parting gesture, Roosevelt also proposed an ambitious World Congress on Conservation to be held at The Hague, and the diplomatic machinery was creaking into motion to advance the idea before he departed the White House on 4 March.[13]

By that time, conservation had come almost to define Progressivism, that broad and sometimes woolly term used to describe reform movements seeking to adjust the United States to the pressures of a newly industrialized society, with all of its corporate power and labor strife. To be sure, generations of historians have defined and debated different strands of Progressivism, so much so that some would abandon the term entirely.[14] A work constantly in the process of becoming, Progressivism was made as a concept and a movement in this period—and by the actors in this story. Conservation became quite central to their hopes and fears, as apprehension over resource depletion peaked from 1906 to 1910. Ultimately, conservation was enshrined in the Progressive Party platform for the 1912 presidential election.

A striking near consensus emerged on the need for conservation as a key Progressive reform in these years, as attested by the statements of the three leading contenders for the presidency in that election. Yet opinions differed over the degree of reform needed. Not everybody agreed with Cronau or like-minded Jeremiahs.[15] The *Literary Digest* asked, "Are We Conservation-Crazy?," while the *Los Angeles Times* recalled the miser who said he was saving for a rainy day but died before that came, thus missing out on "all the comforts and good things of life." The paper openly championed the oil industry, just then developing in Southern California.[16] Many opponents of conservation represented such obvious economic interests and reflected sectional tensions. In the western states, certain grazing, mining, and other groups did not want federal interference in their arrangements to use public lands. They resented what they regarded as collective eastern hand-wringing, particularly when the advocates of conservation called upon the federal government to withdraw those lands from sale, thus locking up resources in developing states. Governor Edwin Norris of Montana said of easterners, "They have eaten their cake, now they want some of ours." Journalist George Knapp of Saint Louis targeted federal control over public lands as the worst of all despotisms, one suffused with the arbitrary authority of petty officials telling the common people whether or not they could farm or mine at all. In practice, many conflicting interests fragmented grassroots opposition in the American West, with larger-scale lumber and cattlemen's operations often supporting federal programs that could stabilize their businesses. Yet Knapp's case expressed the immediate experience of many other farmers, ranchers, and small-scale lumber millers. Opposition grew in strength during Roosevelt's presidency, precisely because of the widening scope of federal intervention.[17]

Underlying and augmenting regionalist responses was legal opinion. For some state governors, conservation within the states was not the business of the federal government, although they acknowledged the need for reform. They cited constitutional arguments that federal power did not extend to nonnavigable rivers, control of waterpower, and related matters.[18] Just as clearly, these constitutional points were aligned with business interests that, Progressive conservationists charged, spearheaded the opposition. The most outspoken champion of untrammeled corporate development was Knapp, writing in the *North American Review*: "That the modern Jeremiahs are as sincere as was the older one, I do not question. But I count their prophecies to be baseless vaporings, and their vaunted remedy worse than the fancied disease. I am one who can see no warrant of law, of justice, nor of necessity for that wholesale reversal of our traditional policy which the advocates of 'conservation' demand. I am one who does not shiver for the future at the sight of a load of coal, nor view a steel mill as the arch robber of posterity."[19] According to Knapp, the entire Washington press had been enchanted by a diametrically opposed position emanating from a federal juggernaut of bureaucrats. The "finest press bureau in the world," he charged, had "labored with a zeal quite unhampered by any considerations of fact or logic." It pandered "not to popular reason, but to popular fears."[20]

Opponents of conservation made a deeper, philosophical case about development as well. Here, nature was malleable. Humans had altered it in ways that embodied value through capital, thus *improving* the prospects of future generations. This position rested on the antebellum Whig political economy of Henry Carey, stated in *The Slave Trade, Domestic and Foreign* (1853): "The earth is a great machine given to man to be fashioned to his purpose. The more he works it, the better it feeds him, because each step is but preparatory to a new one more productive than the last—requiring less labour [*sic*] and yielding larger return."[21] This argument was inherently hostile to alarmism. For example, by building railways, humans increased the comfort of future generations, albeit at the cost of destroying forests to lay railroad ties. Though the implication was rarely stated, nature had actually become incarnate in capital. A *New York Times* reviewer of Cronau wrote: "If we have hewn the forests, we have invented steam. If we have exhausted the mines, we have developed electricity. Assuredly our followers are better off with the reduction of natural resources, accompanied by the inventions which recent generations proffer as a recompense." This critic also rejected guilt over the legacy to future generations by pointing to material progress already achieved: "what right has posterity to expect so much from us? Our ancestors did not do so much for us."[22] And yet the current generation had a higher standard of living than ever before in human history, the disbelievers chanted.

That said, even those who praised the transformation of nature into productive capital still conceded, under a de facto precautionary principle, that conservation should be attempted. When the *New York Times* denied recourse to intergenerational equity, it agreed that there was "need for prudence, but not cause for fright."[23] Careless and unnecessary waste the paper accepted as a sin. Though the *Wall Street Journal* likewise scoffed at apocalyptic conclusions, it stressed that the country was living off its capital by destroying raw materials. Reliance on the free market was "too much like locking the stable after the theft of the horse fully to meet the case." Perhaps the answer was not to stop resource use but to engage in "technical and scientific research."[24] That was, most scholars would argue, a very American response that flowed from the key role of technology in the nation's culture.[25] Nevertheless, the *Wall Street Journal* conceded that technological growth would not simply happen on its own. Government, business, or university research was needed to promote more efficient use and to discover substitutes. The paper praised the prudent policy of resource conservation already applied to forests by Pinchot.

A surprising number of opinion makers across the political spectrum agreed that government needed to do something serious about natural resource waste. Four-fifths of the states created conservation commissions, illustrating in the process the breadth of concern. Manifestly, the enthusiasm for conservation went far beyond partisanship, loyalty to Roosevelt, or purely national action. The acquiescence of so many political figures and media outlets in some degree of conservation sentiment indicated that alarmism had partially won the debate by pushing many of the unconvinced to favor some—though not all—of the action that enthusiasts of conservation advocated.

Previously little used, the term "conservation" became ubiquitous and closely identified with the team of Roosevelt and Pinchot. In fact, Pinchot claimed to have hit upon it while riding in Rock Creek Park, in Washington, DC, one day in 1907,[26] but "conservation" as a concept was used by George Grinnell, editor of *Forest and Stream*, as early as 1884.[27] Grinnell had written of both preservation and conservation, meaning the careful use of resources in the latter instance and the prevention of development of resources in the former, but for forestry, the practice of conservation had already begun in British India, Germany, and other places well before that.

To be sure, the jeremiad on forest depletion was neither new nor exclusively American. Historian Richard Grove has demonstrated that concerns for the disappearing forests in Europe's colonies run back to the late seventeenth century. The German explorer and geographer Alexander von Humboldt expressed a similar alarm regarding Spanish Venezuela in 1800.[28] The United States was not directly affected by such soul-searching, as it lacked overseas colonies—but it did have the West, which served a parallel, quasi-colonial function. In the mid-nineteenth century, Americans already worried about the future of a wantonly destructive nation as the West was rapidly "won." Though prophecies of shortage for other resources were made in that era, the forests were by far the chief worry. In these years

Americans articulated ideas of preserving nature, spurred especially by George Perkins Marsh's *Man and Nature,* the first major American account of the damage humans had done to natural environments. In this 1864 book and later editions, Marsh emphasized lost forest cover that led to soil erosion and floods, and he investigated the impact of deforestation on climate. A quarter of the nation's landmass must be kept in woodland, he argued, to prevent serious problems. Social scientists took up the cause in the 1880s and 1890s, with Richard T. Ely of the University of Wisconsin urging a greater role for the federal government in creating public forests, since only public action could command the "vast scale" needed, as well as comply with the necessity of thinking long term to make forestry profitable.[29] The eminent geologist Nathaniel Shaler of Harvard greatly valued Marsh's work and by the 1890s broached the subject of depleted oil stocks, though his comprehensive indictment of resource misuse was not published until 1905, when he blamed Americans for the "very worst" of the "sinful wastes [*sic*] of man's inheritance in the earth." The book's arrival was splendidly timed to reinforce as well as reflect upon the growing sense of alarm.[30]

Until 1900, little concrete action occurred. The earliest legislative response to the intellectual agitation was the Timber Culture Act of 1873. Ineffective in practice and beset with legal loopholes, the act had been intended to encourage tree planting on the barren plains of the West in the hope that the climate would improve with enhanced vegetation. In another sign of things to come, the US government appointed its first chief of the Division of Forestry, Franklin Hough, in 1881. Concern over the protection of the watersheds in Southern California and elsewhere contributed in 1891–92 to the creation of the nation's first forest reserves, and the Organic Act of 1897 registered the first attempt at administration of such forests.[31] These incremental achievements reveal legal and administrative precedents for Progressive Era action, in which new jeremiads about the environmental failings of the American Republic began to flourish. The survival of forests came to be seen as essential to republican civilization's health.[32] Nevertheless, the scope and content of the response would be startlingly novel after 1900. Professor Ely called the earlier efforts merely "petty measures" that "will never accomplish anything of economic significance."[33]

What appeared fresh in the new century was the breadth of issues discussed, the global scope of concern, the vigorous involvement of the federal government in developing a countervailing conservation policy, and the articulation of intergenerational equity as a serious issue for debate. Within the government and without, conservationists began to assert a crisis with respect to, not merely US forests, but all American resources. Moreover, it came increasingly to notice that other nations were running short of raw materials and that the United States could not necessarily look abroad for these resources essential to power economic growth. The call to action and the ensuing campaigns covered all aspects of "conservation," including minerals, water, soils, forests, and crops. By 1909, even human health was added. Embracing both "preservation" of resources and their wise use, conservation came to be an all-encompassing enthusiasm.

Matching the rapidly expanding discourse over wasted resources was its obverse: the idea of efficiency as the basis of conservation policy. In part, "efficiency" revealed an urge for a stronger state to compete on the international stage. Undertaken by a president "highly conscious of America's developing role in world history," efficiency included far-reaching attempts at administrative reforms to strengthen the American state, such as military reorganization. This drive was also manifest in conservation policy, particularly as pioneered in forestry.[34] But this was no dry-as-dust bureaucratic process. A "gospel of efficiency" and the accompanying jeremiad galvanized Progressive intellectuals and the populace alike as if Armageddon itself was at hand.

This great change from 1900 to 1910 has been portrayed in terms of the passion, enthusiasm, and intellectual force of Roosevelt and those he inspired.[35] Complementing the focus on the president were the intellectual implications of the disappearing "frontier." The idea of a fin-de-siècle crisis that Roosevelt and Pinchot tapped into almost coincided with historian Frederick Jackson Turner's announcement of the end of "free" land at the frontier's official closing in 1890. Few Americans of the time read Turner's essay, but many opinion makers absorbed his and similar ideas welling up from popular culture.[36] To Turner's followers, this momentous change made the American experience unique.

The struggle over the future use of the US public lands of the West broadly correlates with this interpretation. Arable land was becoming scarcer, and this circumstance imposed constraints absent from earlier land-grant policies. Roosevelt created a Public Lands Commission in 1903, tasked to recommend reclassification of land according to its best uses. Thereby he hoped that a rational and planned allocation of land would be undertaken. On the one hand, that suitable for "actual settlement" needed to be maximized to allow homesteads to be built in the yeoman farmer tradition. On the other, crop farming would no longer be allowed where pastoral leasing met the environmental conditions more precisely. This potentially contradictory and politically controversial policy was not successfully implemented. The federal government was unable to introduce general land reform during the Roosevelt years due chiefly to western opposition to interference in the economic affairs of the affected states. But piecemeal reforms were achieved. Roosevelt resorted to executive action, withdrawing, for example, sixty-six thousand acres of suspected coal lands in 1906 from public "entry" in the General Land Office for sale to private interests, and he recommended separating mineral rights from surface rights, as in Australia.[37] Subsequently, the Coal Lands Act of 1909 started the process of implementing this suggestion. Further, the Enlarged Homestead Act of 1909 provided for farming allotments of 320, instead of 160, acres to take account of semiarid conditions, which required larger blocks for fallowing and rotation of crops. Other federal policy areas, such as irrigation, national forests and parks, and restoration of the economic and social viability of rural life, were also consistent with the belief that "free" land, or at least "good" land, was no longer abundant.[38]

Roosevelt himself chose to identify, at least in part, with the effort to protect a wilderness that was receding. He was an advocate of pioneering values and strove to safeguard the nation's natural heritage through the creation of parks, monuments, and wildlife reserves at the very time that the influence of the Turner thesis on the end of the frontier was spreading.[39] Douglas Brinkley documents this achievement amply in his book *Wilderness Warrior*.[40]

Roosevelt did hold the issue of wildlife and its habitat in wild places close to his heart, but he believed that, for political as well as social reasons, a balance must be kept with conservation as efficient use. He cautioned his friend William T. Hornaday, a fellow naturalist, that "preservation of wild life" had to be combined with "forest or other preservation." Moreover, grand schemes for saving wildlife "must always be kept as a mere incident" of the utilitarian movement, "for our success in achieving either movement depends upon our convincing people of the practical beneficial side." Protection of wild places and things needed to be toned down so that an appeal could be made "primarily to the practical business common sense of our people."[41]

Even granting the complete sincerity of Roosevelt's objectives, and even though the frontier's ending contributed to a sense of shrinking resources that combined both land and wildlife, this internally oriented interpretation of American conservation is not entirely satisfying. The ill ease was not purely American but extended to many nations, some facing conditions very different, and others very similar. Newspapers far away quoted Hough and Cronau and took the stirring calls to action from American conservationists seriously.[42] The foreign response encompassed expanding imperial

1.1. Roosevelt at Glacier Point, Yosemite National Park, 1903. Stereograph, LCPP.

frontiers of European settlement rather than contracting or "ending" ones and promoted conservation on a self-consciously international level. When Roosevelt counseled Hornaday to be cautious, he was in the middle of discussing international, not national, conservation politics.[43] The rise to prominence of national conservation capped a decade or more of international debate over efficiency in national life, a debate as prominent in Britain as in the United States.[44] The transnational nature of the Progressive discourse of waste and efficiency has been apparent to scholars for at least four decades, though its implications for Progressive Era historiography and the birth of the conservation movement have been largely ignored. Samuel Hays's *Conservation and the Gospel of Efficiency* long dominated the field, but he neglected this international context in which parallel trends existed in other industrializing countries.[45]

The book that follows begins from the knowledge that the creation of a proconservationist sentiment was not the product of purely American conditions. Historians are familiar with the transatlantic exchanges of Progressive reform, but little has been done to apply this approach to conservation.[46] More important, the transnational context of American conservation was not European or Atlantic but global. The beginnings of "conservation diplomacy" can be detected, and the Roosevelt and succeeding administrations conducted bilateral negotiations with Canada and Britain over boundary waters and inland fishing, and multilateral ones with Canada, Japan, Russia, and Britain over fur sealing. The Migratory Bird Treaty of 1918 with Canada capped this trend.[47] Yet under Roosevelt the ambitions went beyond these specific items to involve comprehensive hemispheric and global conservation.

To understand this new American engagement in all of its depth and span we must connect the conservation debates to another neglected theme in US history. American "empire" and "imperialism" are slippery terms, and a great deal of printer's ink has been spilt upon them. Nonetheless, a consensus is growing that the United States has indeed been an empire with both formal colonies and informal spheres of economic and political control. The boundaries between the two were fuzzy, and it is necessary to specify clearly the changing historical character of that empire at every point. From the 1890s to the 1910s, it was an empire that altered rapidly in response to global circumstances.[48]

More than acquisition of territory, trade expansion backed by financial power and, where necessary, military intervention in other nation-states was important as far as the United States was concerned. Much attention has been paid to exports as a stimulus to this American economic imperialism—but almost none to the issue of a growing American need for certain imports.[49] That reflects the common image of a United States so abundant in resources that it was largely self-sufficient. The evidence is otherwise in two respects. Perceptions of resource shortages grew after 1898, while consumer demand for products that could not be obtained within the continental United States expanded rapidly. The source of those fears of scarcity amid abundance was changing, as tropical areas became more important to American thinking, and as American attention turned to East Asia and Latin America. The growing need for raw materials and the new geographical orientation were linked circumstances. East Asian commercial exchanges with the United States surged ahead in the late nineteenth century, including consumer goods such as silk, tea, and porcelain china. Latin America similarly provided an increasing range and volume of goods. Among the natural resources imported was a surprising variety of tropical forest products. Coffee and rubber from Brazil were already important consumer and industrial items. Fine furniture required wood such as the tropical mahoganies from Central America, and so too did the interior fittings of the homes of the well-to-do. The landmark Arts and Crafts structure of the Gamble House, designed by the firm of Greene and Greene in Pasadena, for example, would have been unthinkable without its extensive embellishment with Central American mahogany, a timber in declining availability. More prosaically, the Wrigley Chewing Gum Company needed chicle from Guatemala as the basis of its masticating products, said to be a necessity "for the girls of America."[50] As the United States began to seek enhanced trade in

Latin America and East Asia, it also sought secure sources of supply for raw materials from these places. After 1900, oil in Mexico and later Venezuela joined the list.

While concern for the natural resources of the wider world can be detected in American activities of the 1890s, both alarm and action intensified after 1898, with the coming of the formal, or "island," empire. In that year the United States acquired its own colonies: the "insular" possessions of the Philippines, Puerto Rico, and Guam. In the same year came Hawaii as a territory and, soon after, Samoa and the Panama Canal Zone. Though not a consideration in these acquisitions, once the deed was done the US government deemed tropical forests useful in helping to fund colonial rule and in developing export markets for the insular possessions.

A cautionary note is needed here. The rise of conservation to national prominence cannot be explained purely by the seizure of a few overseas colonies. Rather, 1898 gave shape and spur to an already-evident American concern with the nation's international connections and its place as a newly important world power. Widespread anti-imperialist sentiment certainly restrained subsequent American actions, but many Americans were proud of the colonies that the United States had acquired and hopeful about the resources they might contain. These Americans saw the colonies as the site of what commentators called a "Greater America," and this term entered common parlance. "Rising to the full splendor of its responsibilities" was a formal empire supplemented by the growing informal empire of American political and economic influence in the Caribbean region and across the Pacific. All this was unintelligible without 1898. Its cultural impact at home has been woefully neglected in the historiography.[51]

Yet the American interest in overseas territory and economic influence was to become more complex still. The availability of resources abroad in the shape of colonies or zones of political and economic influence was not the only source of power for a self-conscious "empire." Equally important was the idea of an explicit *inland empire* to complement the external one. This internal strengthening of American power was more easily acceptable to—and often advocated by—anti-imperialists but was nevertheless transnationally produced in the geopolitical system of "high" imperialism and shared by pro- and anti-imperialists. Because the United States was a latecomer to the worldwide scramble for colonies, both preservation and wise use of internal resources through conservation policies became doubly important. In this respect the United States was less comparable to Britain than to Germany, which had also come late to feast at the colonial table, and which picked up the scraps of Africa and the Asia-Pacific region that other powers initially did not want. Germany had to rely, as Britain did not, on itself and its immediate neighbors for many raw materials and, as a result, cultivated influence before World War I in southeastern Europe and in the decaying Ottoman Empire. Control of *Mitteleuropa* was to be an important German war aim in 1914, if not *the* main war aim.[52]

In a geopolitical sense, this German bid for continental hegemony was similar to nineteenth-century American efforts to build a "continental" empire. Both powers were industrializing rapidly by 1900, and both required raw materials from an expanding hinterland, linked by railroads and canals. For the American political elite of the period 1900–1917, the distribution of resources within this inland empire was just as important as, if not more important than, access to foreign raw materials. In the world system of "high" or "Victorian" imperialism from 1870 to 1914, all the European empires and the United States understood the importance of a neomercantilist hoarding of key resources for the coming industrial struggle, and many applied discriminatory tariffs or other trade practices in their jurisdictions.[53] Because American leaders swam in the same intellectual sea, conservation became inseparable from geopolitical competition in an imperial world. As much as when considering the colonial territories and the informally dominated Caribbean states, the continental United States itself was conceived of as an empire in this geopolitical sense in the era of Theodore Roosevelt.

The neomercantilist streak in American conservation policy will be illustrated in the following chapters, but it is worth noting that such a policy fits nicely with a historiographical critique of the American empire as an "Open Door" empire of free trade.[54] The assumption of an Open Door–controlled American foreign policy has always sat uneasily with the strongly protectionist nature of American trade policy under Republican congresses from the 1890s to 1913 and in the 1920s. Reciprocity deals were certainly negotiated, but these were used to modify protectionism where American resource and export interests were particularly strong. Despite reciprocity policies, American trade was geared toward the creation of surpluses, and it favored internal strategic and economic priorities rather than international commercial exchange under the invisible-hand principle.[55]

The inland-empire idea also helps us to understand how Americans tried to distinguish their own place within the international state system from European colonialism.[56] Certainly, Americans came to argue that their empire abroad was forward-looking rule, one that worked not for imperial overlords but in a modernizing and democratizing way for colonial peoples. Roosevelt formed this view, a stance particularly visible during his visit to Africa and then Europe in 1909–10, when he compared European and

American practices. This empire of modernity, progress, and temporariness is the basis for the enduring idea of American imperialism as exceptional.[57] But Roosevelt explicitly endorsed both British and American colonial rule. Exceptionalism did not entail the repudiation by Roosevelt and his followers of a transnational Anglo-Saxon solidarity. Far from it. He saw the American version of empire as part of a white-settler bloc of broadly Anglo-Saxon nations that would provide global leadership. Americans would have affinity not so much with British hierarchical rule (except where temporarily needed in Africa and Asia) as with the self-rule of settler societies that would become models for nonwhite peoples made subjects of the European empires.[58]

The American empire of those years should, therefore, be seen as having not an essentialist character but rather the characteristics of a complex social formation. Roosevelt articulated this formation in fashioning the strong American nation-state. Conservation was fundamental to this settler vision since it involved the idea of white intruders justifying their demographic takeover and their rule over indigenous peoples by putting down a deeper stake in the land than pioneers had accomplished—by husbanding it. Thereby they established a right to the land and a right to stay *permanently*, wherever environmental conditions would allow the white "race" to thrive in North America, Australasia, and, it was hoped, certain British colonies in Africa. Through conservation of resources, the American nation could lead the Anglo-Saxon peoples across the globe to a moral and material hegemony to replace the purely British imperial version. Though grossly simplifying the realities of settler society diversity in different imperial jurisdictions, this approach figured in many of the conservation reforms that Roosevelt adopted: from irrigation and waterways to human health and agriculture. In this way, the island empire of colonies, de facto Caribbean protectorates, and naval bases, on the one hand, and the inland empire of the US nation, on the other, became inextricably connected to one vision.[59]

Synoptic understanding of this geopolitics did not emerge spontaneously. Translating the changing conception of the power balance across the globe into US policy and practice was the work of individuals. There, accidents do matter. When the serendipity of an assassin's bullet put Roosevelt in the White House in 1901, it helped enormously that the new president had both a deep interest in conservation through his love of natural history and an interest in the relationship of people to land in American settlement experience.[60] From this propitious background, the slow and uncertain process of conservation reform of the 1880s and 1890s was consolidated and galvanized by Roosevelt's precipitation into power. Contributing too

1.2. Roosevelt was fond of being photographed using the globe as a prop to indicate his interests. Here he is shown in the White House, c. 1905. LCPP.

was the personal chemistry that developed between Pinchot and TR. They knew one another in the 1890s, but Pinchot became the president's right-hand man in domestic politics by 1902. Together, they planned to do big things.

Theodore Roosevelt is a much-studied president. His personality has been analyzed and his achievements have been recited numerous times. He is especially well known as the chief executive who put the United States on the map as a political and strategic force. We know also that Roosevelt was a great conservationist who doubled the number of national parks, proclaimed national monuments, and pioneered wildlife refuges. But never have historians put two themes together: the first, Roosevelt's geopolitics, is seen as the outward projection of American power; the second, the conservation movement he helped to fashion, is treated as an inward encounter with a unique American environment. Roosevelt was not an original thinker or the Great Man of History, but he was an adroit and insightful conduit for and interpreter of the imperial conjuncture that brought these two issues into a mutual relationship. Within that reassessment of American power and the nation's role in geopolitics, Americans became increasingly anxious over global pressures and opportunities, none more so than in resource management.[61]

Despite foreign assessments of the American Republic as a future hegemonic power that out-traded Europe and flooded the Old World with "Yankee goods," American political and intellectual strategists remained remarkably unsettled. The source of their anxiety was fundamentally environmental. Brooks Adams, brother of Henry and informal adviser to the American elite, influentially predicted an imminent US bid for global predominance, grounded in America's control of rich natural resources.[62] But imagine the consternation when policy makers began to fear that those resources *within* the nation were in danger of dissipation and those *without* were either falling into the hands of rival powers or otherwise being squandered. Faced with this unprecedented situation, conservationists both in the government and outside assessed not merely the crisis of US forests but the crisis of all resources across the globe.

Roosevelt's last annual message to the US Congress channeled these intellectual currents. He told a moral tale of empires that failed to conserve. For the first time, a presidential address came complete with photographic illustrations for the edification of the nation's legislators. Scenes of Shanxi Province, China, photographed by US Department of Agriculture and US Geological Survey officials, illustrated "Oriental" environmental decay, though the warning was not intended for China. Bare mountainsides, eroded soils, clogged waterways, and debris littering parched plains underlined a plainer message nearer home. In 1908 the Qing dynasty was collapsing, and the cause was unwise rule: an emperor had, centuries before, decreed the removal of timber cover from the mountains of Shanxi. Historian Mark Elvin's *The Retreat of the Elephants* has shown the resource-consuming nature of China's land empire. Roosevelt anticipated that point but extended its geopolitical significance: China appeared as a decrepit state that portended the fall of future empires.[63]

Putting empire and environment in the same frame has both historical and contemporary importance. It enables us to understand better the consolidation of national power under Roosevelt and the intertwining of personal and political objectives in the process of creating a stronger American state. It allows us to see how conservation became the chief vehicle for the assertion of Roosevelt's and Pinchot's program and the bedrock of the president's efforts to leave a lasting legacy as his time in office ebbed. Conservation broadly interpreted enables us to unpack the meaning of the "Progressive Era" and see the role of empire and conservation as central to the search for national consolidation.[64]

1.3. This scene of Shanxi Province, China, 1903, was photographed by USGS official Bailey Willis and illustrated erosion and belated attempts to combat it with terracing. The set of photographs taken by Willis showing erosion in that region appeared in *World's Work* 17 (February 1909): facing p. 11187. Photograph courtesy Huntington Library, San Marino, CA.

Yet the global scope of the conservation crusade also puts Roosevelt in his place, as an agent of a much larger sensibility. It makes us pay attention less to the passing of American wilderness and the end of the frontier as the source of Roosevelt's conservationist stance than to influences from abroad and the impact of great-power politics.[65] At the turn of the twentieth century, large numbers of Americans saw their nation poised on the edge of an environmental catastrophe, and Roosevelt fostered and embodied their hopes for a new path in a country plagued by waste of its natural resources. The story begins with the perceived crisis of the forests—a crisis with roots in nineteenth-century American experience but augmented by anxieties abroad over the threats to forests in Europe and its empires—and the efforts of international experts and reformers to change a persistent pattern of waste.

TWO

American Conservation and the "World Movement": Networks, Personnel, and the International Context

When Theodore Roosevelt and Gifford Pinchot raised the alarm over resource shortage in America, they drew upon a nascent internationalism in conservation thought and practice. Self-styled "international" societies and congresses with official government representation from many nations began to question resource misuse. Meanwhile, ideas about the destruction of nature crossed national boundaries, and non-state actors forged transnational pathways for this expanding knowledge base. The United States was far from being the sole source of the conservation agenda, with Europe taking the lead in many respects. These cross-national contexts are important, because they show that the nation's very substantial achievement in conservation was part of a broader pattern. They contributed to American awareness that conserving natural resources must go beyond the nation's borders, and conservationists gained both intellectual imperatives and élan from these encounters. Transnationally derived ideas allowed Roosevelt and others to position US moves in conservation as necessary to the nation's new standing in the world.[1]

This new sensibility on the international level did not simply waft over Americans like stardust, automatically translating into action on the national, let alone local, level. Ideas had to be articulated within national politics through voluntary organizations, official government channels, and newly vigorous national media. The institutions of the state were not weakened by these broader influences. Rather, Roosevelt used the international context to strengthen the American nation-state. As a strong nation-state, the United States would be able to participate in, indeed lead, wider conservation campaigns for the global benefit. The international and transnational dimensions are both the beginning and the end of the story told in the following chapters. Roosevelt called such global interconnections the "world movement."[2]

The 1870s and 1880s saw the initial flowering of international organizations, congresses, and treaties. The new nongovernmental organizations were vastly eclectic in scope. Subjects canvassed by these NGOs ranged from public health to Christian missions. Some of the latter's moral sentiment would rub off on nascent conservationism. Nevertheless, science was a profoundly connected set of disciplines internationally by the late 1880s, and these fields quickly trespassed upon environmental questions. Zoological meetings beginning in 1889 through to entomological ones beginning in 1910 contributed to the growing scope of internationally oriented science, while technical forums like those of the International Navigation Congress drew the attention of engineers to innovations in the sculpting of rivers and harbors in Europe.[3] All these groups exchanged ideas across national boundaries, contributing to an unprecedented transatlantic flow of information. Secretary of State Elihu Root noted the impact in 1908: "If you look at the international life of the world, you will see that the correspondence between the Nations is continually increasing, not in the letter-writing sense, but in the intercommunication and understanding about the things that they should do in concert for the benefit of all their people. Scores and hundreds of conferences and congresses are being held under Government auspices to regulate the action of the different Nations of the earth."[4]

Zoologists, engineers, and other professional groups did more than exchange ideas in the way that Root suggested. At first glance, scientific groups did not greatly influence any self-conscious and highly mobilized conservation "movement." Yet from the start, they nurtured grassroots activism for nature protection. Among the first to express alarm globally were ornithologists. This group was particularly important because it breached the supposed divide between amateur naturalists and "professional" scientists. This was the case in many parts of Europe and the United States, where networks and alliances were formed that allowed widespread mobilization of conservation opinion.[5] Birds were no respecters of national or state boundaries, and quickly bird lovers and students of bird life raised concern about the depletion of species across many countries. International ornithological congresses began in 1884 and became regular five-yearly events after the Paris meeting of 1900. Threats to migratory bird populations included the pilfering of eggs for food and slaughter for feathers and meat. In the United States, Audubon societies sprang up, mostly from 1896 on, while agitation on an international level to conserve certain species began almost simultaneously.

Over roughly the next two decades, ornithologists coordinated their activities to pass laws in European nations, just as the Audubon supporters did in the various American states. The American groups became affiliated in a National Association of Audubon Societies in 1905, but *Scientific American* acknowledged that the growing campaign for bird protection could already "be called international." The International Ornithological Congress of 1900, held in Paris, had entire sections on the "protection of birds" and their "acclimatization," and American delegates led by C. Hart Merriam of the US Department of Agriculture (USDA) took part.[6] As early as 1902, the National Audubon Society's president William E. Dutcher was networking with like-minded bird lovers for complementary legislation in Canada, Mexico, and Europe.[7] The bird preservationist impulse was a motive behind the first national US legislation to protect species, the Lacey Bird and Game Act,[8] which targeted the interstate trade in protected birds. Roosevelt, as we shall see, developed a strong alliance with the act's sponsor, John F. Lacey.

Bird depletion was viewed as a warning, not unlike the proverbial canary in the coal mine. Avian loss symbolized reckless disregard for the gamut of natural resources—disregard that portended civilizational decline. It was also closely tied to an area of particular concern to Theodore Roosevelt as an advocate of manly yet restrained hunting. In view of the international exchanges in ornithology, it is more than a coincidence that, in 1900, a treaty committed the European powers with colonial possessions in Africa not only to restricting the slaughter of birds and the taking of eggs in their territories but also to stemming the decline of other wild creatures. Though ratification failed internationally, the Society for the Preservation of the Wild Fauna of the Empire (SPWFE), established in London in 1903, put pressure on the British government to fulfill the treaty's purposes and raised awareness of species extinction through publication of a journal. This society was influenced by, among other things, protests over the near extinction of the bison in North America.[9] As an opponent of the bison slaughter and a founder of the Boone and Crockett Club of responsible hunters in 1887, Roosevelt was already sensitized to species extinction, but he also learned about the importance of game reserves from the international work of a British correspondent, Edward North Buxton, the founder of the SPWFE. Roosevelt became an honorary vice-president of that group, and his friend William T. Hornaday (another bison champion and director of the Bronx Zoo) and Dutcher were honorary members. The 1908 volume of the SPWFE's journal carried a message of support from Roosevelt on his "hearty sympathy with all that is being accomplished."[10]

Hunting itself stimulated conservation sentiment, but there is only slender evidence that hunting per se was a key driving force. Hunters were not a cohesive social grouping. While recreational hunting such as Roosevelt and other elite hunters practiced might partially explain creation of game reserves, the wide variety of conservationist concerns, from forests to farms, fossil fuels, and irrigation, were not the preserve of hunters, even those sensitized by class, experience, or scientific knowledge to the depletion of species.[11]

None of this new conservation ethic expressed a modern ecological sensibility; rather, the fear was for the depletion of natural resources. Preservationist currents were developing, as were the first stirrings of ecology as a discipline, but the 1900 fauna treaty distinguished, as did the American state bird acts, between supposedly "good" and "bad" species. The emphasis was on "useful" creatures. For birds, it was often desired to protect those helpful to agriculture because they stopped the spread of insect pests. For mammals, the treaty of 1900 discriminated on both aesthetic and practical grounds. It classified elephants and zebras as warranting efforts to preserve them, while dangerous carnivores (lions) and the aesthetically displeasing hyenas, as well as the reptilian crocodiles, were not protected.[12] Moreover, "conservation" sentiment did not normally consider the interests of the human populations of "natives," in either Africa or North America. This omission stands out in the creation of nature parks. Under various names, the setting aside of land for wildlife refuges (game parks) or for the preservation of certain landforms and tree species typically excluded indigenous people's hunting rights in Europe's empires.[13]

In the United States these actions came under the rubric of national parks, starting with the creation of Yellowstone National Park in 1872, as they did in Canada, New Zealand, and Australia by the late 1880s. But in Europe and the colonies of the European empires, various "nature protection" zones were created. During the Progressive Era, two countries surpassed the pioneering national park record of Americans in one measure by establishing organized systems of national parks: Sweden in 1909 and Canada in 1911. US examples inspired Europeans and others, but no single model was diffused globally.[14] The utilitarian aspect of fauna preservation and the aesthetic judgment of what was desirable to save also characterized the international movement for protection of flora and landforms. Far from being simply an effort to preserve wilderness, in both North America and Europe utilitarian and aesthetic appreciation that derived from European value systems played a large part in nature protection. The National Trust was established in Britain in 1895 to preserve buildings and landscapes, and it soon provided an example to be followed by like-minded people in the United States. American preservationists were influenced by such European precedents and worked with Europeans through the American Scenic and Historic Preservation Society (ASHPS).[15] As part of this movement, the American Civic Association (ACA) became allied, from its inception in 1904, to scenic preservationists advocating national parks. Both the ASHPS and the ACA served as conduits for European-derived views of park aesthetics. Similar European pressure groups such as the Société pour la protection des paysages et de l'esthétique générale de la France sought to save the countryside and implement antiquities legislation. While the French had their distinctive cultural sensibilities,[16] the idea of countryside protection spread throughout Europe and influenced the United States as well. The Congrès international pour la protection des paysages held in Paris in 1909 "was entirely devoted to the matter of protection of nature."[17] By that time, scenic preservationists in the United States had begun to document the work of a host of European groups, and this aesthetic influenced them. The elite journal *Country Life in America* also incorporated this European idea of a cultivated landscape of nature.[18]

Commentators noted that the United States reflected the European interest in preservation of heritage as well. Not just "natural things" but "historic or prehistoric objects, ruins or monuments" won rapidly growing favor.[19] The Antiquities Act of 1906, which allowed the president to declare national monuments, was inspired by European and "Near Eastern" archaeological research. With a number of European countries passing antiquities legislation, American archaeologists, many of whom had trained in Europe or worked on Middle Eastern archaeological digs, wished to close the gap between US and European practice. Roosevelt's policy-minded presidency combined with the Spanish-American War to give strong impetus. As historian Hal Rothman has pointed out, "In the exuberance" that followed the "War of 1898," archaeologists "sought to give American archaeology the intellectual status of European and Middle Eastern antiquities, while simultaneously bemoaning the lack of care American areas received."[20] In this as in almost every other area of conservation policy, US thinking either paralleled or was influenced to varying extents by an international context.[21]

All these conservationist trends developed in a world where personal contacts and exchanges of information on the global impact of humans on flora and fauna were increasing. These ideas were accentuated by and channeled through the activities of specific groups. Operating transnationally, these groups developed a global perspective that fed into US political debates over the nation's resource future. American technical experts and professional societies espoused the importance of efficiency in the development of conservation. Conservationist ideas indeed gained strength and critical mass by flowing through changing professional channels. Three key groups were engineers, geologists, and foresters.

From the 1880s to 1914, American engineers traveled far and wide, gaining an expansive, even global vision for their profession and its impact. They cooperated internationally, especially in the development of engineering standards. They attended international meetings, replete with style and ceremony, that invoked "technical accomplishment, collective interest, and the onward march of human progress." Their professional magazines revealed an emerging cosmopolitan outlook. The *Engineering Magazine* of New York greeted the new century by documenting African railroad building, "engineering opportunities" in the Russian Empire, gold mining in Western Australia, European and American bridge construction, and the possibilities of the telephone for transatlantic communication. American engineers were in great international demand because they were thought to be more practically trained and could best their competitors in the British Empire's far-flung dominions and in Africa.[22] Engineers' conservation interest at this time was generally based not in the narrow scientific-management concept of maximizing profitability through reorganization of labor productivity but in the physical availability of raw materials. For this stance they had practical, business-oriented motives, because it was necessary to supply ample materials for manufacturing to address fears of changing international wage differentials. American competitiveness was based on not only cheap, unskilled labor in mechanized processes but also low-priced natural resources. With American wages for both skilled and unskilled work rising relative to Britain after 1900, the president of the American Society of Mechanical Engineers worried that "high-priced materials," added to "high-priced labor," would "shut" Americans "out of the world's markets." As it stood, he claimed, "other countries" were "under-selling our mills in our own country, by virtue of low-priced skilled labor."[23]

Engineers touring the globe included the future president Herbert Hoover, who amassed a fortune while working in the mining industry in Australia, Burma, and China between 1897 and 1914. Equally impressive but more influential as an engineer was John Hays Hammond, a man who, after adventures on the goldfields of Transvaal, South Africa, in the 1890s, including a period in prison for supporting Cecil Rhodes's infamous Jameson Raid of 1895, advised the Russian government on engineering projects, beginning in 1898. Later, in 1905, he performed a similar service for the United States on the Panama Canal and, in 1918, went to Turkey, where he reported on Armenian and Syrian relief during the refugee crisis of World War I. An equally mercurial globetrotter was the University of California–based irrigation engineer Elwood Mead. He went to Australia in 1907 to head the Victorian Rivers and Water Supply Commission and later toured the world advising on his field in South Africa, Palestine, Mexico, Cuba, and Canada.[24]

The engineers shared this global interest with geologists, with whom they were closely connected because the fields of science were not then so specialized as they would later become. Even as the Geological Survey (USGS) focused its work on the American West, the professional horizons of American geologists were expanding due to university employment and other research opportunities. The lure of the universities and foreign study is illustrated by the case of the USGS's Bailey Willis. He toured China for a Carnegie Institution of Washington research program in 1903, reported on the coal supplies of that country, lectured on the effects of forest depletion, and, on leave from the USGS, advised the Argentinean government from 1911 to 1913 on national parks, irrigation, and conservation. Finally, he moved on to a long career as a geology professor at Stanford University.[25] It was Willis who supplied the original photographs that Roosevelt used to depict Chinese environmental decay, and his work made him a minor celebrity in Washington, in demand not for his geologist's knowledge of rock formations but for his impressions of Chinese forest destruction.[26] Willis's mentor was the older economic geologist Raphael Pumpelly, whose expedition to Central Asia (also funded by the Carnegie Institution) investigated the physical sources and conditions of past empires. Geography supplemented this work through the contribution of Ellsworth Huntington, who wrote *The Pulse of Asia* (1907) and compiled part of Pumpelly's expedition report (1905). Huntington was more than a geographer, however.[27] He was a theorist of racial decay and, in 1917, became president of the fledgling Ecological Society of America.

Given their wide experience and cross-career activity, it is not surprising that these professionals became concerned with the global and regional dimensions of resource waste. They saw the early stages of resource utilization in the colonial world, compared it with US practices, and noted key resources being used up as quickly abroad as at home. They began to consider the planet's life-support systems because of their awareness of global interdependence. The "foundational resources" for "the life of the human race upon this planet" had to be looked at because these were the "elemental necessities of life," stated *Engineering News*.[28] As engineers transforming nature, a conspicuous self-consciousness was present, along with hints of professional guilt. Engineers who had taken such a strong role in "Harnessing

2.1. A Geological Society of America excursion to Harper's Ferry, West Virginia, April 1897, included W J McGee, Bailey W. Willis, Frederick Newell, Charles Van Hise, Joseph A. Holmes, USGS mentor John Wesley Powell, and others. Courtesy USGS.

Nature's great power for the service of Man" through "steam and electricity" in place of muscle had now to contemplate the "profligate waste" that the inventiveness of their machinery allowed. As early as 1901, *Engineering News* urged a profession that had "in the past utilized Nature's gifts for man's service, and whose pride it is to build for the future as well as the present," to "foresee [the] coming scarcity of these gifts." The world would look to engineers "for a warning of scarcity and for guidance as to how future wants may be supplied." Engineers were now expected to serve these "larger interests of humanity."[29]

Yet despite the prominence of "professionals," the rise of conservation was not the product of a fully *professionalized* stratum. Professionalization was far from complete, and experts were often self-trained or self-announced. These engineers and geologists were travelers, polymaths, and social reformers as much as technical experts, and the visions they had for conservation were deeply influenced by their wider experience. Hoover and Hammond were prominent among the many engineers whose personal efforts coincided with a period of American economic expansion in foreign trade, assisting that informal empire to rise. They were businessmen more than technicians. Others were geared toward not only academia but also social and political reform. A common characteristic of these "professionals" was the alacrity with which they moved between disciplines and jobs, thus violating narrow specialization. The coal-testing expert Joseph A. Holmes of the USGS went to Europe twice to study fuel efficiency, especially the use of brown coal (lignite), but he also climbed mountains in the French Alps to ascertain the effects of forest destruction. Back home, he used his brief inspections of French forests and Alpine terrain to moonlight as an "expert" commentator on the likely impact of deforestation in the southern Appalachians.[30] Irrigation engineer Thomas Herbert Means is of interest because he was one of a number of USDA scientists who toured North Africa studying soils and fruits, but he also took the opportunity to report on dams. After nine years in what became the Bureau of Soils "in charge of soil surveys," Means moved to the Bureau of Reclamation and studied "the silt-carrying capacities of western streams." Finally, in 1910, Means entered "private consulting practice . . . specializing in engineering connected with agriculture, irrigation, drainage, reclamation and water supply."[31] His was a highly varied career in interests, expertise, and travel experience.

None better fitted the depiction of the partially professionalized than Gifford Pinchot.[32] Pinchot spent just a year in Europe training at the French forestry school in Nancy in 1889–90, during which time he visited Germany to study the scientific methods used there.[33] Returning to the United States, he worked at the Vanderbilt estate, Biltmore, in western North Carolina, then for the federal government. His meteoric rise to the apex of forestry administration was due to the considerable wealth inherited from his family, his connections as a member of the social elite, and his assertiveness and driving ambition. As the nation's chief forester, Pinchot was to spend twelve years in the federal government working on forest policy, with most of those years as Roosevelt's "de facto adviser."[34]

Expertise was not narrowly defined, nor did it exclude social and political considerations. In truth, Pinchot cared little for experts, except those of the type provided by the Yale Forestry School, which his father endowed. The forestry chief himself substituted personal networks for the political patronage of old. He told Roosevelt, "[I] will not let the civil service examinations stand in the way of our getting good men if I can prevent it."[35] Rather than a colorless bureaucrat, Pinchot was a skillful manipulator of Washington's political power structure. He mobilized support from the USGS and the Cosmos Club, a male-only network of Washington officials that developed a personal esprit de corps. Holmes and Willis were both members. Historian Wallace Stegner later called the club "the closest thing to a social headquarters for Washington's intellectual elite," but Pinchot cultivated even wider

2.2. Gifford Pinchot at work for the US Forest Service. LCPP.

networks. He followed up formal consultations with many a "jolly meeting" or "splendid dinner" at his charming Washington home and exploited his Episcopal Church connections, his support for overseas missionary endeavors, contacts with the General Federation of Women's Clubs, and his mother's role in the Daughters of the American Revolution.[36]

The state structure that Roosevelt and Pinchot inherited at the turn of the twentieth century reinforced these tendencies to rely on informal networking. Federalism and the congressional system meant that old-fashioned pork-barreling could not be eliminated. More to the point, the demands of the Republican Party mandated compromise. Pinchot and Roosevelt had to work with the realities of practical politics, however much they despised the horse-trading process. This meant, not a nonexistent state, but a peculiarly structured one. As historian Bruce Schulman has noted, the development of the American state "reflected the partially successful, personal projects of a few men concentrated in a few agencies." The norms and standard procedures of administration "emerged not as a check on personal power but as an adjunct to it." It was from "the prestige and the personal networks of their top administrators" that the "reputations and influence of government agencies, and much of their success or failure, derived."[37]

In yet another way was expertise heavily qualified. To transmit the ideas that flowed along the transnational networks of engineers, geologists, archaeologists, ornithologists, foresters, and others, it was necessary to reach a broader audience and to translate those ideas into local and national opinion. Some conservation sentiment naturally crossed the boundaries between expert and amateur. This was the case in ornithology, as already noted.[38] But these networks of communication did not work effortlessly. They were consciously directed and had particular circuits and points of concentrated power. Understanding and shaping these patterns were crucial to Roosevelt and Pinchot. They wished to use international comparisons to make Americans feel disturbed at the nation's dismal record in conservation and at the same time to give beacons of hope through schemes shaped in part from abroad but always adapted to American experience. The work of engineers, geologists, water resource people, foresters, and soil scientists was thereby drawn into a public conversation about the future of America, through the use of media. Fortunately for Roosevelt, he had important allies in that field, as the crusading investigative genre of modern American journalism was on the march. As historian Richard Hofstadter wrote, it was "hardly an exaggeration" to say that the "Progressive mind was characteristically a journalistic mind, and that its characteristic contribution was that of the socially responsible reformer."[39]

Declining publishing costs and an expanding, more literate electorate created enormous opportunities for print news. Journalists took advantage of modern technology to portray topics of moral or political outrage to the masses. The "muckraking" investigative journalism that exposed corruption and special privilege in sensationalized fashion is almost synonymous with the Progressive Era and was well adapted to exploit the jeremiad over lost natural resources. It was in the muckraking *McClure's Magazine* that Rudolf Cronau depicted the despoilment of the continent.[40] Though Roosevelt distanced himself from the extreme yellow press journalism by excoriating "the man with the muckrake" and giving the name Muckraker to the cheaper forms of exposé journalism he disliked,[41] Pinchot cultivated the reformist press on behalf of Roosevelt and conservation. Newspaper magnates Frank Munsey and Edward Scripps ran chains of mass circulation papers ideal for spreading alarm over resource use and urging conservation. The wealthy Scripps joined the eucalyptus-planting craze in California in 1908 and supplied authentic conservation support, boasting "the ear" of over a million readers and a budding international news service to advance the national and global cause.[42] His United Press International sent representatives all around the world "to report on news events instead of relying on reporters from other papers for information."[43] When Roosevelt went to Africa in 1909, he abjured journalists officially but took a scribe from United Press International part of the way and provided him and an Associated Press counterpart with "exclusive" material. These mass circulation papers and their press agencies could operate in tandem with the organs of the highly mobilized middle-class Protestant community, such as the *Chautauquan*, which provided an important journalistic outlet. This nineteenth-century self-improvement version of modern adult education was exceedingly popular, with Chautauqua assemblies expanding across rural America from the 1880s. Roosevelt was quoted as saying that the Chautauqua was "the most American thing in America," and it was proconservation. American it was, but it, too, gave space to conservation stories from Europe.[44]

Temperamentally, Roosevelt favored elite media that provided vigorous and informed allies. Lavishly illustrated magazines such as Walter Hines Page's *World's Work* joined the conservation bandwagon along with the Lyman Abbott–edited *Outlook*. The latter publication would become particularly important for the president. W. E. B. Du Bois called it "the leading weekly of the land."[45] Abbott and his son Lawrence enthusiastically endorsed Roosevelt's attack on waste and believed the conservation struggle to be one of the greatest moral endeavors of the era. Lawrence became Roosevelt's postpresidency secretary on his trip to Europe and wrote in praise of the war on waste.[46] A conservationist and friend of Roosevelt, William Bailey Howland, was the well-connected publisher of the *Outlook* and became associated with a suite of magazines,[47] including the *Independent*, a venerable journal of opinion, and the *Outing*, a periodical that publicized the use of the great outdoors for sports, leisure, and wildlife appreciation from 1880 to World War I. Howland was well placed to influence politicians and the public, serving as a vector for the transmission of moderately progressive Republican thinking on conservation. He was also treasurer of two conservation groups, the ASHPS and the ACA, and served as a president of New York State's Niagara Park Commission.[48]

Good-quality black-and-white photography allowed conservation issues to be vividly portrayed as never before. *World's Work* showed the potential with its cosmopolitan coverage and its ubiquitous and often-astonishing photographs of people and places. Under Doubleday Page's management, *Country Life in America*, a sister journal to *World's Work*, even began to introduce "the marvel of color photography." This allowed presentation of pictures "direct from nature, of live water-fowl, of gaily-colored brook trout in the water, of Rocky Mountain peaks with their purpling distances, of flowers, and what not."[49] *World's Work* editor Page, who served as Woodrow Wilson's first ambassador to Britain, championed Roosevelt's "Greater America" that stretched via the Panama Canal from the new Pacific possessions to the Caribbean Islands and encompassed the continental United States in between. Supporting "the reclamation of the deserts" and "the saving of the trees," Page stood against repeating in America the horrors of China's soil-denuded mountains and championed the "calling of an international conference on conservation."[50] Through *Country Life in America*, Page also supported Roosevelt's crusade to help farmers. Contributors there and in the *Ladies' Home Journal* focused on gardens, landscape design, and even "back to the land" ideas.[51] In all these media sources, elite conservationists had allies who could draw their ideas into a wider public conversation, and Roosevelt exploited these avenues fully. In 1909 Roosevelt became a contributing editor of the *Outlook*, thus making the nexus explicit.

As president, Roosevelt seized upon the importance of influencing the press. He played favorites, giving confidential briefings to preferred journalists, whose output was carefully monitored and shaped with the implied or even explicit threat of withdrawal of the privilege if abused.[52] In his use of newsprint he was skillfully aided by Pinchot, "a master and promoter of political publicity," second, if to anyone, to TR himself.[53] As a result of such cultivation, complained anticonservationists bitterly, the press in the nation's capital had mostly capitulated to the administration's emotional pitch.[54] Certainly, the zealous Pinchot forged a link between professional forestry and the wider public discussion of conservation, though Roosevelt solemnly saw Pinchot's role as merely disseminating "facts about forestry."[55] The US Forest Service itself became an important element in the machinery of communication, providing a constant flow of material on the dangers of deforestation and wastage of other resources. The chief forester established a special forestry press section concerned with "publication and education," and it distributed government reports and press releases through numerous Forest Service agents to newspapers across the country.[56]

In yet another way, Pinchot and Roosevelt showed a flair for publicity and skill in mobilizing opinion. Key investigative commissions that the president established were primarily used to spread the message of conservation. The Commission on Country Life, the Inland Waterways Commission, and others discussed in the chapters that follow touched broader opinion—by deliberate design. The NCC circulated hundreds of thousands of questionnaires inviting citizens, often among key interest groups like farmers and lumbermen, to engage with conservation policy and express their thoughts on the issues of efficiency and waste. Through these commissions, Roosevelt and Pinchot sought to create a highly structured public, identify allies (including businessmen, newspapermen, and journalists), and cement critical relationships across the nation. Conservation could be led by an elite, but it still required broad public participation. This complex machinery was deployed to make Americans aware of the destruction of the natural world, not least in forestry.[57]

Whether in a deluded state of alarm or not, foresters under Pinchot's strategic direction campaigned to end the shortage of resources, and they did so by keeping an eye on both European and global experience. As is widely known, the "transnational emergence of forestry as a profession" was centered on German leadership.[58] American foresters fitted this pattern, linked as they were intellectually and professionally to German-trained foresters. But it was not so much the international influence on technical training that was important as the inspiration provided by German forestry. The Prussian-born and -trained Bernhard Fernow was the third chief of the Division of Forestry, 1886–98, immediately preceding Pinchot in that job. As professor at the College of Forestry at Cornell University (1898–1903), and later at the University of Toronto, Fernow remained an invaluable source of international information for Pinchot, even though the German was temperamentally very different. Fernow also edited *Forest Quarterly* (1903–16).[59] German methods of scientific forestry (silviculture) were considered the benchmark for world practice, but the German connection was more complicated than a simple one-way influence. US forestry linkages were global rather than purely transatlantic, and South Asian contacts were particularly important. American knowledge of forestry developments in British India traveled initially through the channel of the German head of the Indian Forestry Service Dietrich Brandis and of the British-based German forester Wilhelm Schlich. The latter also served in India as inspector general of forests before settling in England in 1885 as professor of forestry at the Royal Indian Engineering College.[60] Both Brandis and Schlich mentored Pinchot during his training in Europe and afterward. Pinchot took particular interest in India because of its record in tropical forestry, a field that soon became highly relevant to the United States with the acquisition of tropical colonies. He also believed Indian experience to be useful because it involved improvising on German ideas about scientific forestry within forests vaster and more varied than the neat and uniform systems of Europe. In the light of growing discussions in other countries over conservation, Pinchot hoped to learn how Americans could adapt, rather than adopt, European forestry methods.[61]

Supplementing these transnational educational linkages were international congresses that paralleled those in other professional fields. Although the first such forestry meeting convened in Vienna in 1890,[62] far more important was the one held in Paris at the Exposition universelle, ten years later. A complicated triangular agitation shaped the agenda for the 1900 conference. International exhibitions were opportunities not only to show off the wares of nations but also to gain scientific and technical knowledge in many fields.[63] Pinchot and his allies knew that the United States was participating in the exhibition and called for a comprehensive international forestry meeting to accompany it. Leading lights among the US foresters wanted a census of the world's forests to be undertaken, and they saw the 1900 congress as a way to raise consciousness of the problem of timber shortage and "to bring about the compilation of forest statistics of all the countries in the world, on a uniform basis." German foresters were equally interested because they worked in a newly emerging colonial empire that had annexed part of East Africa in 1885, and they sought better understanding of the supply of tropical forest products. Headed by Secretary of Agriculture James Wilson, the American Forestry Association wished to stop the destruction of US forests with the help of international expert opinion. At the instigation of a German Embassy attaché, the foresters and the American Association for the Advancement of Science successfully petitioned the French government for a comprehensive, broad-ranging forestry meeting.[64]

This global interest represented a change in the discourse over forest destruction. Before 1900, US concern remained episodic and internal, focused on the struggle over the disposal of the vast public domain.[65] So what was different now? Above all, international conditions had shifted. The expansion of European imperialism produced greater desire for an international stocktaking of resources in a world where the European countries competed for military and commercial dominance. The context of colonialism was registered in Wilhelm Schlich's report read in March 1897 at the Imperial Institute. His "The Timber Supply of the British Empire" was translated into French and published in the *Revue des eaux et forêts* and in the *Belgian Forestry Journal*. This diffusion of concern came as part of "rousing the public to the importance and urgency of the cause" of forest conservation.[66] The changing milieu was even more evident at the Paris meeting. There, the French inspector of water and forests Alphonse Mélard's dire warnings for the future of European and world forestry were discussed. To the audience's consternation, he predicted that the European nations had only fifty years' supply left, and he argued for draconian efforts to replant "waste" lands. Though controversial, his views were applauded and appropriated by many.[67] Mélard's work influenced Americans directly through his participation in the 1900 conference and indirectly via the forestry networks of the German and British Empires. The Frenchman's report appeared in the *Indian Forester* and in *Transactions of the Royal Scottish Arboricultural Society*, a move that accelerated the flow of information across the Anglo-Saxon world.[68] Mélard's relevance for the United States was manifest in the 1909 report of the NCC, in which a US Forest Service official cited the Frenchman not only directly but also through the work of Schlich.[69]

In supporting calls for an international stocktaking, US foresters and public opinion makers reflected the growing alarm across European empires about timber shortages. But Americans were also newly aware of their own nation's potential role in resource management in distant climes. This had not been an issue before 1898, but thereafter, the acquisition of the Philippines and Puerto Rico as well as Hawaii had considerable significance for forest protection. In 1899 the *Forester* (the journal of the American Forestry Association) carried several news reports on the trees of the Philippines, and the *Washington Times* wrote of the need to preserve and develop the forests of the new colony in order to make the United States stand out among the empires and, equally important, to encourage conservation at home: "When we learn to preserve our inherited wealth as well as to acquire new riches, we shall be the greatest people on the face of the earth."[70] This was just a year after Gifford Pinchot became the nation's forestry chief. One of Pinchot's urgent tasks was to rise to the challenge presented by this American encounter with the tropics and its potentially abundant natural resources, a new and momentous stage in the development of national and global power. To that discovery of the tropics we must turn.

PART TWO

The New Empire and the Rise of Conservation

THREE

Colonies, Natural Resources, and Geopolitical Thought in the New Empire

Empire is not simply a fact but an idea, an idea that needs nurturing. People live by such ideas. Despite the arguably empire-like wresting of the American West from Indian tribes and the Mexican Republic in the nineteenth century, empire was a notion that did not sit comfortably with many Americans, even when the facts indicated otherwise. "Empire" as an idea needed all the help it could get. As William McKinley's assistant secretary of the navy in 1898, Theodore Roosevelt was in the forefront of those arguing for acceptance of American "expansion" and the white man's burden abroad. But in advancing the cause of formal imperialism, Americans had to overcome hardy intellectual hurdles. Those places that the United States had annexed were in the tropical zone, and the countries to which the nation's leaders looked for resources outside the United States were largely, though not exclusively, tropical or subtropical too. Some key countries, especially China, could be exploited with the advantage of access via the tropical staging posts acquired across the Pacific in Hawaii, Guam, and Manila. One obstacle to the expansion of the kind that Roosevelt and like-minded government officials, statesmen, and intellectuals advocated was a tradition of anti-imperialism, derived from the revolt against the mother country in the War of Independence. Another was environmental. Many anti-imperialists and others feared engaging with the tropics because of the risk that the entire zone posed to health, morals, or politics. Surely the white race would degenerate if exposed either to these nonwhite peoples or to the fetid environments in which they multiplied. But these same anti-imperialists rarely bothered to mention the natural resources that beckoned on those shores.

Such was not the case with pro-imperial strategists of geopolitics, from the well-known Brooks Adams through to the little-known Frank Vrooman.

These writers did not provide a coherent or consistent ideological motivation for empire, to be sure. Rather, they observed trends and showed a capacity for shrewd conjecture and advocacy about outcomes. In the process they contributed a toolbox of ideas that could embolden the ambitious and yet call for self-reflection about the new trajectory of power. As one modern authority puts it, ideas do not supply "sufficient conditions" to explain political decisions, but they can help frame policies and "legitimize policy proposals."[1] Articles by strategists of empire appeared in major periodicals, and their books were reviewed in prominent newspapers and journals of opinion. Politicians, diplomats, and conservation officials echoed their themes in public pronouncements on global resource strategy, as in the Conference of Governors in 1908 and the 1909 report of the NCC.[2] Adams, Vrooman, and like-minded thinkers contributed not only to a pro-imperial moment but also to a moment in which conservation gained momentum. They achieved their aims by revealing the possibilities for global and racial dominance offered by a mostly tropical empire of resources.

The acquisition of colonies in 1898 stimulated not only Americans to think about these issues but also Europeans, who gave the new regime gratuitous advice. *Atlantic Monthly* devoted most of its December 1898 issue to the new colonies, with articles by British civil servant Benjamin Kidd and the journalist W. Alleyne Ireland included.[3] The University of Chicago employed Ireland as a "commissioner" to study comparative colonialism,[4] while the views of the prominent Social Darwinist Kidd on the "control of the tropics" appeared at an opportune moment in the wake of the Spanish-American War. Partly because Kidd visited the United States at that time, he influenced the new American imperialists, but not in quite the way that he hoped. Kidd argued that Europeans could not permanently live in the tropics. Instead of elevating the lesser races in hot lands where the white man "made his unnatural home," Kidd believed, the colonial ruler tended "himself to sink slowly to the level around him."[5] Americans, as we shall see, developed a more positive view.

Kidd's American contacts extended to the corridors of power, where he emphasized the need to use the tropics for their natural resources, not their markets or land for settlement. In Washington he met senators and swapped views with Secretary of State John Hay. This tête-à-tête showed that a strategist like Hay could see Kidd's relevance for the development of the Open Door trading policy. While historians place emphasis on the need for markets for US products as a driving force for empire, Hay was more concerned in his discussions with Kidd about the need to develop complementary, not competitive, trading and therefore access to foreign resources as well as markets. Hay told Kidd that the European and American powers were increasingly engaged in pointless competition because they produced similar things as the Industrial Revolution progressed. Better, Hay thought, for a trade to develop in which the resources of the underdeveloped world were exchanged for the manufacturing products of the developed. This argument focused on the need to eliminate the "waste" and duplication in global commerce, and it was consistent with Roosevelt's developing ideas on efficiency.[6] Kidd returned to Britain to emphasize this point in future writings, but the same idea was found in the popular press. In a statement that Kidd could have written, millionaire Frank Munsey's *Washington Times* proclaimed: "There will come an epoch when the tropics with their vast productiveness must be developed," making it possible for "one people, one continent, one race" to benefit from "the surplus of another."[7]

Kidd's evolutionary determinism was not fully accepted in the United States. Roosevelt, for one, objected to Kidd's *Social Evolution* (1895) on this score,[8] but the Englishman's views on trade rather than biology remained important to a conversation over the future of the United States. He underlined for the American elite the importance of tropical commerce to the coming world order. Europeans had begun a furious contest for dominance in the tropics, where many believed the crucial struggle for control of the world's resources would be played out. With the "filling up of the temperate regions and the continued development of industrialism throughout the civilized world," control of the tropics would become the "permanent underlying fact in the foreign relations of the Western nations in the twentieth century."[9] Not only were Kidd's views respectfully received, but they also achieved an impact indirectly, through another figure, an American whose ideas were closer to Roosevelt's thinking.[10]

On the question of global resources, Kidd influenced a clergyman who spent four years in teaching and missionary work in Japan beginning in 1870. William Elliot Griffis later became a prolific author and advocate of deepening US engagement in the Pacific and an interpreter of Kidd to the American public.[11] His *America in the East* (1899) and kindred articles in the *Outlook* contested Kidd's dismal thesis: "We are very far from accepting his notion . . . that the white man cannot live in the tropics."[12] Kidd's "dogmatism" seemed to Griffis to "rest upon tradition rather than upon thorough knowledge of modern conditions and possibilities."[13] After all, the American upper and middle classes had in his judgment become more cosmopolitan, more willing to travel and experience new cultures. Among these people, the experience forged "an interest and an intelligence concerning foreign countries and races which was unknown two generations ago."[14]

Griffis argued for a more favorable outcome of white "settlement" under the pressures of globalization, especially through improved knowledge of foreign countries and better communication systems. In an epoch of "steam and electricity" the tropical colonies were "but a few days distant." The call to "new duties" had arrived in a timely fashion "when the potencies of science, the harnessed forces of nature, and the printing-press [had] reached a development undreamed of a century ago."[15] Griffis looked to the example of his missionary friends to verify that Europeans could indeed adapt to the torrid regions. The force of their moral conduct gave them an advantage in the struggle for tropical survival. That is, human values rather than the natural environment would influence the outcome of white expansionism, whether European or American. This striving for moral success was a view congenial to Roosevelt.[16]

Far from expressing the polar opposite of Kidd's ideas, Griffis differed from him chiefly on this one question of white people's permanent occupation of "hot" countries. Whereas Kidd believed that Europeans must rely on nonwhites as intermediaries, and control the tropics indirectly, Griffis contended that new technologies allowed direct supervision. On most other matters he agreed with the Briton, especially that the underdeveloped world should not be worked for the gain of individual powers but "governed as a trust for civilization" and supervised by people who demonstrated the "highest" standards of that civilization. Griffis's aim was to legitimize the US role in the Philippines and to advocate a long-term physical occupation by developing a rationale that portrayed Euro-American colonialism as benefiting all through access to raw materials.[17]

Noting the European rush to the Asia-Pacific region in the 1880s and 1890s, Griffis highlighted the changing geography of world power: "The centre of the world's hopes and ambitions has shifted. . . . The Russian is marching seaward, building his railways as he goes, settling the great plains and valleys of southern Siberia, and commanding northern China." This imperial expansion involved a transformative harnessing of nature. "Where thirty years ago forests stood and tigers were shot, stands Vladivostok, a city of fifty thousand people."[18] Griffis drew particular attention to the fact that China was awakening from its Confucian "semi-slumber," and he predicted the entry into world trade of the remarkable natural resources of the Qing Empire. There, "the industrial revolution" had "already begun" and Chinese "exports and imports" were increasing: "She has coal, iron, petroleum, natural gas, sugar-cane, tobacco, indigo, cotton, and all sorts of food supplies. The reign of Confucius will not last forever." Other missionary observers concurred. "Vast and varied are the mineral treasures buried in these mountain masses awaiting the dawn of an enlightened policy," wrote one. Natural resources properly developed would make China "one of the richest nations on earth."[19]

Like Kidd, Griffis emphasized the demand for distant resources, not just the potential for US exports. Both thinkers noted US dependence upon a "surprising" amount and variety of tropical products—surprising, that is, "even to one moderately familiar with the general subject of our dependence upon the earth's middle zone for comforts and necessities." Nearly one-third of "our imports are from tropical regions," Griffis observed. The "modern man" with his "more complex life" was "even more dependent on the products of the tropics" than in earlier epochs, when mostly luxury goods of the East had been coveted by the European rich.[20]

Griffis emphasized that moral values, not physical coercion, must exert influence over modern civilization, and those were to be Christian values in pursuit of righteous empire.[21] This viewpoint, too, was congenial to Roosevelt. A "modern" man for a modern world would, influenced by a Christian civilization and its accouterments, obey the biblical injunction to subdue and replenish the earth. Influenced by Griffis's missionary background, this outlook nevertheless promoted the expansionist nature of Euro-American trade in underdeveloped countries. In the tropics there could be found the "richest vegetable products" opened to human ingenuity, including natural rubber (*Caoutchouc naturel*), exploited already through the technological discoveries of Charles Goodyear, and quinine, which missionaries had introduced to the wider world as a medicine.[22]

It followed that the United States must intervene to protect such valuable sources of supply from other nations scheming to take the tropics for themselves. It followed, too, that the United States must develop a globally oriented policy response. Griffis was a "well-known writer" and, along with TR, was a member of the National Institute of Arts and Letters. His arguments appeared in leading periodicals such as *Harper's Monthly*, but his work reflected not original thinking so much as common ideas in public debate.[23] The *Outlook*, the magazine most closely aligned with Roosevelt, ran seven of Griffis's essays urging the United States to make the most of its opportunities in the "Far East." This is not to say that Griffis directly influenced Roosevelt. Rather, as historian Michael Adas has remarked, Roosevelt "shared" Griffis's view in support of this Anglo-Saxon project to vitalize the tropics and the Asiatic "races." The president also paralleled Griffis's favorable opinion of the Japanese capacity to participate in the modernization of East Asia.[24] That stated, another theorist of empire, Brooks Adams, was of greater significance. He looked more strategically toward the geopolitical

3.1. Brooks Adams, c. 1910. National Park Service.

balance across the entire globe and longer trends in history. He also knew Roosevelt well.

The insightful and controversial Adams was the lesser-known brother of the prominent literary figure Henry Adams. None of Brooks's output attained the literary fame of his brother's autobiographical masterpiece, *The Education of Henry Adams*. Indeed, some thought of Brooks in the mid-1890s as a man possessed of a mind "a little unhinged." So wrote Theodore Roosevelt himself. The future president found it hard to accept Adams's dominant mood of pessimism concerning the future of civilization.[25] Yet Adams's ideas were, within a few years, to become an undeniably important part of the intellectual resources that presidents, congressmen, governors, and reformers mined to recast their ideas on American power.[26] Roosevelt corresponded and dined with Adams during the White House years. The two saw eye to eye on the relationship between domestic reform of the economy and the nation's interests in Asia. Internal and external policies were "parts of a whole and cannot be considered separately," Adams told the president.[27]

The grandson of John Quincy Adams, Brooks Adams taught for a time at his alma mater, Harvard, studied law, and watched with a mixture of nervousness and anticipation the rise of industrial society. In the 1890s, he examined "trade-routes and their influence upon the history of peoples and nations," laying down the principle "that civilization follows exchanges, or commercial growth and decay."[28] Adams first crystallized these ideas in 1893 amid the global impact of the 1890s depression. He published *The Law of Civilization and Decay* (1895), a widely noticed work. His theory was complicated in a way that made the more practical Roosevelt suspicious, however.[29] "Brooks Adams' theories are beautiful," Roosevelt wrote to the Englishman Cecil Spring-Rice, "but in practice they mean a simple dishonesty."[30]

Adams's thesis represented the mind-set of an older social elite alarmed at the rise of American business personified in the enormously powerful tycoons of the robber baron generation. Like his more famous brother, Brooks had become fascinated with the transcendent meaning of mechanized industrial society, its transnational expansion, and its relationship with the natural world. While Henry tried to develop a theory on the application to human societies of the physical laws of thermodynamics, which led him to frame a cosmology of American and world development, Brooks stuck to the slightly more pedestrian terrain of the rise and fall of human civilizations, and so was of more immediate policy relevance than his brother. Brooks had become convinced that "human societies differed among themselves in proportion as they were endowed by nature with energy."[31] Obsessed with the possible decay of civilization allied to the dissipation of energy as predicted in physical laws, he argued that human activity had concentrated energy. But whereas "barbaric" societies had utilized elemental energy through violence and war, the modern commercial world was in danger of frittering its own accumulated energy away.[32] Civilized society had developed a more specialized state in which commercial interests grew and elemental forces waned. Such societies were increasingly unable to tap their energetic potential and liable to fall victim to the more vigorous and unified among them.

To be sure, *The Law of Civilization and Decay* sometimes seems confusing, allusive, and possessed of a mysterious air of serious significance, but it had a "resounding impact" despite the misgivings that Roosevelt and others expressed. With the first edition selling out in three months, Brooks turned this intellectual capital into a journalistic career in the late 1890s.[33] He wrote extensively for the leading American periodicals read by the intellectual and social elite. Soon he hobnobbed with Senator Henry Cabot Lodge, a close ally of Roosevelt in the arguments over acquisition of empire. His later books, *America's Economic Supremacy* (1900) and *The New Empire* (1902), provided a more empirical application of his ideas.[34] Therein he predicted a triumphant US push toward global hegemony based on the control of rich resources. These books spoke directly to the new condition of the nation's power. Though far from universally accepted, they were widely reviewed and praised in magazines sympathetic to the Roosevelt administration, such as the *Outlook*, which proclaimed "expansion" part of the "world-tendency of the epoch."[35] Even the less sympathetic, like the *Chicago Daily Tribune*, called Adams's vision of American supremacy "inspiring and convincing."[36]

Adams wrought his influence precisely because his work clarified the new world order of late Victorian imperialism as a geopolitical chess game. Writing in an 1898 article in *Forum*, he saw the implications of the war with Spain as transforming American power and strategic needs. These implications were the coming struggle for supremacy with European nations in China, especially over resources, and the role of the Philippines as a base for consolidating American power in the East. "The Philippine Islands, rich, coal-bearing, and with fine harbours, seem a predestined base for the United States in a conflict which probably is as inevitable as that with Spain. It is in vain that men talk of keeping free from entanglements. Nature is omnipotent; and nations must float with the tide. Whither the exchanges flow, they must follow."[37]

Not a conservationist in the conventional sense, Adams regarded the relationship between civilization and nature as mediated through the control of resources. In Darwinian terms, *America's Economic Supremacy* explained global history as a struggle for survival of the fittest. Europe's spheres of influence in China, the greatest untapped reserve of minerals, suggested to him that the American nation must muscle up to compete. It must become a more efficient society while at the same time trying to prevent malevolent European powers from gaining more political and economic leverage over East Asia. For this struggle the Anglo-Saxon nations were not fully fit because they were relatively open societies, with small governments and inefficient administration: "Everywhere society tends to become organized in greater and denser masses, the more vigorous and economical mass destroying the less active and the more wasteful."[38] The elimination of waste was associated with the great corporate trusts and especially with German centralization of industry and government. "Under this pressure the people consolidated in a singularly compact mass, developing a corporate administration powerful enough to succeed very generally in subordinating individual to general interests. It is to this quality that Prussia has owed her comparative gain on England." Applying the same measure to the United States produced another sharp contrast. "The national characteristic" of Americans was "waste," and "each year, as the margin of profit narrow[ed]," waste grew "more dangerous."[39]

Though Adams claimed to elucidate the natural law of societies, on a practical level his concern over resources and the dissipation of energy had obvious resonance for the conservation movement and its leadership. His call for a centralization of power that could adequately command the raw materials of the nation and the efficiency arguments that such a nationalist program contained were the same as those underlying the Roosevelt plan for conservation, giving the community "effective command of its resources."[40] After Roosevelt left the presidency, Adams lamented that "Pinchot and conservation" were among the president's favorite people and causes "on their way out." To Adams it seemed as though "an era of darkness were at hand" in those post-Rooseveltian years.[41] Later, Adams wrote that conservation "was contrary to the instinct of greed which dominated the democratic mind and impelled it to insist on the pillage of the public by the private man."[42] There can be little doubt where his sympathies resided on this point during the Roosevelt presidency; they were with Roosevelt's plans to stop that "pillage" and husband the fruits of natural advantage for the struggle between nations. This was both a rationalization of and a motivation for the Roosevelt position.

Brooks Adams was not overtly influential in motivating American imperial expansion. In fact, his major books on the American empire came after that expansion was mostly accomplished. But he was significant in providing an ex post facto rationale for American action—and a convincing case for internal reform through conservation to strengthen the nation's international competitiveness. Adams's work served implicitly as a warning to Americans that they must not dissipate the largesse that they already controlled. He believed that while a fresh and consolidated empire was being signaled and augmented through colonial gains, the true strength of the United States rested not on further external political expansion but on domestic reform to rationalize and consolidate the nation's bounty. "The peculiarity of the present movement" was "its rapidity and intensity, . . . due to the amount of energy developed in the United States, in proportion to the energy developed elsewhere." The shock of the new internal power seemed to him "overwhelming."[43]

Primed by the works of Adams to consider the centrality of resource exploitation, policy makers began to worry that those very resources that had brought the United States to the edge of global supremacy on the economic stage were in danger of slipping away at the climactic moment. For the future, Adams indirectly raised alarm over the disintegration of the Qing Empire in the years after 1895 and European, particularly Russian, meddling in that empire. He praised the efforts of John Hay through the Open Door policy to secure equal access to raw materials and trade. Hay's diplomacy, Adams reasoned, was calculated to advance US interests in this new geopolitical contest. Such ideas did not actually cause political change but became part of the swirl of strategic thinking that shaped and reflected the Roosevelt administration's attitudes and policies.[44]

The significance of Adams for conservation debates was, therefore, not the drive to find markets for surplus manufactures, as commonly assumed, since he merely repeated the conventional wisdom on that score, but a growing neomercantilist acceptance of the nexus between physical resources and the nation's strength within the international state system. Progressive intellectuals, bureaucrats, and politicians were vulnerable to alarm over threats to that strength from perceived insufficiency of raw materials revealed by the obvious pressure upon the forests and the huge increase of demand for iron and coal as manufacturing and mechanized transport multiplied needs.[45]

Adams's anxiety over China's future, the wider question of East Asia's development, and the movement of civilization westward was echoed in the thought of conservationists. We see these calculations repeated by an obscure ally of Pinchot and Roosevelt. Bailey Willis's contribution to the debate over resources went beyond his arresting presentation of the photographic plates of Chinese forest policy mistakes discussed in chapter 1. His report on water resources for the NCC was innocuously titled "Water Circulation and Its Control." What readers got, however, was no technical tedium but an unacknowledged gloss upon the fin-de-siècle fears that Adams stirred. Civilization had begun in the East and moved West, Willis asserted. Was it to return to the "Orient," as Adams suggested it might? "The path of our race" led "from Asia across Europe" to the United States, Willis announced. Would the center of stored energy migrate across the seas and make China or its potential masters, the Russians or Germans, the site and source of power in world civilization? If Americans did not conserve their supplies of natural resources, the nation faced a bleak future of civilizational collapse. In the context of Willis's own interests, that reasoning presented a stark choice of conservation or waste of water, the most basic of the nation's life supports and the underpinning of agricultural and industrial greatness. Americans "occupy the last land that lies toward the setting sun," Willis grimly warned. "Our children and our children's children for countless generations are to enjoy the garden or bewail the deserts we create."[46] Lest the universal applicability of the jeremiad be lost, his report was illustrated with fresh pictures of environmental decay, not in the Chinese empire but the Ottoman.

Other versions of this dire prediction surfaced in conservation discussions. Some concerned Japan. Whereas Adams saw Qing China as the key to world history, the Russo-Japanese War of 1904–5 modified geopolitical predictions to incorporate the rise of Japan and its own exploitation of the disintegrating empire. Chase Osborn was a one-term Republican governor of Michigan and world-traveling mining prospector who saw Japan as the critical strategic threat.[47] When asked to speak at the 1908 Conference of Governors, Osborn honed in on the global resource question and argued that an almighty contest was under way in Asia. "The great menace of Japan in the near future" was "not one of war, but of commercial and manufacturing competition." US global ascendance depended on "natural advantages," and the nation had started with plenty of those. "No country in the world since America [had] gotten [its] good start" had been "able to compete" in so many directions. Americans had a far superior inheritance than the island nation possessed, but all this might change "in the next quarter of a century or less," he argued. "At the doorway of Japan, in Korea, Manchuria, and China," lay "as many untouched raw materials located conveniently with reference to transportation as we have in this country, possibly excepting forests." China had not developed its resources, Osborn derisively judged, because of its tradition-bound customs. "These great deposits of raw materials, consisting of coal and iron ore and other useful materials, have been conserved by what may turn out to be a superstition divinely applied." Confucian torpor had ruled, but no longer was this true. Japan was modernizing and would soon grasp this bounty of nature. "Japan will sweep Feng-Shui aside just as it has its Samurai and its methods of yesterday. It will attack the riches of all countries adjacent to it if possible, and will acquire materials that will enable it effectively to enter directly into manufacturing competition with us."[48]

At the same 1908 conference, Governor Walter Frear of Hawaii seized upon the vital role of the Pacific as the site of trade and industry that would mark the age. Shamelessly boosting the importance of the tropical island territory and its demands for harbor improvements, Frear stressed its location at "the commercial center or cross-roads of the Pacific." The "greatest of oceans, between the richest of continents," was "fast approaching the fulfillment of the long-ago prophecies of von Humboldt, Seward, and others, to the effect that it would eventually be the theater of the world's greatest commerce." Returning to Adams's fascination with the future of East Asia, Frear proclaimed that "one of the most effective methods of conserving the natural resources of the United States" involved taking advantage, through provision for "adequate transportation facilities," of the natural resources "of other countries and especially those of China." Like Adams, he believed these to be, "next to those of the United States, the richest in the world and as yet practically untouched."[49]

What ill would come if the United States fell behind in this race to corral the world's raw materials? Israel C. White, the West Virginia state geologist, warned that the nation might lose its "source of world-wide influence."[50] This fear was reflected in the Roosevelt-led conservation strategy. The case for positive state action to assert American power through conservation was most boldly articulated at the end of Roosevelt's presidency, summed up by a man obscure within his own country. Whereas Brooks Adams, writing at the beginning of Roosevelt's presidency, believed that the material forces of nature shaped historical development in dialectical struggle with human civilization, the Reverend Frank Buffington Vrooman took the world-historical significance of Roosevelt's record to be a more permanently optimistic one of domination. Vrooman called the president a man of destiny who had not simply responded to geographical necessity but who "improved" geography by becoming an active agent in it.

A member of the reform-oriented Vrooman family, Frank grew up on the plains of Kansas. Educated at Washburn College, he later worked for the *Topeka Daily Capital* and the *Emporia News* before going to Beloit College to study theology in 1885.[51] Soon he became one of those restless and peripatetic individuals of the late nineteenth century set loose by the rise of rapid steamship and rail travel. Moving frequently, he sampled many careers, hoping all the while that the encircling iron rails of the age of imperial globalization would lead to a better world for humankind. The depths of the 1890s depression shattered optimism about the pace and direction of economic change, and the Vrooman family registered the kaleidoscope of social and socialist reforms of the era. With his brothers, Frank became committed to applying Christianity to society's problems. As a clergyman in Baltimore, Worcester, and then Chicago, he developed a concept of "dynamic religion" for an "aggressive advancement of human welfare."[52] Resigning his position as a minister of religion, "frozen out" due to theological and political conflict over his radical stances, he headed in 1897 for the Klondike goldfields to make his fortune but soon turned to journalism and became a regional booster for western Canada.[53] In 1907 he participated in a "geographical exploration" of the Canadian Northwest, for which he was designated a fellow of the Royal Geographical Society of London.[54] Vrooman had an Oxford University connection through his brother, the socialist Walter Vrooman, who had aided Progressive historian Charles Beard in forming the workingmen's Ruskin College there in 1899. As a result of this cross-imperial networking, Frank was invited to Oxford to deliver a lecture on the economic geography of Canada's Northwest Territories. The invitation was timely, because it came just when Roosevelt was completing the architecture of his conservation program as a comprehensive plan. "After a decade of wandering Frank had settled squarely in the Progressive camp," historian Ross Evans Paulson writes. He favored a state "strong enough to meet modern problems," armed with the right of government ownership regarding conservation that "empowered it ultimately to control all human necessities." The state could be trusted with such power because high-minded persons (such as himself) would restrain it.[55] Vrooman praised the US Reclamation Service and the US Forest Service, two of Roosevelt's proudest achievements. Of the former, the Kansan stated that it was "another silver-plated screw in the lead coffin of *laissez-faire.*"[56]

Delivered to the Oxford University School of Geography on 8 March 1909, the topic of Vrooman's lecture differed from the original invitation. Instead, Vrooman took the opportunity to identify his Progressive creed, for "[h]e found the perfect embodiment of his theories in Theodore Roosevelt."[57] He cites Roosevelt as "the first American President" to combine the "practical statesman" and "ethical philosopher" with "scientific" geography. For Vrooman, the Gilded Age was a time of "arid mediocrity, when few public men were aware of an issue other than the tariff or graft." Blazing "new trails for American destiny," Roosevelt's rise marked the close of these sterile years.[58] So far, the assessment was not unusual, but it was the way Roosevelt implemented his plans that impressed Vrooman most. "Roosevelt is a new kind of geographer," he announced. History's public men could be divided into "static geographers" and a more dynamic type. The president was in the latter camp: "One studies and describes that geography which man helps to make; the other helps to make that geography which other men describe."[59] Vrooman rejoiced at Roosevelt's conservation program, which he saw as the epitome of accomplishment for what London's *Daily Mail* called the "man of action."[60] Nowhere could Vrooman find this characteristic clearer than in the administration's initiatives to make water use efficient. Rivers would be sculpted under Roosevelt's plans, to bring the American West into closer contact with international commerce. Because it was in Vrooman's estimation the nexus of the internal and the external in American life, no case was more relevant for Roosevelt's greatness than the Panama Canal: "The perfecting" of the Panama and the Lake Michigan Canals, the canalization of the Illinois River, and the enhancement of the "channel of the Mississippi itself" would be the conservation scheme's backbone. As far as transport was concerned, cargo would speed "direct from the wharves of Minneapolis or Chicago, Pittsburg [*sic*] or Omaha, to Bombay, Liverpool or Hong-Kong."[61] The inside and the outside of American imperial expansion were inextricably linked in Vrooman's mind, as in Roosevelt's.

This channeling of the by-then ex-president's strategies provided a bookend to the theorizing that began in the 1890s in the work of Benjamin Kidd and Brooks Adams. Both Adams and Vrooman were theorists of American empire, but Vrooman was a theorist of Roosevelt's accomplishments as the epitome of empire in action. Vrooman put human will at the center and captured the synergy between US imperialism and TR's conservation program. The significance of Roosevelt's schemes was neither their external thrust nor their internal political deployment but the link between the two. Whereas Adams saw the nation responding to environmental constraints, seeking to stave off inevitable decay at as distant a point as possible, Vrooman detected a preemptive approach to the nature of imperial power. If Adams described empire as lying beyond US borders, Vrooman depicted Roosevelt in 1909 as reasserting the material platform of American abundance, thereby linking the new geography of power within the United States to global networks of resource development and conservation.

This flood of understanding came in the context of geopolitical concerns created by the march to formal empire. Of the Philippines, Vrooman stated that President McKinley saved those islands from Kaiser Wilhelm's grasp, overruling in the process the "amiable" American people's suspicion of imperial conquest.[62] The Oxford lecturer conveniently sped past the brutality of regime change from 1898 to 1902 to praise the "entire liberty" that the people of the Philippines had attained under American rule. Looking back from 1909, Vrooman believed that Roosevelt acted not just for "America" but for the wider Anglo-Saxon race. Like other imperialists, Vrooman pronounced the Panama Canal the central move in Roosevelt's strategic vision, and one with significance for the future union of Anglo-Saxon settler peoples through ties of commerce and communication. Ever anxious since he walked out on his church career to blend personal advantage and geopolitical curiosity, Vrooman became editor of the *British Columbia Magazine*, a periodical devoted to the boosting of the western province of Canada for land development through its coming links with Anglo-dominated Pacific commerce. Opened to traffic in late 1914, the canal would be central to this dream.[63] In Vrooman's mind, this bright future would be impossible without Roosevelt.

Vrooman's Oxford lecture remains an obscure piece, yet he reflected the drift of Progressive reform thought. The idea of Vrooman toiling away intellectually to articulate the union of the Anglo-Saxon world around dynamic geography would likely have pleased the recently retired president when it was delivered, if he had not been preoccupied with his imminent visit to Africa. The lecture aligned well with the president's own high valuation of the white settlement colonies of the British Empire as fraternal racial states. The message was relevant to that empire's leaders and opinion makers, if newspapers provide any guide. The *Times of India* found entrancing the "clarifying elemental force" of the president's character.[64]

By the time Vrooman wrote, of course, Americans' own idea of empire had congealed. It was to be a new kind of empire, based on a strong nation-state, internal conservation of resources, tropical colonies, informal economic influence abroad, and American leadership of an Anglo-Saxon bloc of settler societies. This new American empire gained in both material power and moral prestige during Roosevelt's presidency. Vrooman did not influence that empire, but he did reflect its characteristics. The empire was taking root not through theorizing, whether by Kidd, or even Adams or Griffis, but action. The actors in this drama were the politicians and government officials who drew upon these diverse armchair thinkers to forge ideas in the fire of experience and to make policy. Together they worked with Americans who sought to test the tropics for their true value to science and economics. Their actions helped change American attitudes toward resources and spurred conservation. These actions would also change the nature of the world—and change nature in the process.

FOUR

Encountering the Tropical World: The Impact of Empire

The day of 1 July 1898 was highly significant for the future of conservation in the United States. Two men far separated in distance experienced that day very differently, but in ways that would ultimately bring them together in the service of the nation. Gifford Pinchot was in Washington, assuming the office of chief of the Division of Forestry in the US government. That very same day, Colonel Theodore Roosevelt found the fame that projected him to national renown. Roosevelt was in Cuba, fighting the Spanish-American War with the regiment of cavalry volunteers colloquially called the Rough Riders. It was the day remembered in American history for the charge up Kettle Hill, where TR gained kudos for his bravery. In the version given in Joseph Pulitzer's *New York World*, Roosevelt inspired his men to charge despite withering Spanish fire. To the *World*, the Rough Riders faced the dire task of mounting the hill "in blistering heat," without cover, except from three Gatling guns. Roosevelt saw the soldiers waver under the impact of shrapnel and, "riding far out in advance, yelled for his men to follow him." His regiment and the Tenth Cavalry, with its "colored," or "Buffalo," soldiers, responded and "were right alongside. They went up on the double quick, yelling and shouting. Bullets and shrapnel rained upon them, stretching many a poor fellow dead with the cheering of his comrades still in his ears. . . . The men had not gone thirty yards in the open when Roosevelt's horse went down. The Colonel landed on his feet unhurt" but nevertheless proceeded to run along "to keep in the van."[1] The charge and the larger battle to take Santiago de Cuba were both successful. The role of the "Colored Infantry" of the Tenth, whose men showed equal bravery, was sidelined, but the still-youngish colonel anxious to engage in manly pursuits had turned himself into a hero of the yellow press. The story of the charge became the subject of a Frederic Remington painting and marked an important stage in Roosevelt's rise to national prominence.

Gifford Pinchot seized no such fame when he assumed the position of Division of Forestry chief in the Department of Agriculture on that 1 July. True, he was already well known in Washington circles and among the social elite. From the well informed he received a round of "hearty and widespread applause."[2] But few outsiders were aware of this "young man" said to be "worth half a million, who is handsome, highly educated, and determined to carve out a brilliant future for himself."[3] If his name was not yet on the lips of every American, it was not for want of self-promotion. He came, he told everyone who would listen, with foreign training, boasting of his forest studies directed by the venerated Dietrich Brandis—thus credentialing himself through the name of a man with whom he had actually spent little time and under whom he had done no formal coursework. Press reports wrote of Pinchot as "an expert," a "professor,"[4] and a man who had "thorough training"[5] because he had "studied forestry abroad." With "such experts as Mr. Pinchot, the possibilities" of the Division of Forestry were said to be "almost limitless."[6] It was a bureaucratic job that this heir of a business fortune took on, but like TR, Pinchot aimed to be an activist. His ambition was "to carry on the investigations in the forest rather than in the office."[7] He toured the western states in the following six months, conversing with cattlemen, sheepmen, and others who used the public lands; he cultivated the company of lumbermen, attended seemingly endless forestry and irrigation conferences, and told all and sundry "how to save the forests."[8] Like Roosevelt, Pinchot was clearly a man in a hurry. His central concern was to make a rational and planned national forestry policy. But he assured listeners that he did not seek to impose foreign methods. Rather, he would apply his experience of private forestry gained at the Biltmore Estate, in North Carolina, in the mid-1890s in such a way that forestry would be shaped, in its "present conditions of development," by "practical suggestions" for better utilization of resources.[9]

During these first months in which Pinchot became acquainted with the nation's forests, and the nation's lumbermen and westerners got to know this wealthy scion of the eastern establishment, the nation was learning about its potential colonial possessions. In December 1898, the possibility became reality. The acquisition of colonies came as a shock. The Philippine Islands "abound with minerals," said the *Los Angeles Times*.[10] More accurately, *Gunton's Magazine* stated that the former Spanish possession was "an unknown empire." Except for Japan, that magazine of political economy claimed, the "face of the United States" had "never been turned towards

Asia." This statement coyly neglected the nation's fascination with China, but *Gunton's* correctly understood that "[l]east of all have the Philippine Islands figured in our outlook upon the world," and Americans had "only the vaguest ideas" of where they were and what they contained. Americans could ill afford to leave these islands to the international intrigue that would mean seizure by another power and so must keep them for the time being. "What a rich prize it is that has fallen into our hands," *Gunton's* rejoiced.[11] This was just how President McKinley and his cabinet saw it. Though the president agonized, the United States assumed the mantle of a formal empire.

Aside from taking up the white man's burden despite the country's self-image as an enemy of colonialism, the most salient fact about the new possessions was their tropical character. The battle for Santiago de Cuba had given Americans a warning lesson. To the *Outlook*, the US Army fought "under the fierce rays of a tropical sun" and "the density of the forests placed the severest physical strain upon the men."[12] This was only the beginning of a profound set of tropical encounters. The colonial expansion of 1898 had a disproportionate effect on American relations with the rest of the world from 1898 to 1917. Only US entry to World War I would eclipse it in importance in this respect. As part of this extensive tropical engagement, Americans reassessed their relations with all natural resources, including the world of plants.

However decisive plants might be in the survival of ecosystems, the choices humans make about plant use also tell much about cultural and intellectual life. So it was for American colonialism. In boosting US interest in the world of plants, especially in the tropical world, the USDA played a central role through its Seed and Plant Introduction Section (1898). Incorporated in the newly established Bureau of Plant Industry (BPI) in 1901, the section was known from 1908 as the Office of Foreign Seed and Plant Introduction, but it had already become a vehicle for disseminating temperate, tropical, and subtropical plants.[13]

It is significant that only in 1898 did the USDA begin "wholesale importation of plants for cultivation in the United States."[14] To be sure, for much of the nineteenth century, the federal and some state governments, as well as private individuals, had practiced plant introduction known as "acclimatization."[15] In the case of California, plant exchanges with Pacific societies in the 1880s, particularly Australia, had been undertaken on a large scale. Acclimatization activities directed toward both plant and animal transfers were especially fashionable in the British and French Empires.[16] But US interest in such exchanges was, with the exception of California, minimal until

4.1. There was curiosity about how American troops would fare in the tropics, along with an interest in the exotica of tropical vegetation. Here, US troops rest under a banyan tree in Manila's Botanical Garden, c. 1902. LCPP.

the 1890s. Federal plant acclimatization effectively began in that decade. Accelerating markedly from 1898 on, plant acclimatization became closely associated with the growth of national power and global engagement. As W. H. Hodge and C. O. Erlanson state, experts went abroad under the "aegis of the new [Seed and Plant Introduction] Section." They explored species that could contribute to American economic growth, and the program "received immediate prestige."[17]

Though the nation's new territorial acquisitions were tropical, this work was far from exclusively concerned with the tropics per se. Rather, it reflected a greater global awareness and a new spirit of global competitiveness that accompanied the nation's enhanced international presence. The first explorations of the 1890s were in Russia and northern China, for durum wheat in aid of Great Plains farmers, who constantly faced strong competition from other countries and problems of drought. Other crops, such as soya bean from China, were able to be adapted to both temperate and subtropical climates.[18] The tropical and subtropical aspect quickly grew in visibility in the BPI's work, however, and it reflected the acquisition of an empire in distant climes.

A crucial first step was the Federal Plant Introduction Garden, also established in 1898. At first a "modest 6-acre tract" in present-day Miami, it soon expanded its allotments until transferred to Coconut Grove in 1923.[19] The Florida site was chosen because it opened possibilities for introducing, in "a relatively frost-free testing ground," many tropical and subtropical species suitable for "peninsular Florida, the warmer parts of the Gulf Coast, southern California, Puerto Rico," and Hawaii. The tropical importations were considerable. By the 1950s, hundreds of species of "palms, various insecticide plants, miscellaneous latex-producing species, mangoes, lychees and papayas as well as many lesser-known subtropical fruits" had been introduced to the United States. In 1904 enthusiastic citizens of Chico, California, lobbied successfully for a duplicate plant introduction garden for temperate stone fruit, and further USDA gardens were soon established.[20]

At the time of its formation in 1901, the BPI took encouragement from the growing scientific belief in the adaptability of plants and the possibilities of modifying them for new environments. Crucially, plants adapted from the tropics might, in some cases, become crops for more temperate climates. Cotton was tendered in evidence, but plant experts expected many others. The widely published clergyman William Elliot Griffis embraced this idea,[21] but the BPI was also influenced by Swiss French agronomist Alphonse de Candolle, whose *Origine des plantes cultivées* (1883) was considered a foundational text in the study of plant improvement through cultivation and global transfer. The NCC's report of 1909 contained in its studies of food crops a distillation of Candolle's work.[22] Departmental officials spoke of the thrill of mixing wild seed gathered in Asia and Africa that could be "tamed by breeding with others now in cultivation, thus contributing to the creation of fruits and vegetables that the world has never seen." This they saw as an emerging process of cosmopolitanism, because the seeds produced "excellent" foods from other countries of which "we are ignorant."[23] The American palate was changing under the impact of globalization and empire, as historian Kristen Hoganson has shown, and the BPI both contributed to this process and responded to a greater dietary openness.[24]

The tropical and subtropical interest gained greatly from the tireless work of David Fairchild, who led a group within the USDA labeled "agricultural explorers." Fairchild (1869–1954) later ran the Office of Foreign Seed and Plant Introduction under its various names from 1904 to 1928 and gave the work of the BPI a sharper foreign orientation (rather than focusing on plant swapping within the continental United States). Despite, or perhaps because of, his chilly Michigan origins and Kansas schooling, Fairchild developed a special interest in tropical plants, but that was not all. The plant lover acquired wanderlust from association with the wealthy globe-trotter Barbour Lathrop, another plant enthusiast, who personally financed the young USDA officer for his first foreign scouting trips for the US government in the mid-1890s.[25] Equally important as the ability to tap private funding sources was Fairchild's capacities as a publicist. He penned several books later in his life and technical publications in the first decade of the twentieth century. Prolific in spreading his views to Washington's intellectual and scientific elite, he was also capable of outreach to the general public through *National Geographic*, which was expanding its popular coverage in highly successful ways in the aftermath of the tropical acquisitions of 1898–99. Fairchild told *National Geographic*'s audience that the United States had the most varied climatic range of any country. It thus needed to source plants from all areas of the globe.[26]

If a flair for publicity was something Fairchild shared with the conservation-oriented president and Gifford Pinchot, the new "agricultural explorer" had other important attributes that made him one of those typically Rooseveltian Progressives in spirit and action (he praised Roosevelt as a leader who was "the first and last president . . . to have a biological sense of proportions").[27] Fairchild's affinity went beyond enthusiasm for the world of the naturalist: he was an exponent of Roosevelt's plans for a vigorous expansion of government power, in this case to help farmers. Thus, Fairchild envisioned the BPI as a "government enterprise" to introduce "as many of the valuable crops of the world as can be grown here." Government leadership would do what "private enterprise will not naturally undertake," in order to strengthen the economic and social fabric of the nation through a healthy agriculture that would tap the "wealth-producing power of American soil."[28]

Fairchild was an equally emphatic supporter of the idea that American power required commodity security, a view expressed often in the 1908 Conference of Governors and the report of the NCC. For timber, this meant establishing more forest reserves, while for fruits, vegetables, and the major staple crops, self-sufficiency mandated strengthening American plant resources with virile stock from abroad. Government leadership and international engagement came together with commodity security when Fairchild urged measures to "encourage the production of food and other products that we now import from other lands" and to establish new farms to support what he claimed would be many thousands, if not millions, of extra people lured back to agriculture.[29]

Though seed borrowings were often for temperate zone crops such as wheat, they were also undertaken to meet projected demand for tropical products. Fairchild recalled: "the Spanish War created a wave of interest in tropical agriculture when our soldiers returned from Cuba and Puerto Rico with tales of new and delicious fruit."[30] Even as plant enthusiasts in the BPI were overly sanguine about the possibilities for transforming the southern tier of states around the Gulf Coast into a subtropical cornucopia, the acquisition of colonies within the tropics encouraged two concrete activities: plant transfers to southern Florida and similar introductions from tropical places around the world to the nation's new possessions abroad, where they could be grown for the American market. Thus, the BPI planted bamboo from southern China in Florida; and the growing of mangoes, principally for export to New York from Puerto Rico, was aided when Fairchild visited Bombay in 1901 and sent home cuttings. In 1902 another officer went to San Juan to survey fruit prospects and make recommendations for farmers, and the BPI subsequently promoted what Fairchild called a "mango craze."[31]

Fairchild's botanical interests were broad, in both plant types and the areas that he incorporated in his schemes. He noted, "From the tropical regions of Porto Rico, Hawaii, the Philippines, and the Panama Canal Zone, there are hosts of new possibilities open."[32] Dr. Seaman A. Knapp, another departmental agent, wrote that his tour of the Philippine Islands in 1901 "confirmed the opinion formed during a visit in 1898, that from an agricultural standpoint these islands are among the most valuable territories of all Asia."[33] A plant introduction program within the Philippine Bureau of Agriculture was established in the islands to test a wide variety of crops. Effort went into staples such as coconut palms for oil production and gutta-percha trees from the Dutch East Indies and the Malay States, which could supply the cable linings for the world's rapidly expanding telegraph routes. Much of this work concerned cross-colonial fertilization between empires, especially the Dutch, British, and American. Some was basic scientific discovery. A colonial botanist in the Philippine Islands Bureau of Forestry visited the famed Dutch Botanic Gardens at Buitenzorg, Java, and the British equivalent in Singapore, exchanging plant specimens and studying the classification of the region's flora to help botanical projects in the Philippines.[34] Other work was narrowly utilitarian but quixotic. The BPI imported rubber trees from British Borneo and the Straits Settlements to the island of Luzon, without much success in demonstrating commercial yields. Fairchild believed that research into rubber would combat the already-apparent pest problems facing the rubber industry in its native Brazil.[35] In 1904 Fairchild also drew attention to hybridizing coffee bushes to improve "the run-down coffee varieties" of Central and South America that needed "new strains to invigorate them." This could be achieved by introduction of the "wild coffees of Abyssinia." An American consular official, Robert P. Skinner, secured these in a trip to the East African kingdom.[36]

Along with other USDA officials, Fairchild had an expansive view of the "tropical," since it included the "hot" desert lands of North Africa, a region to which USDA officials with links to California and Arizona were often drawn. Fairchild's fellow agricultural explorer Walter Tennyson Swingle took a trip to French Algeria in 1898 "to attempt importation of superior varieties" of palms. The possibilities of date production especially interested Fairchild. Convinced, due to his racially influenced prejudices, that the production of dates would remain substandard and "dirty" in the Arab world,[37] Fairchild promoted North African dates (and the Smyrna fig) for the southwestern United States. Under Robert J. Forbes, the Bureau of Agriculture Station at Tempe, Arizona, carried out experiments in acclimatizing these plants. From the Arizona station and the efforts of the plant introduction specialists came palm plantings across parts of the American Southwest that "yielded thousands of pounds of fruit in the early years of the twentieth century."[38]

Plant introductions contributed materially, Fairchild boasted, to the prosperity of America. To be sure, the entire history of plant introductions in the Progressive Era is a large issue deserving of greater attention, but its significance was contradictory. On the one hand, plant importation highlighted deficiencies within the American resource base. On the other, it posited the entire world as a cornucopia newly open to American quasi-imperial development through the USDA's global engagement. In this sense, plant importation did not mix perfectly with the growing mood of pessimism over scarcity of resources. Rather, Fairchild gave a message of hope, provided that plant exchanges built up the networks of tropical production sites on American-controlled soil. Another aspect of the tropical world encounter did fit that gloomier mood far better, however, and that concerned the rainforests of the Philippines, Puerto Rico, and Hawaii. Within the Washington bureaucracy and Roosevelt's circle of advisers, this encounter was influential for conservation policies.

At the end of 1902, the United States was of its own accord granting Cuba its "freedom," subject to the strictures of the Platt Amendment. Simultaneously, Gifford Pinchot found himself in another colony, which the United States had no intention of renouncing, the Philippine Islands. There, 40 to 55 million acres of tropical forest had fallen into American hands. It was almost all publicly controlled land.[39] Tropical forests were not a consideration in the acquisition of 1898, but subsequently, the archipelago's forests seemed useful in helping to pay the colony's way, if export markets could be developed and if experiments on unfamiliar tree species could reveal new forest products.[40] The sheer size of the forest acquisition was a factor for the United States, where foresters, politicians, and reformers increasingly lamented that mainland forests were being carelessly frittered away. Stateside, the remainder was in private hands or in mostly "unreserved public forests" of 442 million acres. In 1898 just 47 million acres had been named forest reserves. Thus, while the Philippine woodlands equaled less than 10 percent of those potentially available in public lands on the US mainland and Alaska, they could double the forest reserves currently under the control of the US government, a highly significant figure. Even a decade later, with huge increases in mainland reservations, the Philippine forests still represented more than a quarter of all timber by area in US government reserves.[41] Nobody knew exactly how much forest there was in the islands—a matter of anxiety in itself—but the extent was not the only consideration. Were the Philippine timbers a mixture of hardwoods and softwoods (a desirable combination to maximize use)? Were there particular species that could be easily worked, and were these finer timber trees accessible? And how durable would Philippines timber be for uses such as wharf pilings and other exterior construction? These were only some of the many questions that remained to be answered.

Seeing is believing, or so it is said. In August 1902, Gifford Pinchot decided to see for himself. Traveling via Liverpool and Saint Petersburg, he ventured by train across the huge expanses of Russia's temperate forests and tundra in Siberia. Passing through China, he proceeded to Hong Kong and on to Manila, eagerly noting the details of Russian and other forests as he went on his epic journey. Pinchot was already testing the hypothesis that the world was running out of wood and was ascertaining in which countries sustainable production could be expected. Pinchot's tour of the Philippines lasted only from late October to the year's end. He was back in time for the January Committee of Agriculture meetings and strode triumphantly into the Capitol Hill committee room soon after reaching home. In some respects the whirlwind tour of the islands resembled a view from the bow of a ship (a metaphor famously deployed in Southeast Asian historiography by Jacob van Leur). Conveyed by a government steamer, Pinchot surveyed the mighty prospects and plunged eagerly but superficially into the surrounding forest margin at each island he visited. Undoubtedly, he did not fully understand the tropical forests or its inhabitants and entered colonial forest management with North American, Progressive preconceptions.[42]

4.2. Americans found tropical forest lush and commercially alluring but exotic, peopled by "uncivilized" tribes, as in this stereograph picture where four "natives" are visible in the forests of Mindanao, Philippine Islands, 1919. LCPP.

Nonetheless, Pinchot's views came from something other than a ship's foredeck, and the Philippines was not a tabula rasa for the introduction of his views "virtually unopposed." The foresters there were not under his control, and he had to work informally through forestry networks and his position in the Roosevelt administration to influence the Philippine Commission and to shape the drafts of the colonial forestry laws.[43] Nor was he entirely unacquainted with tropical forests, having learned from his German mentor Brandis of their complex composition. From that experience he thought not of American forestry as the model but the Indian Forestry Service and its forest school at Dehra Dun.[44] Even this approach needed to be critically appraised, he understood, and adaptations would have to be made, just as they had in the United States. He also had intelligence from George Ahern, his right-hand man on the spot and head of the Philippine Islands Bureau of Forestry, whose appointment he engineered in 1900.[45] Ahern emphasized the need to learn from local conditions and from the surrounding empires.[46]

There was potential for scientific forestry exhibited everywhere Pinchot went, as he reported on his return to Washington. Speaking at the National Wholesale Lumber Dealers Association meeting in March 1903, Pinchot was effusive about the future of the colonial possession: "We are going to have in the islands one of the most productive forest regions of the globe, both for our own markets and for all the markets of the east, all of it conserved by practical forestry."[47] Journalists embraced the enthusiasm. In the *World's Work*, Roy Crandall boasted of "great tracts [of forest] awaiting American capital and transportation improvements."[48] At the instigation of the Philippine Islands Bureau of Forestry, timber concessions were handed out to the lumber companies from the United States, while exotic woods such as "Narra" (*Pterocarpus indicus*) went on exhibit at the 1904 Saint Louis World's Fair.[49]

Pinchot indeed believed that the islands would provide the foundation for a great industry, but only when social, economic, environmental, and political conditions improved. With the rinderpest viral infection killing up to 90 percent of the Asian buffalo used in the movement of lumber, this infestation needed to be eradicated. More telling was Pinchot's reference to the happy future time "when the Filipinos have learned to work, as they will readily learn under the instruction of the Americans."[50] This paternalistic statement revealed the more significant and immense obstacles to scientific forestry management.

The social structure of the countryside made swift progress impossible. Shifting swidden agriculture practiced by peasants and tribal groups led to forests being replaced with useless and almost impenetrable masses of cogon grass (*Imperata cylindrica*).[51] These farming zones were known as *caingins*.[52] Peasant fires had certainly changed the land, yet this was not new. They had done so since before European occupation. Scientists and anthropologists still dispute just how damaging *caingins* can be, since swidden agriculture's effects vary with local ecological conditions and with the intensity of modern capitalist development and population pressures,[53] but the perception of irrevocable deterioration prevailed in US colonial circles. Oftentimes blame was attached to landless peasants obstructing the path of material progress under colonial tutelage. To the Philippine Commission these were "the indolent and irresponsible . . . throughout the provinces."[54]

Of greater perceptiveness was the analysis by forester Barrington Moore Sr., who identified the social and economic roots of the problem. Serving in the Philippines from 1908 to 1910, Moore observed that the forest regulations governing lumber concessions encouraged exploitation in order to gain for the Philippine government self-sufficiency in revenue through royalties to the Bureau of Forestry. These "concessions must be given on ruinously favorable terms: at extremely low stumpage rates (unfortunately fixed in the forest act), and for long periods of years (twenty at least). Otherwise, nobody will embark on such a new and untried venture."[55] Thus, commercial exploitation tended to be harsh in its effects. The Insular Lumber Company, which had eight hundred employees and the most advanced machinery imported by its owners in an "exact copy" of their Washington State business, clear-cut much of its concession in the search for usable wood. The land was later converted to sugar plantations.[56]

In addition to noting the early impact of commercial American logging, Moore was the first to tie the origin of fire damage to the socioeconomic condition of the peasantry. Plantation agriculture and the land claims of the *cacique* (prominent Filipino villagers) had driven swidden cultivators into more desperate practices on the margins of the plains, where "virgin" tracts could be burned without interference. Cogon-affected areas were abandoned because if fully cultivated they were likely to induce land claims from the powerful *cacique*.[57] But even Moore remained somewhat derisive in his description of the landless peasants. "On account of his aversion to the harder kinds of manual labor, he prefers making a [new] Caingin in the forest to cultivating the Cogan [*sic*] by hand."[58] Certainly, Bureau of Forestry officials could not afford to pour too much blame on timber concessionaires, who provided the colonial government with needed revenue, and the forest fires of the landless provided an obvious scapegoat.[59]

The Philippine-American War reinforced an American perception of trouble in the forests. Though the war was declared over in 1902, fighting continued until 1906 and even beyond in the south, where, in Mindanao, the most prodigious forests beckoned. The danger of disturbing domestic tranquillity through ill-conceived forest policies was a real one. In 1904 the new Philippine Bureau of Forestry rules banned "unauthorized" firing and clearing of forest, and the felling of trees and removal of stone and earth for local, subsistence use became subject to licenses granted under strict controls, requiring payments. (Taking small quantities of inferior timber for personal use required a "gratuitous" license subject to close bureaucratic supervision.) "Licensed" *caingins* were allowed, but only for land on which there was grass or little timber of "inferior" species, and only after written application and upon physical inspection by forestry officials. Violations for *caingin* offenses included both fines and imprisonment, and fines also applied for other transgressions. These regulations were patently unworkable, stretching forest personnel so thinly that normal tasks of assessing reserves and tracts of forest for commercial development proceeded at an unacceptably slow pace. Just over a year later, the bureau had to revise the rules to extend traditional peasant rights to firewood, timber, earth, or stone for personal use, without any license for five years, while prosecutions were assigned to the municipal authorities. In 1910 special communal forests were established for subsistence use, and soon after that, individual "homesteads" were reserved to encourage shifting agriculturalists to become "settlers," as Americans wanted.[60] Even then, resistance continued to frustrate foresters,

4.3. Two wood-sellers on their way to market on the island of Luzon, c. 1901. The insular government had to accommodate traditional peasant rights to firewood, and the extensive but small-scale trade in domestic wood for fuel use continued. LCPP.

and the implementation of the law remained inadequate. The 1910 report of the Philippine secretary of the interior Dean Worcester explained: "The continued making of unauthorized caingins, or forest clearings, is, and will doubtless long remain, a great menace to the forests of the Philippines."[61] Yet as late as 1914, under a new "aggressive" enforcement policy, municipal authorities had acted upon only a third of complaints.[62]

These reports make clear that US colonial forest policy was compromised from the start by underfunded budgets and the needs of revenue raising, needs that conflicted with scientific forestry when applied to a tropical environment; the record makes equally plain that forest policy faced severe opposition from both peasants and larger landowners, resistance that was effective before World War I. It was not only the poor who offended in the eyes of the Philippine Commission: "a class consisting of the wealthy few . . . deliberately encourage the ignorant and the poor to violate the law in order that they may participate in the profits resulting from such violation." In practice, the *cacique* induced the poor to burn forest, then submitted claims to the land for themselves. Given uncertainties and confusions of land tenure, this tactic was often successful because it was "difficult to secure evidence."[63]

In the light of experience, Pinchot, Ahern, and their subordinates became less sanguine about the prospects for quick resource development. The effect was twofold. On the one hand, Americans set about achieving the longer-term objective by forging a new model of efficient colonial development for global imperialism, using forestry as a key component. A forestry school was established at Los Baños in 1910, forest reserves began to be mapped, and mechanized logging was encouraged in concession areas. By 1911 a vigorous marketing campaign was under way to make the islands the center of an East Asia and Pacific commerce. The expansion was based upon research into the useful dipterocarp hardwoods, of which Red Lauan (*Shorea sp.*) was the most prominent.[64]

This process of a modernizing US colonialism is well understood in the historiography, even as it must be seen not as an American imposition but as an adaptation to the circumstances of the Philippines and the surrounding countries. American foresters learned much from forestry in India, the British Malay States, and Dutch Java, yet they sought at the same time to outdo these competitors in terms of resource management.[65] Forestry played an important part in this articulation of modernity, and even before US entry into World War I, the forestry service in the Philippines acquired a reputation for economic efficiency and scientific prowess that resulted in the Republic of China calling on Philippine- and Yale-trained foresters for their own fledgling forestry unit. Experts from within the Philippine Bureau of Forestry and the Bureau of Agriculture were also hired in Brazil, Venezuela, and Argentina to help deal with tropical and subtropical forest problems in those countries. Whereas British India's foresters advised Ahern in the early days, now they "began visiting the Philippines to study steam-logging techniques." Meanwhile, bureau officials were asked to "develop working plans" for the Sumatran forests.[66]

In these personnel and plant exchanges, the influence of the US Forest Service was never contained within the borders of the United States and its dependencies. It operated transnationally, but the US colonial possessions were the lynchpin of this extended influence, where American forestry precedents were put on show for the rest of the tropical world. From this basis, an international forestry expertise developed in the 1920s, yet it was largely separated from internal American practice. This was precisely because tropical forestry did not, a few exceptions aside, apply to US domestic circumstances of climate and vegetation types. What the Philippine experience did do was increase US interest in any future cooperative international inventory of tropical forestry to see how American discoveries might be applied to other forest resources of the tropical zone.[67]

On the other hand, the short-term impact was more significant for domestic practice. US foresters and scientists quickly registered certain dismal facts about the global significance of the archipelago's forests. A simple European or American sustained-yield approach did not work well in the colonial setting. Forestry development necessarily involved an accommodation to local Filipino elites as well as peasant practices. Preserving colonial power took precedence over efficiency.[68] Foresters concluded that the Philippines could not provide itself, let alone the world, with adequate timber in the immediate future. For the more distant prospect, the growth of population and industry, the inaccessibility of the bulk of the forests, and the needs of the colony for its own commercial development would "prevent the timber export from ever being an important factor in supplying the United States," reported forestry official and Pinchot ally Raphael Zon in the *NCC Report* of 1909. Indeed, the initial material impact of commercial forest ventures was negligible. Expansion of lumber production occurring during American rule came mostly post-1917, when wartime conditions and the economic prosperity of the 1920s put pressure upon these tropical forests and accelerated deforestation under Bureau of Forestry leases.[69]

Less positive still were the prospects for colonial forestry in Puerto Rico. According to the 1904 survey conducted by German-trained John Clayton Gifford for the Bureau of Forestry in Washington, much of the land was in private hands, and many of the most valuable trees had already been removed. Impenetrable weeds infested cutover landscapes. The remaining wooded area centered on an existing Spanish reserve. Partly because the reserve was far too small for a forestry industry capable of providing significant imports to the United States, Gifford favored continuation and extension of the reserve, though he also envisaged scenic tourism in line with Progressive Era conservation's nascent multiple-use approach.[70] The most important of these multiple uses was watershed protection for farms and villages. Underlying Gifford's thinking was the need to study tropical forests and their impact on water supply, erosion, and economic activity. He urged the creation of a "center of forestry" with "educational value" like the already-established agricultural experiment station on the island. In 1907 Roosevelt was delighted to make the area, on Gifford's recommendation, the Luquillo National Forest, the first US tropical forest reserve.[71] Under this model, Luquillo eventually became valuable for transnational research and exchanges of information on tropical plants, but not until 1939 was an official Tropical Forest Experiment Station established, and the fact remained that the vast majority of the island had already been deforested.[72]

The recently annexed Hawaii likewise promised little bounty for tropical foresters. In 1903 Pinchot sent William Logan Hall, then in charge of forestry extension services in the USDA, to scout the islands for usable timber. In a paper "widely noted by the press,"[73] Hall stressed the destruction of vegetation at the hands of the indigenous people through "Cutting and fire" and lamented the introduced pigs and cattle, which were "a menace" to trees.[74] It could not be "claimed that these forests have great commercial value," Hall concluded, but they did have other uses, such as protecting the slopes from erosion and providing steadier supplies of water for the cane fields below. A territorial forest service under a board of agriculture and forestry was established and a trained forester appointed as superintendent.[75]

These largely unfavorable experiences of Puerto Rico, Hawaii, and the Philippines fitted into the wider imperial discourse that Alphonse Mélard had generated. Though German foresters contested his opinion within the global forestry community, the examples confirmed Mélard's pessimistic view in US eyes.[76] Yet the key problem was not the scarcity of wood but the composition of the forests. As John Gifford observed of Puerto Rico, the "essential tropical character of the forest, in which a great number of species contend with one another for possession, makes the problem of management a very difficult one."[77] This model derived from the island empire was not exceptional in European encounters with the tropics. In "nearly all of the tropical forests," Zon concluded, the "few trees known at present as commercially valuable are invariably found scattered among a great many unutilizable ones."[78] From a utilitarian viewpoint, the floral jumble that American foresters encountered compared poorly with the more uniform pine and spruce forests of Europe. Perceived "insufficiency" backed Mélard's alarmist arguments that one could not rely upon tropical colonial possessions to meet metropolitan timber demand.

The impact of this evaluation was widely felt in American conservationist circles, with the idea applied indiscriminately to global shortages. In his work in Argentina, the geologist and Pinchot ally Bailey Willis drew upon tropical forestry assessments. There, a similarly dispersed pattern of forest species was manifest across the Southern Cone (principally Argentina and Chile). Also, much subtropical land in northern Argentina had passed out of the control of the government, and Willis argued that what remained must be preserved from logging.[79] This recipe for preservation was exactly the same as Gifford's for Puerto Rico and Hall's for Hawaii.

Discovery of the apparent inadequacy of these distant forests paralleled efforts to find domestic sources of tropical timber through a Florida land boom. These assessments were far from uniformly scientific. Gifford argued (in a way similar to David Fairchild) that the vegetation of the southern part of the state was "distinctively West Indian in character." Southern Florida, he contended, could supply the nation's tropical forest needs and awaited only "American capital and enterprise."[80] As the recent acquirer of real estate in the region, his views were potentially self-interested. He became a booster of development and an advocate of "conservation" through the filling in of "useless" swamps.[81] Though much of southern Florida ultimately became cities and suburbs rather than timberland, and exotic trees, notably the Australian *Melaleuca* species that Gifford recommended,[82] became pests rather than assets, Gifford's views show how US forestry experts inclined to look *inward* for forest development, even at the cost of alienating rich ecosystems through afforestation of swampland. Roosevelt himself gave "implicit support," though the idea of draining swamps "never got off the ground" during his presidency.[83]

Instead of emphasizing either formal or informal colonial ventures, Roosevelt and Pinchot took their own experience and that of their underlings and international experts seriously; they argued for conservation of US resources. The experience enabled Pinchot to drive home the message that American forests must be conserved by a strong national institution under his direct control and subject to minimal political interference from Congress. This was essentially the message that the encounter with the tropics gave, though there is little doubt that this was the message Pinchot wanted to hear in the first place.

The creation of a strong national forestry authority had been on Pinchot's mind in the 1890s,[84] but now he was in a political position, with Roosevelt's support, to have legislation pass Congress. This, together with executive orders, created a capable national bureaucracy not only to direct forestry policy but also to explore wider plans for conservation of resources.[85] Previously, the forestry head had charge of a small division within the USDA, while the forest reserves themselves were under the Department of the Interior. Except by exhortation and informal arrangements, Pinchot had no direct say over what actually went on in the forests. Late in 1904, he and Roosevelt lobbied Congress successfully to have foresters and forests put under the same roof, as it were, in the USDA. Through the Forest Transfer Act of 1905, the US Forest Service, with Pinchot as chief, was established. Pinchot's creation obtained a significant degree of independence because the revenues from timber sales and leasing of forestland would go into a war chest for forest improvement, the hiring of trained foresters, prosecution of illegal users, and cultivation of local political support with promises of spending in the relevant local communities. The Yale Forestry School, whose networks he influenced, supplied many of the foresters. In 1907 the forest reserves were renamed "national forests," a change that underscored their new significance and their management in the long-term interests of the nation.[86]

Even as Pinchot's approach valued expertise exerted through top-down control, its deployment was highly political and infused with a

4.4. Gifford Pinchot, chief of the US Forest Service, 1907. LCPP.

world-historical agenda for social and cultural advance. Abroad, he vigorously endorsed American rule in the Philippines as a moral imperative—a benevolent, patriotic duty.[87] The only risks to the nation's anticipated hegemony in the world of nations were internal, he averred. Readers of the *North American Review* learned that the young United States had inherited the best resource endowment of any nation, an advantage threatened only by the profligate wasters of the nineteenth century. Pinchot's aim was to ensure that the "white race" of "settlers" could continue its unparalleled ascent. In this process, Pinchot viewed forests as key among the "foundations of [American] prosperity."[88] Supporters of the Forest Transfer Act also adopted a broad view of forestry's importance to the international standing of the United States. In arguing for legislative changes, the American Forestry Association evidenced foreign examples of forest retention.[89] Legislators such as John F. Lacey, the influential chair of the Committee on Public Lands in the House of Representatives, similarly stressed the international benchmarks of forest conservation and the contrary example of fallen empires that failed to heed the call.[90]

If Pinchot was as much an imperialist as a man expressing scientific expertise, his on-the-ground practice was also pragmatic. Though he took trained men to be forest inspectors in the United States and secured the same for Ahern in the Philippines, he cultivated western grazing interests and lumbermen by promising controlled access to forests and other public lands. He recognized that western interests were not monolithic, and he assiduously sought local power brokers, submitting, for example, detailed plans for reserve extensions to interest groups and sympathetic politicians, courting the latter to frame workable executive orders through prior grassroots feedback.[91] This political cultivation was vital for the Forest Transfer Act's original passage through Congress. Thereby he persuaded western senators and congressmen that he would do a better job than the Department of the Interior's fraud-ridden and inexpert General Land Office and, more compelling for parsimonious legislators, that he would convert the Bureau of Forestry into a paying proposition through timber sales and leasing fees for grazing.[92]

Under Pinchot's highly political leadership and with Roosevelt's backing, the growth of the US Forest Service was phenomenal, reaching 2,500 staff by 1910 from 10 in 1898. Moreover, many millions of acres were added to Pinchot's domain from the public lands. Historian Brian Balogh has noted that by the time he was forced from office in 1910, Pinchot ruled "an empire of reserves" approaching "200 million acres."[93] This was a 400 percent increase since Roosevelt became president. In addition, the Philippine and the American western experience indicated the need to educate forest users carefully on the rules and purposes of this new forest "empire." The publication and education office (the de facto press bureau) that was established within the Forest Service grew in importance as a vital outlet for Pinchot's ideas. It distributed huge quantities of proconservation publicity releases to newspapers across the country with a combined circulation of some ten million readers. Pinchot assembled a mailing list of 670,000 people and a slate of institutions to which he could dispatch Forest Service bulletins and other material. His officers monitored the results through press surveillance. With a version already introduced in the Philippine Islands Bureau of Forestry in 1904,[94] Pinchot had a new manual, *The Use Book*, printed in the following year to guide his growing army of employees back home. Strikingly, this *Use Book* differed from the earlier American forestry manual of 1902 and the Philippine one. While the latter was a book of laws, rules, and regulations for the foresters, the *Use Book* of 1905 addressed the users as well and included explanations of the rationale for scientific forestry. More succinctly written and forcefully presented, it was a tool of communication with the public.[95] The version broadcast in the Philippines envisaged no such dialogue, and the *Use Book* did not perpetuate this mistake at home.

While the Philippine Islands' own Bureau of Forestry "anticipated" in its actions "similar changes" within the United States in "scientific forestry and executive control,"[96] the direct impact of the one upon the other is difficult to disentangle completely, since national and colonial forestry policy evolved together. Pinchot certainly viewed his advice to the Philippine Commission as an aid in establishing his credentials as suitable to reorganize government forestry in the United States, and so too did Governor-General Taft. It became a valuable part of Pinchot's résumé for this purpose. Taft told the secretary of war that Pinchot was "of great use with respect to the forests of these Islands."[97] From the American colony Pinchot learned much about the importance of communicating with forest users, the imperative of controlling fire, and the desirability of centralizing decision making within his office, but he already favored the latter outcome.[98] Pinchot also learned what to avoid in enforcement regimes. Though Ahern wished to prosecute offenders burning forests for *caingins*, he lacked the authority. The insular government's fears of renewed outright rebellion against American rule combined with budgetary restraints to inhibit his and Pinchot's plans. Repeatedly complaining that his hands were tied, Ahern had to report, through the Philippines' secretary of the interior, to the Bureau of Insular Affairs in the War Department. Missing was the political and financial freedom that Pinchot obtained in 1905. This different bureaucratic organization was important for enforcing forestry rules. Pinchot appointed a legal office within the Forest Service and skillfully reinterpreted the latent powers within the provisions of the 1897 Organic Act (the original forest management act) to assume authority for prosecution of offenders taking timber illegally or otherwise disturbing forests. This was an innovative use of executive power, based upon experience, rather than the product of new legislation on the part of Congress. If the Philippines tested and confirmed the need for rigorous state action at the national level and gave Pinchot inspiration and propaganda for national forest protection, it also emphasized the value of efficient, if devious, enforcement and a global vision that he did not ignore at home.[99]

The reinterpretation of the 1897 law was not the only creative manipulation of existing statutes. Roosevelt also used executive power to supply the most essential ingredient of all, forests. Since 1891 the president had been able to establish, by "proclamation," forest reserves from the public lands, withdrawing them from entry for sale through the General Land Office. But Roosevelt used this power far more than his predecessors, proclaiming some 150 million acres.[100] By 1907, western discontent over the removal of such lands from private development for mining, lumber, and other uses was growing fast. Pinchot's forestry extension policy met a stern political challenge. Under the Agricultural Appropriation Bill of 1907, congressmen proposed an amendment to stop the president from creating further forest reserves in a cluster of western states without explicit congressional approval. In response, Roosevelt had Pinchot and his aides worked at top speed on a plan to protect all the forested or partly forested land that Pinchot could quickly identify in the public domain. Though the work continued day and night for a week, to his opponents the areas nominated became, in effect, the "midnight forests" stolen from citizens' access by a nimble president and his scheming forest czar, Pinchot. The thirty-two reserves created or enlarged totaled sixteen million additional acres in the six states named in the 1907 bill. After presenting his executive proclamations, Roosevelt approved the bill, thus leaving Congress with a hollow victory.[101] The bitter implications of this maneuver for TR's conservation drive and the struggle with Congress soon became apparent. In the meantime, Roosevelt and Pinchot used the strengthened Forest Service to advocate bringing many different resources into the field of conservation.

Preparing notes for presidential candidate William Howard Taft's speeches in 1908, Pinchot observed that a "great awakening" had occurred as the United States assumed the role of a great power. The nation was "becoming acquainted with the rest of the world." After 1898, claimed Pinchot,

Americans undertook the fastest crash-learning course in human history. "With better knowledge of the world came a better understanding of the relations between our Nation and others—and naturally and necessarily . . . people began to grasp, as they never had before, the conditions which control the strength and perpetuity of our own Nation."[102] Leaving aside a certain excusable element of politically induced hyperbole, this statement contained the foundation of conservation policy.

Of these experiences—with tropical plant introductions and forests—discussed in this chapter, the latter would be more influential in the rapidly congealing views on conservation. A nationally controlled scientific forestry, based on adaptation of European principles and colonial experience, became the goal. Yet the utilitarian concept of maximum sustained yield, first advanced by the Danish fisheries expert C. G. Johannes Petersen in 1903, was not yet fully articulated for American conditions. Formulations such as "intermittent" or "periodic" harvesting, "sustained productiveness," and "annual yield" appeared in forestry periodicals and manuals, derived often from German and French publications, while public justifications regarding forests at the Conference of Governors remained emotive as much as scientific. Harvesting the "natural yield" of the forests for "the benefit of the people" required "proper regard for forest protection" against "avarice" and must be conducted under "careful superintendence."[103] This ethic drew deeply from elite revulsion against what had been done wrongly in the past and from the fear that those wrongs were being duplicated across the world. The world mattered because, without access to global supplies, the United States would run short of certain timbers, American foresters believed. Yet more important was the cautionary tale. Humankind's ignorance of forest destruction's consequences seemed duplicated everywhere, and attention to global examples served the trope of declension that Roosevelt and Pinchot favored.

The issue with forests was the need for renewable practices, but what about the nonrenewable? In theory, renewable and nonrenewable resources were conceptually distinct. Forests could be replaced, while coal, copper, iron, and oil could not. If the cutting of trees matched the plantings, and if savings could be made and waste reduced, supply and demand might be brought into balance in the arboreal world. Productivity could be sustainably managed. What worried Pinchot and TR most was the import of forest destruction for the future of nonrenewable resources on a global level. Throughout the nineteenth century, forests had, in the United States and in many other countries, been essentially treated as a kind of free resource within a commons.[104] Only feeble efforts had been made to replace trees felled. Pinchot and the president felt that the threat of imminent timber catastrophe supplied a warning for the likely outcome of all natural resource use and a clarion call applying even more firmly to resources that truly were nonrenewable. That led to the question of the fossil fuel revolution and its consequences.

FIVE

Energy and Empire: Shadows of the Fossil Fuel Revolution

"During the last decade the world has traversed one of those periodic crises which attend an alteration in the social equilibrium. The seat of energy has migrated from Europe to America. . . . A change of equilibrium has heretofore occupied at least the span of a human lifetime, so that a new generation has gradually become habituated to the novel environment. In this instance the revolution came so suddenly that few realized its presence before it ended." So went the ominous ruminations of Brooks Adams in *The New Empire* (1902).[1] We have seen how the geopolitics of this direct descendant of two American presidents both captured the mind-set of opinion makers and influenced it. Yet these ideas had no purchase without material changes occurring in American society and environment. A sense of national disquiet was enhanced by a rapid diminution of forested land. Of course, the grim scenario for forests that Pinchot and collaborators documented might not apply to seemingly abundant fossil fuels. But, as Roosevelt stated, "The Conservation movement was a direct outgrowth of the forest movement. It was nothing more than the application to our other natural resources of the principles which had been worked out in connection with the forests."[2] In the case of fossil fuels (and most other minerals) the supply was hardly lacking, but the theorem that waste leads to shortage, and shortage to civilizational decline, was duly applied. Indeed, authorities such as the future chief of the Bureau of Mines Joseph Holmes went further. Mineral fuels deserved "special consideration" because their use involved "immediate and complete destruction." The "barbarous waste" must be stopped.[3]

Though the alarmist model dealt with oil and gas as well as coal, because all were fossil fuels, the greatest concern was with coal. Even though supplies of oil were periodically uncertain, neither oil nor natural gas yet held the key to the running of the American economy. Coal did. Factories used

5.1. The railroad nation and its smoky consequences: Delaware, Lackawanna, and Western Railroad yards, Scranton, Pennsylvania, c. 1900–1910. Detroit Publishing Company Collection, LCPP.

it, transport needed it through voracious demand from steam engines, and homes were heated with it from the eastern cities to the far-flung West. Coal was available in abundant quantities in the United States, no doubt. In fact, there was so much mined that its grimy imprint registered on the denizens of cities all over the Northeast and Midwest and in the eyes of any schoolchild silly enough to stick a head out the window while riding in a train anywhere across the countryside. The nation that once relied upon wood had become the fossil fuel nation.

It was this sudden change from wood to coal that provoked a sizable portion of the anxiety over energy resources. The revolution hit the United States later than Britain due to the former's abundant primeval forest that even some railroad companies had plundered for years as fuel rather than resort to coal. Thanks to this generous endowment of nature, Americans resisted the use of coal as long as they could. In particular, householders preferred the smell and low cost of wood to the dirty, stinking black mineral. Yet not only did the use of fossil fuels exceed that of wood from 1880 on,

but demand also rose exponentially in the prosperous years from 1896 to 1907. The anxiety over coal shortages reflected complex misgivings, particularly among the social elite and political leaders, about the threats that the nation's expanding appetite for the substance posed. Smoke was good for business, and it propelled American power in an increasingly competitive world. Ever since the world-historical changes of 1898 and indeed for some years before, the central role of fossil fuels in American economic power had been plain to all. In 1898 the anti-imperialist businessman Edward Atkinson stressed the nation's abundant cheap coal as the basis of an Anglo-American commercial prosperity that would spread the benefits of commerce all over the globe without the need for formal empire. But within just a few years, it was precisely this reliance on coal for productivity that made entrepreneurs and politicians fear running out of fuel.[4]

Principally, concern about coal registered the staggering rate of increase in use. This boded ill for the future of America, and Roosevelt's view of America was all about the future and the empire of material expansion that must never end. In 1907 Roosevelt argued for withdrawal of public lands from entry and for the separation of the surface rights from underlying mineral deposits because of the need to preserve western coal in the light of prodigious use.[5] At the Conference of Governors in 1908, the president explained: "The mere increase in our consumption of coal during 1907 over 1906 exceeded the total consumption in 1876, the Centennial year. This is a striking fact."[6] Here Roosevelt implicitly tied American progress and the entire republican form of government since the Declaration of Independence to energy security. At the same meeting, Andrew Carnegie chorused: "Coal consumption is increasing at an astonishing rate. During the period for which statistics have been gathered, it has doubled in each decade; of late it has more than doubled. In 1907 the production was about 450,000,000 tons."[7]

Predictions of *near-term* fossil fuel exhaustion were few. In fact, many people discounted alarm on that score. Frederick E. Saward, the editor of the *Coal Trade Journal*, argued that the mineral's supply in the United Kingdom was certainly dropping, but not in the United States. Moreover, if one's own country burned coal with abandon, it could be replaced with that of others. A "broadened understanding of national and international relations has drawn the whole world closer and is ever enforcing the concept that the great storehouse of the world is a common one," Saward soothingly assured readers.[8] More than outright shortage, experts and politicians predicted *future* "exhaustion" caused by accelerating use, made worse by unbelievably wasteful practices. The words "exhaustion" and "waste" occurred frequently in the *NCC Report* of 1909 and during the Conference of Governors meeting of 1908.

The rhetoric and the reality of shortage were quite different, but the alarm persisted. This is not understandable without realizing that "the future" was, in the minds of Roosevelt and his followers, not merely the next few years but hundreds, if not thousands, of years. For Roosevelt, "the patriotic duty" was not only "insuring the safety" but also the "continuance of the Nation."[9] Charles Van Hise, president of the University of Wisconsin, was one of the chief theoreticians of conservation, Roosevelt style. He specified what "continuance" meant: "It is only within the past two centuries that the lands of the country have been subject to agriculture upon an extensive scale, and the main drafts upon the soil of this country have been within the last century."[10] Yet what was "one or even several centuries compared with the expected future life of the Nation"?[11] Mines expert Holmes agreed. Conservation would be undertaken "for the hundreds of millions of American Citizens who will people this country through the future centuries."[12]

Why should businessmen and professional and political spokesmen think this way? Geologists' acceptance of the long accretion of the earth's physical changes and biologists' comprehension of species evolution gave confidence that humans could, if well adapted to their environments, persist over millennia.[13] Thus did Van Hise put it: "There is no reason, from a geological point of view, why human beings may not live upon this earth for millions of years to come, perhaps many millions of years, and, so far as we are concerned, such periods are practically infinite."[14] This long-term perspective clashed with the rapidity of modern change in energy use. Reformers' embrace of a long time span reflected also their sudden awareness of the nation's arrival as a world power and the realization that American power was founded on abundant, good-quality natural resources. Israel C. White, the state geologist for West Virginia, was an international traveler who advised the government of Brazil on coal deposits and assessed the mineral fields of Mexico. He specifically linked the dependence of the American economy and its role in the world to fossil fuel power. "The nations that have coal and iron will rule the world," he stated. "Bountiful" nature's heritage gave the United States "both coal and iron richer by far than that of any other political division of the earth." This was the "source of world-wide influence" that could be threatened only by "criminal waste and wanton destruction."[15]

Ruling the world was indeed a key issue explaining the change in attitude toward fossil fuels. Roosevelt was in the process of building a strong navy and came to regard this and the Panama Canal, which would facilitate a two-ocean naval policy, as among his greatest achievements. One argument for the acquisition of the island empire had been that these new territories would serve as coaling stations for the fleet, but the fuel experts of the USGS were advocating, as early as 1904, more efficient ways to use coal to power naval vessels, so that they could travel twice as far without refueling. The substitution of oil for coal in this process was already another possibility for the future that made the conservation of all fossil fuels ever more important.[16]

Allegations of profligate use came from many people and concerned many other minerals, to be sure. Participants at the Conference of Governors made disquieting assertions about a range of metals under a "history of . . . exploitation."[17] The eminent mining engineer John Hays Hammond recalled South Africa and the example of the goldfields, where he had worked in the 1890s: "It has been, unfortunately, the popular custom to refer to large deposits of ore as illimitable and inexhaustible. Such hyperbole characterizes the description of the famous gold deposits of the Transvaal. As a matter of fact, we mining engineers know that even these exceptionally extensive deposits will be practically exhausted within a couple of decades [and] certainly within a generation."[18] Though "for some of the metals we may illustrate the marvelous increase" in production since 1850, Van Hise claimed, "even the most sanguine calculations can not hold out the hope that the available high-grade ores of iron, copper, lead, zinc, gold, and silver, at the present rate of exploitation, will last for many centuries into the future."[19] On iron ore the NCC contributors agreed, but their assessment was rosier for the precious metals, pointing out that increased use had thus far been matched by new discoveries or better utilization of old deposits. Because future uses of metals were unknown, it was impossible "to strike a balance between us and our descendants," the *NCC Report* concluded. In retrospect, we know that this judgment was more accurate than Van Hise's. Metal prices relative to wages generally dropped through the twentieth century, including the period 1900–1914.[20] Nevertheless, Carnegie could point out that "reduction of copper ores" for over fifty years meant that the industry was struggling to cope with the increased demand. The future seemed unnecessarily mortgaged to the present.[21]

Fossil fuels presented a more extreme case, because they were deployed in all the other processes. For many minerals, there would be no shortages for at least the medium term precisely because coal could be used to run machinery that could dig deeper for—and power the transport of—new but less conveniently situated supplies. Moreover, the parallel between fossil fuels and forests could most easily be made because both were carbon-based power sources. Commentators understood the profound switch in energy usage going on and perceived the organic connection between forest and fossil fuel use: the decline of one was in symbiotic relationship with the other's rise. The lessons of the forests provided a self-evident model for the abuse of coal and oil. Israel White charged: "not content with destroying our magnificent forests, the only fuel and supply of carbon known to our forefathers, we are with ruthless hands and regardless of the future applying both torch and dynamite to the vastly greater resources of this precious carbon which provident Nature had stored for us in the buried forests of the distant past."[22] This was future-eating of the worst kind.

The bearers of bad news came with the imprimatur of the government bureaucracy in Washington. Taking the lead was Edward W. Parker, the government coal statistician within the USGS, 1899–1908, and head of the Division of Mineral Resources, 1908–1915.[23] Quite optimistic about coal's longevity as late as 1898,[24] Parker drew attention in his 1904 forecast to growing demand and rising prices of both anthracite and bituminous coal over the previous five years,[25] then hardened his position by 1907, when he took aim at the depletion of anthracite: "Our best and cheaply mined coals will, at the present rate of drain upon them, be largely depleted by the end of the next century." Yes, there would still be plenty left but Americans were "taking the cream and leaving the skim-milk." Anthracite producers in the eastern states had already begun to feel the effects "of the lessening supply." Parker challenged them to "heed the warnings" and "practice greater economy in the mining and utilization of our fuel supplies."[26] Even though coal was not scarce, the costs of extracting the less desirable seams would stifle business expansion.

Perhaps because he was speaking to mining engineers, Parker praised the "wonderful development in the coal mining industry." Yet he went beyond conservation as efficiency, introducing moral and nonutilitarian reasons for his fears. Referring to the "wickedness of men who were stealing from the Lord," Parker stated that he was "sufficiently uncommercial" to seek the preservation of nature itself. Parker's text indicated that he allowed anxieties over the pumping of unprecedented amounts of "carbonic acid" (carbon dioxide) into the atmosphere—a process that seemed to him, as to other conservationists, unnatural—to overwhelm his technical assessments and impart a tone of moral and cultural criticism as a citizen rather than a geologist. "What becomes of it?" he asked of this cloud of industry. Noting how a scientist assured him that "the corn crop of Kansas will take up as much carbon dioxide as all the trees cut in a year," Parker was nevertheless skeptical: "I am not in a position to deny it, but I am inclined to doubt it."[27]

Delivered at the American Mining Congress in Joplin, Missouri, in November 1907, Parker's assessments of coal supply and demand created a storm of controversy. USGS Director George Otis Smith asked Bailey Willis, then head of the Division of Areal Geology, to write a clarification, in order to scotch what he termed misunderstandings of Parker's position. Officials of the USGS had, Willis argued, been wrongly identified in the press with "sensational items predicting the early exhaustion of our fuel supplies." The Parker report had been misquoted and the press "omitted [the author's] essential qualifying statements." But Willis gave virtually the same figures for the shortage of anthracite and the predictions of exhaustion as Parker, merely trying to put a better gloss on them. Willis cited "uncertainty in predicting for the future," which depended "upon many unknown factors, chief among which is the future rate of increase in consumption." Nevertheless, production had "increased enormously" in the previous fifty years, Willis admitted, and if continued, the trend would mean severe depletion. Such concern persisted despite technical improvements in the international science of geology for measuring available coal supplies.[28] This was because US officials assessed for the NCC the likelihood of coal availability on the basis of a continued increase in use at existing rates of successful extraction, and because the required perspective was long term. It mattered little that there was no objective measure of shortage. The NCC's report on "future coal production" by another USGS official, Henry Gannett, reemphasized that all viable supplies would be used up within 240 years, a time frame that attendees at the Conference of Governors believed was far too short for an exceptional republic.[29] Gannett was a statistician as well as a founding member of the National Geographical Society and its president in 1909. He was also involved in the US imperial project as assistant director of the census of the Philippines (1902) and author of gazetteers on Cuba and Puerto Rico.[30] It is notable that, like Parker, Gannett had changed his tune from the 1890s, when he had judged these resources inexhaustible.[31]

Could the United States look to foreign lands if these dire economic predictions proved correct? English political economist Stanley Jevons had already in the 1860s pointed to a growing American dominance over Britain in coal, so it seemed impossible that Britain could supply the growing need. China, with its vastness and its unknown and untapped resources, seemed the best bet. Israel White noted that the USGS's Willis had gone to Shanxi in 1903, financed by the Carnegie Institution of Washington (on whose board Roosevelt sat), to investigate alternative sources of coal for the United

States and the world: "It was formerly supposed that China would prove the great store-house from which the other nations could draw their supplies of carbon when their own had become exhausted, but the recent studies of a brilliant American geologist in that far-off land, rendered possible by the generosity of the world's greatest philanthropist, tell a different story. The fuel resources of China, great as they undoubtedly are, have been largely overestimated, and Mr Willis reports that they will practically all be required by China herself, and that the other nations can not look to her for this all-important element in modern industrial life."[32] There was nowhere left to turn but homeward, to reexamine American patterns of use.

Though the alarm over fossil fuels was centered on coal, concern over its impacts was infectious and far-reaching. Petroleum was seen, not as a substitute or savior, but as the next ninepin to fall. Geologist David Talbot Day wrote the chapter on petroleum for the *NCC Report* of 1909. Appearing also as an article for the *American Review of Reviews*, his assessment moved "from technical to public discourse, in step with other Progressive federal officials" in their efforts to popularize the Roosevelt agenda. Surveying production region by region, Day "concluded that the United States had between 10 and 24.5 billion barrels of oil left, inclining to 15 billion barrels as the likeliest figure."[33] Day's reckoning was based on a very narrow reading of the evidence and made no allowance for new discoveries.

This call for restriction did not please the oil industry. It flew in the face of commonsense observation, as well as self-interest. The overproduction of oil in some areas encouraged anti-alarmism. Gluts occurred whenever new wells were struck. Wild fluctuations in prices and supplies followed, sometimes due to the monopoly activities of Standard Oil in attempting to squeeze out small producers.[34] For Day, these gluts were only such in the eyes of fools. The resultant feverish competition when prospectors struck gushers in new fields was an aspect of the petroleum industry that Day particularly wished to curb. It led, he argued, to unnecessary waste through underpricing in an effort to off-load oil quickly and maintain cash flows. The result would be longer-term shortage.[35] Whatever the cause of market instability, the *Los Angeles Times* was not at all sympathetic to Day's argument. Decrying "the alarmist theory so common with government experts," the paper pointed to the promising growth of Californian prospecting. "Oil fuel has made California great," the boosters shouted, and "every Californian" would oppose restriction on "immediate production."[36]

Despite the scoffing and gloating from the industry and its apologists, Day's position was not indefensible at the time. When Day published his conclusions in article form in January 1909,[37] further oil discoveries of the

1910s in California, Oklahoma, and other places were still to come, and industry surveys also predicted potential decline.[38] Day wrote in the context of vastly increased use of oil in the immediately preceding period of 1900–1908 and in full knowledge of the dwindling oil reserves in some formerly important states, such as Texas. To be sure, US prospectors were already crossing the border in search of oil in Mexico. The Texas Oil Company struck its first gusher in Tampico in 1901, at a time when US fields were "declining, and competition among the oilmen was fierce."[39] Yet foreign supplies were not what Roosevelt's advisers wanted. They adhered to the idea of national self-sufficiency in a resource necessary to maintain "civilized progress" and world leadership.[40] As with coal, Day was alarmed by the exponential increase in petroleum use that had occurred in just a few short years. The expansion of industry's needs for machine oil and the beginning of automobile use had raised demand beyond all expectations since 1900.[41] Moreover, again in parallel with coal, the emphasis was not on immediate catastrophe. Rather, Day stressed the need for conservation to prevent waste in a way that would promote American international competitiveness and social stability.[42]

Day made many suggestions for better use of existing resources to achieve this end. Most of all, in Rooseveltian fashion he called for removal of oil lands from public entry to preserve the black gold for future generations and restriction of existing supplies to essential usage such as machine oil. Such a plan would require extensive market intervention by the government and industry-government cooperation. Some of the oil withdrawn from general use would be reserved for the navy, if oil-burning ships were to replace coal. Already under Roosevelt's orders, given upon USGS advice, the first oil lands were withdrawn from public entry in late 1907, and Taft expanded this policy in 1909. From this action would come the first naval oil reserves, established in 1912.[43] Once more, the demands of empire influenced the jeremiad over waste.

Under the precautionary principle, research should also be conducted, Day believed, into "artificial petroleum from various vegetable and animal waste."[44] This paralleled calls from the public to develop ethanol fuels and wave power, calls that came to nothing before World War I, but that revealed broader community anxiety over future energy shortages. In 1908 Pinchot ordered a review of an unsolicited proposal for wave power, but the bureaucracy deemed it insufficiently developed. Hydroelectric power was an entirely different matter, and the search to secure sites for this fossil fuel substitute was central to the administration's plans to deal with the anticipated crisis of expensive power. Water was abundant but Pinchot feared that the same forces that corralled coal would control the price of the hydroelectric alternative, commonly called "white coal." The struggle over public control of waterpower sites soon reflected this anxiety over the energetic power of empire.[45]

Another substitute for coal and oil under consideration was natural gas (methane). Not scarce, it was nevertheless wasted in ways that even those who benefited from the oil industry's expansion could not deny.[46] Oil-prospecting adviser Israel White stressed the gas escaping during petroleum extraction. Here was "a fuel of which nature has given us a practical monopoly," White bemoaned. In his view, natural gas was one of God's greatest gifts for the future. Though "lavish in abundance, already transmuted into the gaseous stage and stored under vast pressure to be released at our bidding when and where we will," this "best and purest fuel" was wasted in what amounted to "a national disgrace."[47] The report on minerals to the Joint Conservation Conference in December 1908 by Republican Senator Frank Flint of California backed this assessment with "startling facts" that showed how escaping gas from oil and gas wells would, if retained, annually light all US cities with over 100,000 people.[48] Former governor Newton Blanchard told how a German expert had visited the Caddo field in Louisiana and found that high-pressure ("wild") gas wells were often drilled accidentally when searching for oil. Wasting seventy million cubic feet a day, these were typically fired rather than controlled.[49] In response to such evidence, conservationists urged capping high-pressure wells until they were able to be used.[50] Oil and gas were not the chief concerns in the age of mechanization, but the prediction of future famine was highly significant because it indicated that substitution within the range of nonrenewable resources could not be relied upon to save the nation from a gloomy fate. It seemed that the United States must horde both its unpredictable oil and gas stocks and its most desirable coal supplies.

A key implication of the alarm over fossil fuels was the effect of soaring exploitation of coal upon the prices of almost everything. As the costs of extraction rose, they would feed into inflation and then wages, commentators feared. As early as 1904, the *New York Times* ran Parker's story that coal prices had risen more than 39 percent in just five years since the war with Spain.[51] The reality of resource shortage and consequent spikes in costs was self-evident to participants in these debates, but it is not self-evident to modern economists.[52] The latter have shown that, in the late nineteenth and early twentieth centuries, the real costs of basic mineral resources were declining, because supply was increasing. However, this generalization does not apply to US anthracite, the most valued form of coal. This, the preferred source for household use, was subject to modestly rising prices from 1897 to 1913. To be sure, the increase of 42 percent was only slightly above the general increase in living costs of 37 percent, but it was more significantly above the average nonfarm income increase of 25 percent from 1900 to 1913.[53] On the other hand, the cost of the inferior and dirty bituminous (soft) coal (which people did not prefer for home fuel) remained steady. Increasing numbers of industrial processes were utilizing this type of coal because it was cheaper and more bountiful, with no such midterm shortages as could be predicted for anthracite.[54]

Nonetheless, the modest trend of coal prices relative to inflation raises questions that cannot be easily squared with alarmism, unless we analyze the widespread *perception* of rises that contemporary observers associated with scarcity. The debate concerned real-time fluctuations in monetary values, not the retrospective analysis of economists with price series indexed for inflation. Myopic as it may now appear, newspaper and periodical headlines were full of panic that the cost of coal was heading up.[55] Commentators claimed that if energy prices increased, social unrest would too. At the Conference of Governors, railroad magnate James J. Hill was as adamant as White that "upon iron and coal our industrial civilization is built. When fuel and iron become scarce and high-priced, civilization, so far as we can now foresee, will suffer as man would suffer by the gradual withdrawal of the air he breathes." Hill added that these "facts were pointed out not in the spirit of the alarmist, but in order that attention might be directed to the way by which the nation may escape future disaster."[56] Nevertheless, such unsettling forecasts fueled anxieties.

The perception of rising prices can be unpacked to reveal the impact of real and imagined pressures upon the social fabric of the nation. Contemporary commentators generally agreed that the "high cost of living" in the United States from 1896 to 1914 was due in some measure to fuel prices. Yet, as to the causes, wildly different explanations proliferated. One was the holding back of production in the case of pig iron and coal to maximize price, together with other forms of monopoly "pooling." Closely connected were the restrictive haulage and rate practices of railroads. It was, stated one irritated observer, "touch and go" whether coal would get through in the winter of 1907, as freighters did not use enough cars, due in part to the restrictive "combination of coal carrying railroads."[57] The role of middlemen (distributors and retailers) also figured in the debate,[58] alongside high tariffs, with the *New York Times* concluding in a 1912 survey of the controversy that there had been "no world-wide rise" of prices for all goods, as defenders of high tariffs argued, and that "American-made goods" sold "for less abroad" than at home.[59] In this scenario, the tariff seemed the villain in the inflation story. Then there were the strikes and rumors of strikes that affected distribution and, in a variation on the theme, alleged collusion between labor and capital to keep coal scarce.[60]

All of this—real, imagined, or exaggerated—fed into the debates over the existence and growth of monopolies and combinations against which Roosevelt contended in outspoken ways after 1902. The launch of legal proceedings against the Northern Securities Company, a railroad holding company organized by none other than J. P. Morgan, gave Roosevelt his reputation as a "trust-buster." In 1904 the Supreme Court ordered the breakup of the company. Exaggerated images of the trusts as the enemy and of the trust-busting Roosevelt as a white knight became closely bound to conservation politics. On the one hand, conservation functioned to channel concern over price-fixing by commodity-manipulating trusts. On the other, conservation allowed those accused of monopolistic practices, such as Hill, to deflect public irritation and rebuild their reputations. Through conservation, capital could cooperate with government to reorganize the basis of American finances to ensure the orderly supply of resources to industry. In this way, conservation was a vehicle to transcend nascent intraclass division. Hill had been affected by the Northern Securities case, and his Great Northern Railroad continued to endure censure for its haulage practices. It was condemned in the press for not carrying enough winter coal, despite the fact that Hill had acquired Canadian mines to supply his system.[61] By focusing on the larger issue of resource misuse, Hill could discharge his frustrations over what he regarded as unwarranted attacks on his haulage record.[62]

In these debates, coal had become a social, not an economic, commodity, with a central role in industrial society and of fundamental importance to society's tranquillity. To the *Outlook*, the "exorbitant" fuel prices of 1903 made urgent the need to preserve "the people's coal rights."[63] This position was closely connected to social reformers' advocacy of measures to stem wider unrest. Underpinning these assessments was the desire to mediate between capital and labor. The Anthracite Coal Strike of 1902 had been a key event in US labor history, during which Roosevelt had forced a modicum of arbitration upon the contending parties to prevent what he called "untold misery . . . with the certainty of riots which might develop into social war."[64] Newspapers continued to report the rumblings of union discontent in the coalfields in 1906 and 1907, further but more sporadic striking, and the fear that consumers would suffer.[65] The winter of 1906–7 brought sensational press stories of people shivering and even freezing through lack of household coal. In response, Roosevelt's supporters returned to the argument that coal should "play an important part in cheapening the cost of living in the United States."[66] To the architects of empire and social stability, the corporations and others wasting resources stood condemned for fueling class conflict as seriously as "the wildest anarchists determined to destroy and overturn the foundations of government." For Israel White, no one could "act in a more irrational and thoughtless manner than have our people in permitting such fearful destruction of the very sources of our power and greatness."[67]

When Roosevelt linked conservation with that "certainty of riots which might develop into social war," he touched upon what many social reformers, including quite conservative ones from the mugwump tradition, regarded as the central problem facing the American Republic. The United States did not have a powerful socialist movement, but it did have labor unrest, interclass hostility, and industrial violence aplenty. The presence of the blue-collar working class, which reached its peak in proportion of population in this period, and the pressing need to prevent class tensions were the oft-unstated assumptions behind moderate social reform and much of Progressivism. Key people involved in the coal scare also participated in organizations such as the League for Social Service of New York City, formed in 1898 with the Reverend Josiah Strong as its president. Strong is better remembered as the provider in the 1880s of rationalizations for future American imperialism through Social Darwinism, but his concerns after 1898 connected external commercial and demographic expansion in a globalizing world to the internal stability of American society achievable through social reform.

Reorganized in 1902 as the Institute for Social Service, the organization that Strong founded was intended to address the severe social problems of the present. This "transition period" was the age of the fossil fuel revolution. "Industrial, economic and social conditions inherited from an old civilization" were, according to the institute, "slowly and painfully undergoing readjustment to the new civilization created by steam, electricity and machinery." The institute saw its work as "international" in "all civilized countries" because, in those places, the mechanized world that fossil fuels drove was increasingly ascendant.[68] The *Outlook*'s William B. Howland, ever a supporter of Roosevelt's conservation policies, was on the organizing committee promoting the institute's activities. In a similar way, others connected to the Institute for Social Service, such as the prominent civic reformer and advocate of worker housing E. R. L. Gould and government statistician Carroll

D. Wright, were committed to moderately Progressive social change steered by sections of the political and intellectual elite. Roosevelt had appointed Wright to be a member of the Committee on Arbitration to settle the 1902 strike.[69] The institute's secretary, publisher William H. Tolman, coined the term "social engineering" to describe supportive "industrial betterment" of workers' conditions through the "mutuality" of capital-labor cooperation.[70] As Andrew Carnegie wrote in praise of Tolman's scheme, "It is by the efforts of individual firms that the right solution of the problem will be furnished, and not thru [*sic*] Socialism, which can only talk speculatively, while individuals can work practically, curing evils that Socialists point out."[71]

Concerns about the social question of class spoke to matters of life and death, morbidity and mortality. Carnegie decried the coal gases that were not consumed but escaped during coal coking: "Much of our coke-making is still extravagant; some ovens use the gases, and all should do so without delay if necessary, under State regulation, since the people have some rights both in the preservation of their heritage and in maintaining the purity of the air they breathe."[72] A movement for cleaning up the smog and grime of the cities had simultaneously sprung up through "smoke abatement" societies, as in Britain.[73] The ACA under horticulturalist and journalist Horace McFarland's leadership vigorously campaigned during the first decade of the new century for city beautification and urban and regional parks, partly to allow citizens to breathe cleaner air and have spaces for recreation essential to well-being.[74] Historic preservation societies in the eastern states similarly condemned the lack of public space to which citizens might escape for a time from toxic city tenements. State and national park proposals from both McFarland and the ASHPS drew upon this theme. Championing Adirondack preservation in New York against lower-class woodcutters and hunters, the ASHPS nevertheless saw state and urban parks as the lungs of the cities of the east and a safety valve for urban discontent.[75] As an admirer of the ASHPS, Roosevelt aligned himself with these urban, elite concerns.[76]

However unlikely, the deleterious impacts of fossil fuels on the working class struck a chord among certain rich and powerful people. They, too, had to breathe polluted air from steam trains and witness the unsightly effects of mines and quarries, sometimes from the windows of their residences, such as for those living in New York opposite stone quarries on the Jersey-side cliff faces of the Hudson River. Howland and J. P. Morgan were office bearers in the ASHPS, a group that deplored such effects of a machine- and steam-dominated civilization, even as its members could hardly be said to oppose the commercial development of society itself. Indeed, although Morgan and his allies funded and owned railroads, they regretted certain

5.2. A Smoky Hudson Terminal, New York City, 1909. Public Domain.

aesthetic impacts of the nation's industrialization. Of railroads, the ASHPS could argue that these "avenues of steam traffic through the cities and villages . . . are, in the nature of things, almost invariably bordered by unattractive conditions,"[77] and they campaigned for tree plantings and the provision of pleasant garden landscaping at train stations. While entrepreneurs from Andrew Carnegie down understood the essential role of steam in the creation of American industrial wealth, they were now convinced that its polluting consequences must be mitigated in the interest of social stability and public health.[78]

Aesthetics melded into moral concerns. Charles Van Hise proclaimed: "It is good morals to insist" that nonrenewable products such as coal and mineral phosphate "shall not be wasted," because "we should think not merely of the next fifty years but of the future centuries."[79] According to labor leader John Mitchell, "The present generation has no moral right to destroy these resources which were not created by Man or given solely to us."[80] At the 1908 Conference of Governors, participants used the word "moral" quite freely. Roosevelt, whom Pinchot called the "preacher of righteousness," invoked morality four times in his speech to convey the responsibility of the present to the future concerning "conservation of our resources in the interests of our people."[81] The moral course of action was to provide for the future through a de facto precautionary principle, coded in what Roosevelt called "foresight." For the first time this idea became explicit in the political discourse over conservation, and mechanisms for calculating intergenerational equity were laid out. As Roosevelt stated, "One distinguishing characteristic of really civilized men is foresight; we have to, as a nation, exercise foresight for this nation in the future; and if we do not exercise that foresight, dark will be the future!"[82] The alarm over the increased use of fossil fuels and about price hikes only makes sense if this long-term moral vision of the president and his supporters is recalled.

Enhancing this widespread perception of intergenerational imbalance was the sheer pace at which social and economic changes seemed to be occurring, thanks to fossil fuels. Whether alarmist or not, the underlying reason for the concern was the perception that time had sped up, due to the communication changes associated with late nineteenth-century globalization. As James J. Hill put it, "The changes of a single generation have brought the Nations of the Earth closer together than were the States of this Union at the close of the Civil War." This material shift affected ideas and impelled newly conceptualized solutions to the problem of possible resource shortage: the United States had to think globally and plan carefully over a longer period of time. There were transnational, as well as national, implications to this insight. The "movement of modern times" had "made the world commercially a small place," Hill continued, thus producing "solidarity of the race such as never before existed." The human "race" had "come to the point where we must to a certain extent regard the natural resources of this planet as a common asset."[83] This sentiment revealed a growing sense of interdependence and cosmopolitanism in conservation policies and attitudes. The fossil fuel scare lay behind these impulses.

Yet despite invocations to treat energy conservation on a global level, Progressive thought was contradictory on the implications. On the one hand, professions of internationalist sentiment spread, and unexpected concerns for the welfare of foreign nations were expressed. "Newer" countries like Canada and Mexico were warned not to follow the American path, while "older" ones were cited as examples of past mistakes to avoid.[84] On the other hand, because fossil fuel energy was increasingly seen as the foundation of American greatness,[85] a persistent desire to hoard these resources at home was also revealed. Neomercantilist ideas flourished, as they did in the European empires, and the USGS increasingly emphasized before World War I "how to make America industrially independent" in fuels and other minerals.[86]

And what of the planet itself? While the debate over fossil fuels centered on supply and demand, social class and social stability, and the struggle for supremacy between nations, it also concerned possibilities for genuine global cooperation. The implications of booming inanimate-fuel usage and the need for "modes of economizing supplies of coal and iron" had echoes in other countries, especially Britain and parts of its empire, as Roosevelt himself pointed out.[87] To be sure, few scholars, let alone politicians, raised commodity exploitation's effects on the natural world across the globe. Such insights came almost entirely in an indirect form through criticism of industrial pollution and demands for wildlife and national park creation or protection, but an opening for a fundamental debate over the fossil fuels' systemic impacts did appear, due to the work of a faraway Swede. Svante Arrhenius first described the "hot-house theory" of the atmosphere in German in 1903, and he summed up the implications of the greenhouse effect for the American public in his *Worlds in the Making: The Evolution of the Universe* (published in New York in 1908).[88] Intended for a popular audience, this work followed upon the "startling scientific prediction" of William Thomson (Lord Kelvin) that the earth had only four centuries of oxygen supply left, a statement made in Toronto in 1898.[89] American geologists had already proposed theories of climate change, but this earlier discussion was based on changes in the physical properties of the earth, not the human-induced action that some conservationists now questioned.[90]

These European sages stirred the fears of some American scientists over the scope of fossil fuel use. When Edward Parker expressed concern about the atmospheric impact of oxygen depletion, he cited "no less an authority" than Lord Kelvin. The possibility of keeping the carbon dioxide cycle in balance was receding, Parker concluded. "One of the great consumers of carbonic acid is the forests," Parker stated, and these were "being used up even faster than the coal." Could farm production take up the excess carbon dioxide? While enormous farm cultivation could "provide for the consumption of this product of fuel combustion" through carbon sequestration, Parker doubted whether farms would "be able to do so" if Americans continued to "increase the production and consumption of coal as indicated" by the projected expansion of steam power. If a balance of carbon dioxide emissions and sequestration were not achieved, "this would produce a condition that would make the earth uninhabitable by man."[91]

In his influential text *The Conservation of Natural Resources*, Van Hise analyzed parallel anxieties over excessive fuel burning. Carbon dioxide was already being pumped into the atmosphere at an alarming rate, he understood, and this gigantic experiment with the earth's atmosphere could not fail to have dramatic effects. He believed that the climate would "become milder" as a result. Extrapolating from Arrhenius's work, Van Hise speculated that vast regions of the Northern Hemisphere might support a denser population because of balmy future conditions. Conservation, too, might benefit as the need for heating homes and offices fell: "For the temperate regions, in which coal is located in the United States, less coal will be used for fuel."[92] Thus, for the northern United States and Canada, economic development might be more successfully perpetuated. Still, Van Hise channeled Progressive anxieties and was not as sanguine as Arrhenius about the likely results. Whereas Arrhenius speculated that colder countries like Sweden would gain a better climate from the spewing out of greenhouse gases, Van Hise considered the interregional-equity implications of a warming planet: "if this were the case," it was "probable that the equatorial and temperate regions would be warmer than now, and perhaps less favorable to high civilization."[93] All the speculation over the future of the tropics as a field for colonial development might, in this circumstance, be settled in the negative.

In the end, however, Van Hise stepped back from the brink of gloom, and so did Parker. Despite uncertainties concerning climate, Van Hise judged that humankind was "able to an amazing degree to adapt . . . to changing conditions."[94] Because of the perception that humans were changing their environment through fossil fuel use, gone was the assumption that climate was merely the product of natural conditions. Humans could indeed contribute, a possibility Willis raised as early as 1901.[95] Parker similarly invoked technological optimism, even though the assertion seemed to be made in order not to leave his audience of practical miners and engineers in open revolt. Yes, oxygen might be used up and life rendered extinct, but "[b]efore that time arrives, we may rest assured that man's genius will have so subdued and utilized the forces of nature that the need for the combustion of fuel in the production of heat, light and power, will have passed."[96] This invocation of technological optimism might seem quintessentially American, but neither Parker nor Van Hise believed society could right itself through panaceas. For Van Hise, the efficient use of resources under public control of new technology to prevent waste provided the only possible answer to the consequences of disequilibrium in energy economics. Wise regulation in the public interest would alone suffice, not the resort to technological fixes. This judgment was compatible with Roosevelt's approach to conservation as a problem of enlightened personal leadership.[97]

The debates over fossil fuel energy that peaked between 1906 and 1910 were intrinsic to the reassessment of the nation's role in the world. They registered an increasing openness to global conversations about the impact of the new energy systems on the planet and its peoples, but the impacts on ecosystems would have to be pioneered by others, much later. This anxiety did not become linked to nascent ecological thought, nor did it provide a precursor to Earth Day, 1970. Meanwhile, American scientists, bureaucrats, politicians, and businessmen worried about more than fossil fuels and forests. Along with farmers, they cared about water resources and soils. These were anxieties stimulated by the new relationship to the wider world, but each made more pressing a strong national response to the crisis of waste in rural America. For this, many conservationists would advocate an inland empire.

SIX

Dynamic Geography: Irrigation, Waterways, and the Inland Empire

Less than a year into his presidency, Roosevelt put his political weight behind a signally important piece of "conservation" law. The law did not mention the word "conservation"—or "efficiency." Its purpose was to "reclaim" the arid American West with irrigation works to be financed from the sale of public lands. The act was closely tied to the homestead idea. It was sold to Congress and the American public as a way of extending nineteenth-century American opportunity by making the creation of family farms possible in the arid sections of the public lands. Such a result would be achieved through the application of government-supplied water from irrigation channels. This National Reclamation Act, popularly called the Newlands Act after its congressional sponsor, would spur further developments in what later became known as conservation. Francis Newlands, a representative and then senator from Nevada, made conservationist claims influenced by the changing debate over forestry that had animated farmers in parts of the American West as far back as the 1880s. Newlands drew upon the concept of "an entire watershed" in which to "conserve the snow and the floodwaters" by retaining forest cover. Forests would help regulate streamflows and provide the steadier water supplies needed by irrigators.[1] Watersheds, and the role of the federal government in their conservation, were topics that became important under Roosevelt, and the National Reclamation Act started the process, helping to shape the wider conservation movement as it emerged in the second half of Roosevelt's presidency.

Unlike forest policy, the Reclamation Act ostensibly reflected pressures far from the sites of foreign empires or colonial possessions and stemmed from internal dynamics, both economic and political. Irrigation policy has typically seemed to historians to be the product either of local interests and the competing claims of pork-barrel politics or of engineers and water experts seeking to impose order on the sprawling American West through scientific development. Irrigation policy came to embed these contradictory drives: at the top, expert intervention and, at the bottom, grassroots agitation that reflected the diversity of American federalism.[2]

These practical pressures for irrigation from interest groups cannot be denied, but the ideology of irrigation and the development of a strong nationalist rationale for irrigation were also critical elements. Men and women do not live by pragmatic local interests or by the drives of disinterested experts alone; they live by dreams as well. Irrigation's promoters sold such a dream, and many believed in it. From its formation in 1891, the National Irrigation Congress, led by William Ellsworth Smythe, embodied the hopes of this movement for western land development based on the concept of the garden landscape through supply of water to deserts. Thus would ordinary Americans gain extended opportunities to become prosperous, independent, small-scale agriculturalists. This egalitarian dream of the "yeoman" farmer had wide appeal, with both national and transnational implications. Demands for government intervention to aid irrigation and international exchanges of irrigation practices and technologies developed in tandem and were in full view at the early meetings of the national irrigation congresses.[3] The nation-state was a vital part of the drive for irrigation because irrigators increasingly looked to the federal government for assistance, and because the public lands that could use water were in the hands of the Department of the Interior. But the national focus after 1901 also reflected Roosevelt's attempts to forge American national power amid a global struggle for empire. The connection with empire was threefold. First, there was the obvious emulation of techniques from the British Empire and the desire to surpass British standards of technological achievement in irrigation. The second link was, paradoxically, a reaction against formal imperialism. Anti-imperialists argued for internal reform through a consolidation of the United States as a continental or inland empire. Water would be vital to this reform. The third aspect—Roosevelt's determination to strengthen the nation within an imperial system rather than to provide an alternative to imperialism—blurred the contradictions between these two perspectives. Both anti-imperial and pro-imperial forces could unite after 1900 on the virtues of developing a strong nation through irrigation of the arid West on a "settler" society model.[4]

American irrigation promoters and experts displayed a growing global awareness in the late nineteenth century. The periodical *Irrigation Age*, which Smythe founded in 1891, registered from 1898 to 1901 the significance of the imperial turn just as everybody else did and included items dealing with the Philippines' own irrigation potential and the plants that could be acclimatized from the tropical world. Aided by USDA technical advice, farmers competed for global markets but frequently took inspiration from the achievements of other nations and borrowed technology and biota.[5] When USDA personnel traveled to foreign places in search of plants, they brought back news of irrigation development as well. Based at Arizona's Agricultural Experiment Station at Tempe, USDA official Thomas H. Means toured North Africa in 1902 studying plants suitable for arid locations. Yet he found equally enchanting the record of irrigation in Egypt, which, he stated, was "now one of the most prosperous agricultural countries in the world, with crops superior to any raised elsewhere." He attributed this success to "the magnificent and untiring efforts of the English engineers in charge of irrigation and similar work."[6] "In many ways," he concluded, Egypt was "teaching the world."[7] This imperial and global context spurred the irrigation debate and produced a toolbox of ideas with which to improvise on the national level.

Revealing US national ambitions for a heightened international status was the representation of dam technology. In the first decade of the new century, irrigation engineers and other experts saw global prestige and national pride in the construction of impressive dams. Frederick Newell, the first director of the US Reclamation Service, had a "concern for monumentality in dam design," favoring "huge structures that would visually testify to the power of his vision."[8] For Americans, the search for "monumentality" was not simply a psychological quirk but a means of expressing imperial parity with the British. The Aswan barrages and dams on the Nile River generated inordinate interest in conservation circles.[9] Apart from mere curiosity, irrigationists as much as imperialists of every stripe and interest desired to learn from the British administration in the Egypt of Lord Cromer, who was generally thought by pro-imperialist Americans to have devised the ideal form of indirect rule and forward-looking empire. The fascination also expressed a desire to do better. American dams were compared with the Aswan on this basis, with the projected Roosevelt Dam on the Salt River in Arizona touted as the "World's Highest Dam and Largest Artificial Lake." Pundits noted that not only would the Roosevelt Dam in Arizona and the Elephant Butte Dam in New Mexico be bigger than comparable works in India and Egypt, but the American models would also do more for the advancement of humankind than European dams.[10]

The Roosevelt administration did more than display a competitive interest in imperial irrigation; it fostered broader international exchanges of ideas on irrigation too. Secretary of the Interior James Garfield encouraged the National Irrigation Congress to invite foreign delegates to its meetings

6.1. Roosevelt Dam, Salt River, Arizona, completed 1911. US Bureau of Reclamation.

and aided the dissemination within the United States of knowledge about foreign irrigation plans. Due to the nation's geography and size, the diversity of conditions in the different applications of irrigation made the United States especially attractive to foreign experts. Because US irrigation needs varied within the United States, it also made domestic sense to cultivate as broad a set of links with foreign irrigators as possible.[11] During Roosevelt's second term, Edward McQueen Gray, an Episcopal clergyman and president of the University of New Mexico, conducted research that emphasized the key role of national governments around the world in providing a stimulus to farming. The son of a British diplomat, Gray was in touch with the Canadian and British governments on agricultural development and irrigation through Britain's ambassador Lord Bryce, who spread news of American irrigation precedents to Canadian provincial governments.[12] In 1909 Gray published *Government Reclamation Work in Foreign Countries* for the Department of the Interior.[13] Gray traveled all over South America and Africa and, according to the press, knew "the continent of Europe thoroughly." Irrigators reciprocated by making Gray their official "foreign secretary" in 1908 to conduct what might be termed irrigation diplomacy.[14] In the eyes of leading irrigators, Gray's international contacts and frame of reference "made him particularly valuable." Gray addressed the eighteenth

National Irrigation Congress in Albuquerque in 1910 and reviewed the progress that had been made since 1902. Looking back at the work of the irrigation congresses, he stressed that they were "a potent force in the comity of nations." Gray boasted that "officials of this government at Washington" had approved of his efforts "toward establishing a nice balance between this and foreign countries." A good deal of conviviality had evidently accompanied the comity of nations at previous gatherings, and "distinguished" foreign attendees expressed the appreciation of "nations great and small in all parts of the globe."[15]

These international interests were not new in the first decade of the twentieth century. American irrigation schemes had long been part of an international discourse, both drawing on foreign examples and inspiring like-minded people in other nations. Because the yeoman farmer ideal and its garden landscape concept underpinned irrigation promotion in the British colony of Victoria, the Australian experience with irrigation had feedback effects in the United States. The Canadian American George Chaffey, who organized the irrigation settlement of Ontario, California, had gone to Australia in 1886, lured by offers of land to develop irrigation at Mildura, Victoria. He returned to the United States in the mid-1890s, where he continued the exchanges of information and sought to apply his Australian experience in further land development. The government of Victoria eyed with interest the National Reclamation Act and, in 1902, wrote Senator Newlands commending its passage. It soon embarked upon its own plans to put the state's watercourses under government control.[16]

Irrigation transfers were multilateral as well as bilateral. By 1900, irrigation technology and policy exchanges were closely integrated with the peripatetic nature of the engineering community, with Americans taking their social ideas and technical expertise from one country to another. Irrigation technology was, as Jessica Teisch shows in her study of American mining and hydraulic engineers, fundamentally transnational. The migration of irrigators also contributed to grassroots support for transnational exchanges. Afrikaner settlers discontented with their loss to Britain in the Boer War (1899–1902) came to New Mexico to settle in the Mesilla Valley. They were drawn there by prior knowledge of the irrigation reputation of California and the wider American West. The South Africa sojourns of American engineers such as John Hays Hammond had established that reputation.[17]

International exchanges in irrigation were institutionally represented in the national irrigation congresses. The 1893 congress in Los Angeles was formally billed as an "International Congress"—and had delegates from Australia, France, Russia, Mexico, and Ecuador. Later conferences had

Canadian, Brazilian, Chinese, German, Mexican, Peruvian,[18] French, South African, and Australian delegates.[19] These conferences demonstrated support for irrigation by an international fraternity of like-minded workers. Benefiting from cheaper and speedier international travel, foreign governments in the expanding settler societies of the Americas, Australasia, and South Africa sought ideas and expertise abroad for the development of their agriculture. Though India, Spain, Egypt, and Italy had long provided examples of pioneering irrigation work and inspiration concerning the technologies and legal frameworks of irrigation, from 1890 to 1910 foreign governments frequently sent emissaries to the American West to learn from the implementation of irrigation systems in a specifically democratic political culture.

At the elite level of politics, more immediate diplomatic pressures also drew American attention to the international dimensions of irrigation. In 1894 the irrigation congress in Denver called not only for a national commission to supervise irrigation works but also a permanent international body to adjudicate differences between New Mexico and Mexico over the diversion of waters from the Rio Grande upstream from the Mexican border. As the Mexican delegate stated, his people were already "being deprived of some of their water rights" and claimed "such rights as prior appropriation," while the Yankee population of New Mexico and Texas was "constantly looking on the waters of that river with longing eyes."[20] A boundary commission had been at work since 1889 under an 1884 treaty, but it was not empowered to consider irrigation diversions at all.[21] The Canadian delegates joined the lobbying during the succeeding congress at Albuquerque in 1895. As a result of concern over the Rio Grande's fate, the irrigation congress called for "appointment of an international commission to act in conjunction with the authorities of Mexico and Canada in adjudicating the conflicting rights which have arisen, or may hereafter arise, on streams of an international character."[22] But the United States rebuffed this initiative in 1896. Caught in the midst of a depression and in a presidential election year, the Cleveland administration had enough on its plate already, and the matter lapsed. It was also held up by internal dissension and litigation over the location of any dam built upon the Rio Grande in American territory.[23] Only upon Roosevelt's ascension was progress on the international front achieved. Mexico and the United States negotiated a border treaty in 1906 that focused on irrigation diversions in the upper Rio Grande. Though the result satisfied both nations' immediate needs, the treaty was still weighted toward US interests, and Mexico pushed for further bilateral discussions on the lower Rio Grande. Initiated in 1908, these were disrupted after 1910 by the Mexican Revolution. A new treaty eventually had to be negotiated in 1940.[24]

The 1895 Albuquerque congress represented a high point for international action on irrigation, and for several years thereafter, American political attention and that of the constituent states shifted to the internal aspects of the irrigation question. A solution to social and economic problems in rural and industrial America became pressing, even as foreign examples continued to provide inspiration. The opening of the major sections of western arable lands was mostly complete after the Oklahoma land rush of 1889, but less desirable public land continued to be entered for settlement.[25] In the 1890s many parts of the prairies faced extreme drought, and together with widespread urban unemployment in the depression of 1893–96, climatic conditions dashed farmers' dreams and spurred reformers' renewed interest in the use of irrigation to clear the cities of their huddled masses. In 1893 the *Los Angeles Times* had drawn attention to the unemployed and their distress when arguing for irrigation to create viable farming land. It drew upon the predictions of Britain's Lord Macaulay and Thomas Carlyle that American troubles "would begin when [the] public lands are exhausted."[26] This fear became part of the growing late nineteenth-century jeremiad of the American Republic that questioned American exceptionalism in the light of the economic depression, industrial unrest, and the official "closing" of the frontier. Favoring government help to restore progress and prosperity, the national irrigation congresses responded to such perceived threats. Now it was hoped that the federal state would help to prevent this potential fall from grace. Attention therefore shifted from the ideology of irrigation to its practical implementation. The 1900 congress was described as one that would "get down to business" rather than continue "generalizing on the stupendous possibilities."[27] Such an approach made the national sphere more important than the international.

Faced with persistent lobbying from the irrigation congresses for action on dams, both Democratic and Republican platforms adopted "pro-irrigation planks" for the 1900 presidential election campaign. Preliminary federal appropriations for western irrigation work began in that year, but no act specifically addressed the issue until 1902. Enterprising congressmen sponsored legislation to advance this cause and their own careers or business interests. Senator Newlands had extensive landholdings in Nevada, a state that he hoped would benefit from the harnessing of the region's rivers, to his political and perhaps economic advantage.[28] Initially, eastern interests continued a long-standing tradition of opposition to legislation that seemed unconstitutional intervention in local matters, and those interests blocked western irrigation appropriations, but key congressional opponents changed their minds after westerners filibustered a bill containing river and harbor projects desired by those same eastern interests.

Here, Roosevelt entered the fray. He claimed conversion to irrigation in the 1880s, while ranching in the West. In his *Autobiography* he recalled, "While I had lived in the West I had come to realize the vital need of irrigation to the country, and I had been both amused and irritated by the attitude of Eastern men who obtained from Congress grants of National money to develop harbors and yet fought the use of the Nation's power to develop the irrigation work of the West."[29] Immediately after his election in 1900 as vice president, Roosevelt promised the National Irrigation Congress support for government control of water storage facilities. The national government should give "generous aid to the movement," and "at considerable length," he stressed "the necessity for preserving the forests" as essential to the conservation and steady flow of irrigation water.[30] By 1902 the lobbying reached a climax with the National Reclamation Bill that proposed federal seed money for western irrigation projects. These projects would be financed over time by the sale of public land to small-scale settlers, who would then pay for allocated water from the irrigation projects as well. Roosevelt was a strong supporter of the bill, and he organized its passage with Republican Party arm-twisting of congressmen and senators. The provisions for prior surveying of the land for suitability and other supervisory and fiduciary powers given to the secretary of the interior were designed to ensure prudent and proper use of this US government intervention.[31]

Obtaining passage of the law required compromise. As historians note, Roosevelt toned down the "nationalizing" rhetoric of the original bill to placate powerful western interests.[32] Skillfully aiding him was the Democrat Newlands, whose grassroots political skills were considerable, and who worked for a compromise to counter the objections of western states to the accretion of national power.[33] As a result, the scheme catered to parochialism by proposing projects in proportion to the public lands of the respective states (an important concession to federalism), and this condition undermined the prospect for efficiency in irrigation expenditure.[34] Partly because of its modification, but also due to "lax enforcement" of the act, the small farmer's chances to benefit diminished.[35] Also influencing the act's implementation was Roosevelt's perceived need to demonstrate executive vigor. This prompted premature action, action that catered to political imperatives to shore up western constituencies for the Republican Party prior to the 1904 election. Roosevelt, a leading authority has stated, went "further" than he needed to go "to accommodate" these state interests.[36] The rushed and fragmented mishmash of dam and canal projects led to poor choices of soils and settlers and to "chronic problems with drainage."[37] This was an American story associated with federalism's effects on the structural and functional composition of the state, but it was not unique. In another federal system, Australia's, attempts to supply water for the benefit of small farmers were only partly implemented due to interstate wrangling over that nation's major river basin, and with comparable technical difficulties. Nor did alternative methods of farming in the West, such as dry-farming techniques, achieve consistently better results than irrigation farming.[38]

Despite the pork-barreling that was inevitable in achieving passage of the National Reclamation Act and the shortcomings in its implementation, experts gained an important platform to advance their objectives in irrigation politics. The act established a Reclamation Service headed by chief engineer Frederick Newell, which quickly had "little to do with the parent agency," the USGS within the Department of the Interior. At the national level, Newell worked hard to align the civil engineering profession with the aims of "permanence" and foresight in American conservation efforts.[39] This Pennsylvania-born ally of Pinchot had the job of vetting and running the projects. Newell emerged from the same milieu of the USGS as William John McGee (appointed secretary of the newly formed Inland Waterways Commission in 1907), Bailey Willis, and Joseph Holmes. Newell was active in the Cosmos Club and the American Geographical Society, where he, together with Pinchot, networked for conservation measures. Through Pinchot, Newell had met Roosevelt in 1900, and the three formed a vital triangle of executive government influence for the development of national irrigation policy.[40]

Though at times it seemed otherwise, this group was not just a collection of experts. Progressives regarded the American irrigation effort as a mark of civilization, comparable to the achievements of older empires. As much as politicians did, engineers emphasized social, rather than purely technocratic, outcomes. William A. Follett, who served the government as a consulting engineer in the international negotiations over the Rio Grande and the division of its waters with Mexico, informed the American Society of Civil Engineers that the US government provided not only farms, which the British did in India, but democracy, as British imperial precedents there did not. Speaking to the nation's civil engineers in March 1909, Follett charged that "Stupendous" as those British projects were in the number of people affected, "they do not mean so much in proportion to size as do those in the United States in adding to human advancement." Whereas the Chenab irrigation area in the Punjab had provided for one million people—"two inhabitants for every three acres irrigated"—and "ameliorated the condition of the natives," it had "not brought in so high a civilization as do smaller enterprises in the United States, where homes are made, universities founded, churches built, and all the complexities of a higher civilization set under way." Such reclaimed American lands constituted "a vast addition to the happiness of mankind."[41] These were the comments, not of a pure technocrat, but of a man who mixed social commentary on empire and nation with irrigation expertise.

The idea of purely economic efficiency achieved through a regime of technocrats was also compromised in the way that irrigation became linked to a highly politicized and racialized version of nation building. This was rooted in the idea of the white settler society as providing, through human labor working upon the raw materials of nature, a "settler theory of value" justifying occupation and development, with the United States as the preeminent example of what postcolonial theorists call "settler colonialism." That ethic endorsed the premises of an expansive Anglo-Saxon race.[42] The Reclamation Act itself evinced such a racial stance. It included a provision, advanced by Newlands, that no "Mongolian labor" be used in construction, an explicit legislative endorsement of a racialized settler landscape. Roosevelt's own version of the race argument did not require such extreme exclusionist machinery, since his worldview was Lamarckian rather than eugenic at the time of his presidency. He had "a divided heart" on the possibilities for uplift of the "lesser" races. But the more negative view was widely supported in the American West, especially among Democrats and pro-labor people, and was therefore politically necessary. The almost simultaneous 1902 extension of the Chinese Exclusion Act was a move that Senator Newlands recognized as working in tandem with the Reclamation Act. Both were ways of strengthening the racialized nation. And, though a Democrat, Newlands became to an extraordinary extent another of Roosevelt's right-hand men in the legislative implementation of his conservation program.[43]

Roosevelt viewed a nationalism embodied in an activist state as vital for this racially constructed landscape. Private enterprise had already taken on the relatively easy projects, leaving the most difficult and costly ones of broader significance to the "National government." "These great works" of irrigation were what Roosevelt called "an essential part in bringing the Nation to its full development" through the enhanced "well-being" of "a dense and vigorous population" that fertile, well-watered land would encourage.[44] As historian Laura Lovett has argued, "[i]ntertwined" with Roosevelt's "advocacy of masculinity and the 'strenuous life' " was concern for maintaining the farm population. Support for a racially white nation became centered on the reinvigoration of farm life, a move in which Roosevelt "equated the status of the nation with the status of the rural family."[45]

This was not a purely inward-looking nationalism, however. The concept was stated with an eye to momentous international implications. In 1903 Roosevelt proclaimed in a message to the irrigation congress that the National Reclamation Act was "one of the greatest steps, not only in the forward progress" of the United States and its constituent parts, but in "that of all mankind." His statement went much further than his endorsement sent to the same body in 1900. Roosevelt now transformed the legislation from a matter of regional politics to an imperial issue of world-historical development. The change was "so great," he averred, "that we hesitate to predict the outcome."[46] Boosters affirmed irrigation's potential to change the international economy, and irrigation experts fully expected the effect to be international as well as national. By drawing Europeans and the urban easterners to the American West to farm the formerly arid land, the region would become a critical food bowl for the world. The president's statement reflected such commonplace beliefs.[47] Implicitly for Roosevelt, American irrigation would lay down the platform for human progress, providing examples of good practice in irrigation to replace its subordination in older countries to nondemocratic governments, and it would underpin his idea of a hegemonic racial union of the Anglo-Saxon "settler" peoples. In Washington, pro-irrigation politicians shared these ideals. Though uneasy about formal empire, Senator Newlands saw the United States joining the ranks of those older civilizations that had practiced extensive state-controlled irrigation since time immemorial—so much so that irrigation was used "in more than half the world." The European powers, including Britain in Egypt and India, had deployed irrigation to enhance their agricultural strength, and Newlands believed that there was no reason why Americans should not undertake similar yet superior measures.[48]

For all that, in its affinities with imperialism, American irrigation differed from other aspects of Progressive conservation policy. It encouraged commitment to an internal colonialism that could be equated with the rule over foreign peoples that European powers had undertaken. Implicit in Roosevelt's position on irrigation was the belief that it would have the same transforming effect on internal policies as the American acquisition of overseas territory in 1898. Supporters of irrigation emphasized this internal colonialism. The Indian-born William Booth-Tucker, commander of the Salvation Army in the United States, spoke in 1903 of settling the West through "scientific colonization" and applauded Roosevelt's warning on "the dangers of race extinction" in the absence of ever-extending quantities of arable land. "[T]he well-being of the nation" would be promoted through an invigorated (white) settler family structure that such internal colonialism would sustain.[49]

This internal colonialism did not marginalize quasi-imperial thinking but incorporated it through its treatment of Native Americans.[50] The Bureau of Indian Affairs continued the allotment policies of the Dawes Act of 1887, but now with the aid of irrigation, especially after the passage of the National Reclamation Act. An Indian Irrigation Division developed "reclaimed" acreages in western states, with Native Americans relinquishing extensive tribal land in return for small, irrigated plots. Irrigation allotments were seen as a way to protect these supposedly helpless people by giving them a livelihood. In practice, this process released the surplus of large, formerly tribal territories to white settlers. In addition, Native Americans were pressed into labor to build aqueducts and dams, just as the British did with their own subject peoples in Egypt and India.[51]

In a self-conscious way, the Reclamation Service compared this "settlement" of Native Americans on individual plots to external colonizing. Frederick Newell explained how and why the irrigation program benefited the nation's indigenous population, and he drew an analogy with the subject peoples of the British Raj. For Newell, irrigation enhanced the development of "the neglected or little considered natives of the United States and of other countries where water conservation has been wisely practiced." He evidenced the "improved conditions . . . of India and Egypt, through the work" of that paragon of racial progress, "the British engineer." In an analogous way, American Indians were the beneficiaries of American "improvement" through reclamation projects. In their "native" state, Indians lived wild and could not devote time to the pursuits of social life because they had to compensate for lack of water by incessant labor and nomadism. Where storage reservoirs were built, however, these indigenous tribes could be "lifted" in "the scale of civilization," making them less dependent on hand-to-mouth existence and able to "practice better agriculture."[52] No longer need they maintain a "slavish dependence upon the fluctuation of water supply."[53] As Newell went on, he appeared almost to dream of himself as a pith-helmeted, khaki-clad colonial official. Whether applied externally through formal imperialism or internally through surrogate imperialism over Native Americans, irrigation produced a higher stage of civilization. Apaches were "members of a tribe reputed to be among the most bloodthirsty in the world," according to Newell, but "conservation and utilization of the water" made them "self-supporting citizens."[54]

While in some sense the irrigation policy realized a domination over indigenous people intrinsic to imperial state structures, the process was more complicated. The outlook was informed by more than the desire to reshape nature and people with the aid of engineering.[55] Newell gave his comments in a lecture well after the Reclamation Service had been established, and they described an imagined world rather than the real world of Native American land loss and labor exploitation. In practice, the two relevant bureaus, Reclamation and Indian Affairs, fought in Washington over protecting Indians from fraud and loss of land or water rights. The impact of irrigation on Indian tribes required especially careful management. Reclamation Service diversions of water threatened Indian economies where tribes continued to assert collective riparian rights. Roosevelt and his successive Department of the Interior secretaries sought to retain such rights for Fort Belknap Reservation Indians against upstream diversions by white settlers on the Milk River in Montana, for example, although the tribe's legal position deteriorated through further legal wrangling after the president passed from office.[56]

Roosevelt also stepped back from the brink of total assimilation to be achieved through the long-standing allotment policy.[57] Whereas the Dawes Act coerced Indians into becoming individualistic farmers as a prerequisite for citizenship and deprived them of vast acreages, Roosevelt preferred paternalistic intervention, not the ruthless and impersonal logic of bureaucratic expropriation by a domineering state. He did so, not out of respect for Indian culture nor entirely out of a nostalgic desire to preserve the ancient hunting ways of the West's first inhabitants as part of a pioneer milieu, but because of the noblesse oblige he considered inherent in imperial responsibility. With his "absolute honesty" and "devotion to the needs of the public," James R. Garfield became Roosevelt's mainstay in public land policy, and he represented for the president the high-minded ideals of service in this interest. Taking over as secretary of the interior in 1907, Garfield immediately saw the danger of making Indians subordinate to market forces, and he used the provisions of the Burke Act of 1907 to delay completion of land transfers to private ownership (and thus full citizenship under the act). On a western tour to familiarize himself with the situation, Garfield conferred with Newell and Indian representatives, paying particular attention to the alignment of allotment policy with irrigation. Garfield agreed with Roosevelt that Indians were "not competent to perform properly the duties of citizens." For this reason, administrators needed to supervise the public entry of land tightly, in order to prevent widespread fraud against the tribes (just as they sought a parallel policy for the entire community on oil and coal lands). Officials "must not agree to a line of action which would make it possible for the Indians to easily alienate their property," Garfield proclaimed. The imperialist mind-set underpinned the need for such intervention, since Indians were, in Roosevelt's view, a parallel case to "wards" of the European empires, where many years would pass before civilization was achieved, just as generations would be needed for Filipinos to be ready for independence. The negative side of this policy to protect Indians was, for many, a two-decade delay for Indian citizenship.[58]

Enthusiasm for this quasi-colonial task at home did not necessarily translate into popular support for formal US imperialism abroad, however. Committed anti-imperialists urged the irrigated empire as a clear alternative to the formal one. In the debates over annexation of colonies in 1898, Senator Richard Pettigrew (South Dakota) had argued: "Instead of spending hundreds of millions in conquering the Philippines, it would have been far better economy and better business judgment to spend it in reclaiming the arid lands of the west."[59] At the meeting of the American Forestry Association in 1899, Abbott Kinney, the president of the California State Bureau of Forestry, spoke eloquently. Significantly he was a Democrat, in which party opposition to Philippine annexation was strong. Kinney, who was also a real estate promoter and better known as the developer of Venice, California, put the case for creating national forests in the watersheds of the West (and their corollary, the steady supply of water in a dry land). His proposal was a "matter of vital importance to California, a question of expansion," but it was "expansion *under our own flag*. A great arid country lies at our door, . . . and we have the opportunity to conquer a great empire, to be taken up by people of our own blood and language, who will do a great deal to build up our trade and commerce." Contrast that with the search for an empire abroad among people not "of our own blood." In Kinney's anti-imperialism, racial objections to formal imperialism and support of settler colonialism at home were closely aligned.[60]

The theme "annex arid America" became a rallying cry at the 1900 National Irrigation Congress. Reflecting the anti-imperialist demands of Smythe, Kinney, and others, this "annexation" nevertheless demanded a national focus on irrigation programs and policies that had not existed before, and it was aimed at strengthening the American nation-state within the imperial system of states. Though irrigation appeared to be a purely domestic solution for social ills by 1902, the wider context was the promotion of an American empire rooted in the efficient harnessing of all natural resources, including water.[61]

For Roosevelt, the importance of irrigation lay in how it would invigorate the nation and foster the growth of American world power. In his first annual message (December 1901), he argued for the beneficent multiplier effects of irrigation projects, developments that could counter ideas of a stalled American sense of individual opportunity in the wake of the frontier's end. Irrigation would make the nation stronger and nurture trade:

> The increased demand for manufactured articles will stimulate industrial production, while wider home markets and the trade of Asia will consume the larger food supplies and effectually prevent Western competition with Eastern agriculture. Indeed, the products of irrigation will be consumed chiefly in upbuilding [*sic*] local centers of mining and other industries, which would otherwise not come into existence at all. Our people as a whole will profit, for successful home-making is but another name for the upbuilding of the nation.[62]

Whatever the actual economic impact of irrigation at the local level, the political and cultural promotion of irrigation suggested conservationist ambitions of creating a "big" state, as a part of the Republican "large policy" that would, in empire-like terms, dispense welfare to the ordinary farmer at home and paternalistic modernization of agriculture abroad in the island dependencies.[63] Despite the Jeffersonian rhetoric of the empowerment of small farmers, the aim was to use a strengthened rural America based on white settlement as a foundation for national power and the global projection of that power.[64]

Of course, the significance of irrigation itself mattered mainly to the West, an ideal place for the fantasies of internal empire that Newell articulated. But in Roosevelt's mind (and in that of Senator Newlands) irrigation was not a local panacea.[65] For Roosevelt especially, the Reclamation Service's work was a step only toward a coordinated national response to the program of industrialization in an increasingly global political economy. For this reason, irrigation must have spin-offs for other programs. By 1907, irrigation had become linked to a broader policy for the reordering of waterways in the interests of the nation's newfound economic position. In his 1907 message to Congress, Roosevelt stated, "Irrigation should be far more extensively developed than at present, not only in the States of the Great Plains and the Rocky Mountains, but in many others." Swaths of the South Atlantic and Gulf states should have irrigation too, "hand in hand with the reclamation of swamp land." More important, the president declared this rearrangement of nature to be a US government responsibility. The national government would be committed to the task of shaping nature with the knowledge "that utilization of waterways and water-power, forestry, irrigation, and the reclamation of lands threatened with overflow, are all interdependent parts of the same problem."[66]

By this time, Progressive conservation was adding the concept of multiple-use to its repertoire. Water would be harnessed for more than just irrigation, and its applications would now include hydroelectric power and flood control, using "the water resources of the public lands for the ultimate greatest good of the greatest number." This statement bore the indelible marks of Pinchot and of the multitalented USGS official William John McGee, who saw waterways as the foundation of a broad-ranging conservation policy for the final years of the Roosevelt presidency. But the marks of Roosevelt's own thoughts on American racial destiny through a prolifically expanding Anglo-Saxon settler society were present too. Only with that settler society approach would all the pieces of the conservation jigsaw come together. The federal government's responsibility was to "put upon the land permanent home-makers, to use and develop it for themselves and for their children and children's children."[67] This the control of water could do.

Events external to the United States encouraged such a universalistic view of irrigation's significance. While local and sectional interests demanded regional water policies, international contexts spurred interest in a nationwide plan. The development of national water policy was more than a simple response to parochial pork-barreling from below and to the need for sectional compromise. It also reflected external pressures from above, articulated through Roosevelt and Pinchot. The demands of both Canada and Mexico for treaties covering boundary waters, first exposed in the national irrigation congresses of 1894 and 1895 and renewed after 1902 by Canada, raised issues of not only national but also international significance.[68] Shared international waterways could not be adjudicated in line with efficient resource allocation without linking them to a coordinated national strategy on conservation.

A second pressure favoring a more nationally focused and comprehensive water policy was the organizational and technical model of an explicitly imperial venture. With Roosevelt's conniving, the United States acquired the Panama Canal Zone in 1903 to implement the long-desired objective of a sea route through the isthmus. A canal would enhance the nation's ability to defend its growing strategic interests in the Pacific, but the economic and social implications of such a canal were also substantial. As the planning for the ambitious project proceeded, economic interest groups and federal officials saw how the canal could spur the development of waterways within the United States. These changing international circumstances determined quite precisely the timing of the Inland Waterways initiative.[69]

The exercise would center on the very heart of the country and involve the redesigning of "nature" in the Mississippi Valley. With construction under way in 1906, the Panama Canal became central to this strategy. In the late nineteenth century, farm production across the upper Midwest and Great Plains increased, but so too did international competition for markets. Whether because of discriminatory railroad rates, as often alleged by Progressives, or not, rail transport failed Mississippi Valley farmers seeking to sell their crops at competitive rates nationally and internationally. As railroads acquired canal and riverboat interests in the 1880s, water alternatives to rail almost disappeared. From Chicago to New Orleans, trade groups dreamed of using engineering to refashion rivers, thus allowing farmers "to ship crops entirely" through the Mississippi Valley river system by the inexpensive, if slower, means of a rejuvenated water traffic.[70] Midwestern interests that looked to the Latin American and China trades for the answer to farm overproduction sensed the opportunity, with the prospect that the region's exports could be cheaply carried to distant Pacific as much as Atlantic ports.[71] The waterways project was hatched in regional lobby groups that thought along these lines; they wished to extend federal river and harbor projects. It was, federal officials admitted, "largely in response to a petition" from such groups that the Roosevelt administration's campaign for the improvement of the Mississippi Valley waterways began.[72]

Farming was not the only interest. The Panama Canal "raised the possibility" that, along with western wheat, Pittsburgh steel "could be delivered to the West Coast or to the Orient more cheaply" than from the Atlantic seaboard.[73] Though railroads were bugbear number one for Progressives, track congestion and the boxcar shortages encouraged even some rail companies to seek relief via water transport redevelopment. Railroad baron James J. Hill championed the "improvement" of rivers. In part, this stance was economically rational, since if agriculture were more strongly profitable, Hill's railroad interests that served the agrarian Northwest would become more lucrative. Though the produce of the upper Midwest and Great Plains had doubled in the ten years from 1897, railroad capacity reputedly increased only an eighth. Due to his own experience, Hill was also painfully aware of rail's shortcomings.[74] His Great Northern operations had incurred the wrath of Progressives for failing to supply coal during 1907's vicious winter, and Hill now agreed with Roosevelt that railroads were no longer "able to move crops and manufactures rapidly enough to secure the prompt transaction of the business of the Nation."[75] The "canalization" of the Mississippi would provide relief from rail competition, natural obstacles, and "transportation congestion," Hill argued. In future, both Roosevelt and Hill believed, rail and water networks should be coordinated. To the *Outlook*, it was Hill's railroad position that made the argument for reform most convincing, as his case seemed to transcend the obvious special pleading of port interests.[76]

The leadership for action to canalize rivers in the Mississippi Valley came from sectional economic groups. W. K. Kavanaugh, president of the Lakes to Gulf Deep Waterway Association, pushed the case strongly. The owner of cross-Mississippi ferries and a major coal transportation firm in Saint Louis, among other interests, he proposed a National Department of Public Works "just as the government is building the Panama Canal." He thought comparable river work of national importance ought to be speedily and efficiently completed under government auspices. In Kavanaugh's opinion, reduced railroad rates to the Atlantic had followed the opening of the Eads Jetty system, which had cleared a siltation bottleneck at the mouth of the Mississippi River in the late 1870s. When presented with waterborne competition throughout the valley, Kavanaugh suggested, the railroads would reduce rates to save their trade and thereby make life easier for farmers.[77]

Though initiated at the local level, at the national and international levels the architects of American national power were deeply involved in turning potentially pork-barrel schemes into an overarching plan to strengthen the American empire. The figure behind this particular branch of the new empire was McGee. Trained as a lawyer, not a geologist or engineer, he first rose from a position in the USGS to prominence in the Bureau of Ethnology within the Smithsonian Institution, taking charge of the Ethnological Exhibit at the Saint Louis World's Fair.[78] Subsequently, he moved to the Bureau of Soils, from which he was tapped to join the Inland Waterways Commission in 1907. A pretentious man who did not use periods for his initials and went only by the moniker "W J," McGee came to be known as "no points McGee."[79] McGee certainly had a high opinion of himself. Before dying prematurely from cancer of the jaw (probably from incessant pipe smoking), he arranged to have his brain dissected so that the entire scientific community could be shown the wonders of its working. There was, the sage proclaimed, a "need for studying the brain of intelligent persons."[80] Sometimes discounted retrospectively by historians, in life McGee was certainly influential, especially when working with Pinchot and Roosevelt. Pinchot even claimed that McGee was "the scientific brains of the Inland Waterways Commission, as well as of the Conservation movement in its early stages."[81] Nevertheless, Newell of the Reclamation Service also pressed for a comprehensive water policy. Irrigation expertise was thereby put into service to plan and execute the inland waterways strategy. An "excess" of water could be handled as the obverse of a deficiency, and the engineering experience of the American West applied to the humid East. In both cases, waste would be stopped.[82]

Federal officials came to see water transport as the lynchpin in a campaign to coordinate the nation's resources. To McGee it was not so much that the nation faced a crisis of resources in any absolute sense. Rather, it needed to order these treasures more efficiently through superior brainpower. When viewed against the surging growth of mining, power usage, and population, the case for comprehensive national planning in water policy seemed to McGee convincing. Refining the mantra of efficiency, McGee noted that production and consumption would hum in harmony if the entire spectrum of interests were harnessed in a national plan. Irrigation, electric power, flood control, and water transport must be managed together, with water "better utilized in order to derive full benefit from lands and forests and mines," he argued.[83] With his wide government experience in the USGS and the Bureau of Soils, McGee brought together different aspects of conservation policy, enabling him to see the whole. The logic of geography and hydrology drove McGee to favor such a comprehensive multiple-use scheme, quite apart from the need to harness pork-barrel interests at the local level. Forest and soil conservation had, McGee believed, an enormous bearing on flood control and navigation, due to the thousands of tons of valuable soil deposited as silt in the nation's major rivers each year. To McGee this was yet another element of waste. Spurred on by such anxieties, McGee began to turn conservation into a secular religion, and under his influence, the Inland Waterways Commission gave the conservation movement "its first great impetus" as a multi-issue movement.[84]

If geography, geology, soil science, and hydrology drove McGee's ideas, he also linked the scientific analysis to an argument deeply rooted in American exceptionalism and the nation's providential role. With the "opening of the Panama Canal," wrote McGee in a sweeping geopolitical brief presented to Roosevelt, "a new era will dawn in exchange between our eastern and western coasts, and between these and the rest of the world."[85] More prosaically, Roosevelt, along with Pinchot and McGee, envisaged that the giant ditch across the Isthmus of Panama could enhance the American empire by strengthening the national economy. Navigable rivers would move resources and food more cheaply than rail and thus, coordinated with the canal, develop certain food and fossil fuel resources more efficiently, while saving others for the nation's future.[86] Conveyance by water would lower the cost of transport and hence prices to "producers and consumers," leading to "cheapened power," an issue at the core of Progressive conservation aspirations. The scheme seemed to McGee a masterstroke of a masterful mind—his own. Though the issue of the national economy was central, the political implications of a society in which the working class were increasingly hostile to the accumulation of great wealth by capitalists underlay this thinking.[87]

When it came to "expansion" and "large policy," McGee was as overtly imperialist as Roosevelt. He rejoiced at US territorial acquisitions in the Pacific and saw these processes as linked to America's moral leadership through innovation in resource use.[88] In "The Growth of the United States" (1898), McGee had depicted territorial enlargement across the continent and abroad as "natural actions by an 'enlightened' nation, whose purpose was 'the elevation of humanity [and] the ultimate peace and welfare of the world.' "[89] His "National Growth and National Character" (1899) referred in Social Darwinist terms to the "Law of Human Progress." Thereby, humanity moved in as "orderly" a way as "planetary orbits." From "savagery into barbarism, thence into civilization, and finally into enlightenment," human progress across its "vital stages" was "never dropping backward save by extinction."[90] In McGee's stages-theory interpretation, "Americans had surpassed the level of achievement of their European colonial predecessors," becoming the "offspring of the strongest stock" that the world had ever seen.[91] Americans represented "enlightenment," the pinnacle of evolutionary hierarchy. They had evolved as "fit representatives of humanity, invincible in war yet generous to fallen foes, subjugators of lower nature, and conquerors of the powers of primal darkness."[92] Adapting Rudyard Kipling's exhortation for Americans to "Take Up the White Man's Burden," McGee concluded that Americans must assume "the Strong Man's burden" and develop a more muscular political economy to lead the world's less fortunate peoples to universal progress.[93] The Bureau of Soils official had modified his Social Darwinist roots and, like Roosevelt, moved toward a Lamarckian evolutionary position in which better knowledge would propel the fittest peoples upward. By representing the highest stage of humanity, the bulked-up United States was duty bound to take on world leadership.[94]

On 4 February 1907, McGee wrote a remarkable draft letter for Roosevelt to send back to him as the president's own invitation. Therein, it was envisaged that Roosevelt would appoint McGee and four others to an Inland Waterways Commission. His associates would be Pinchot, Newell, General Alexander Mackenzie, chief of the Army Corps of Engineers, and Lawrence

O. Murray, the assistant secretary of labor. In the draft, McGee stated that the United States had great resources in minerals and agriculture, but he believed that the nation could better utilize its vast mineral stores, about which there was so much anxiety at the time, if the problem of the waterways were first solved. This would occur through conservation and development of the entire Mississippi Valley.[95]

The waterways initiative was a project of "The Valley" not as a mere catchment area but as an imagined space, seen as the centerpiece of future American greatness. Academics applauded how the role of the Mississippi Valley seemed to be rising in presidential estimation. American historians organized their regional historical groups into the Mississippi Valley Historical Association at the same time in 1907, studied river and harbor improvements, and strongly supported the idea of national conservation. McGee even published one of his key articles, "The Conservation of Natural Resources," in the pages of the historians' proceedings. He was acutely aware of the world-historical significance of the movement with which he had become associated in the Roosevelt administration. This movement he sought to document historically and scientifically.[96]

The Inland Waterways Commission came under McGee's guidance as secretary, with four extra members added to garner political support for the bureaucratically led initiative: Representative Theodore E. Burton (R-Ohio), as chairman; and Senators Newlands, William Warner (R-Missouri), and John H. Bankhead (D-Alabama). The commission was expected to seek out international benchmarks that, when surpassed, would cement the nation's place in the geopolitical firmament. McGee was concerned to "take account of recent progress by other nations" in hydraulic engineering such as shown in the Dutch draining of the Zuider Zee. Especially he pointed to the achievements of the dominant empires then contesting for world power. German advances in water management, especially on the Rhine, were well known in American engineering circles. McGee reminded Roosevelt that "several of the leading rivers of the German Empire have been opened up to deep-water navigations [*sic*] so that many of her interior cities are seaports." This work reputedly made the German Empire industrially strong and efficient, but Chicago and Saint Louis could be equivalents for an American empire. In their far-flung possessions, the British were not neglectful of rivers either. The Nile had been "brought under control," McGee rejoiced, "by British enterprise," and to repeat the cliché common in US irrigation circles, Britain had in its empire undertaken "some of the most notable engineering works in the world." McGee understood that the Inland Waterways Commission would not follow any single international benchmark, but he warned Roosevelt that "the example of the mother-countries should be understood and intelligently applied."[97] This was hardly a lesson the president needed to be taught.

The "conservation" adopted here had, of course, nothing to do with wilderness and much to do with human manipulation of nature. Social reformer Frank Vrooman saw McGee's plans as the sculpting process that qualified the master Roosevelt to be justly called the "Dynamic Geographer." In this interpretation, the genius of the proposed waterways model was not so much that it overturned "nature" as that it improved and completed nature as part of the evolutionary process. The proposed canalization of the whole Mississippi Valley would enhance the drainage that nature had already provided courtesy of the deep waters of the Mississippi River, "whose channels Nature has digged [*sic*] so conveniently in their ramifications for the uses of this great area."[98]

Reorganization of nature in this way was indeed McGee's aim. "While rivers rank among the important resources of any country, they are liable to become destructive agencies endangering life and impairing the value of considerable districts," McGee opined.[99] Certainly, there had been great floods, as the recent inundations of 1906 in the lower Colorado illustrated, but the 1906 flooding had happened because a private company had recklessly diverted the bountiful river's water to a faultily constructed irrigation channel. In the United States, such disasters, whether "natural" or not, had been a spur to action, especially federal action, and McGee noted that "some of our most notable engineering enterprises have grown out of efforts to control the destructive tendencies of streams." In McGee's view, the "ultimate end" should be "the entire control of our running waters and the complete artificialization of our waterways in such a manner that this great natural resource may be turned wholly to the benefit of our people for the benefit of humanity."[100] This last concept McGee seized as "the keynote of my personal idea,"[101] yet the idea was far from original. It was fundamental to the way the engineers of the Suez Canal and the Aswan Dam had worked. McGee himself agreed that water "utilized by artificial means in any country" was "the measure of the civilization and progress of that country."[102] Civilization was, in turn, the sine qua non of Rooseveltian Progressive thinking about the progress of the American nation and empire.

McGee divined alchemistic significance in all this, defining progress as mastery of the elements of fire, earth (minerals), air, and water. Of these, he claimed, water was the last to be conquered. This step would be "no less sweeping and cosmic than the Conquest of Fire." This process could be perfected only when water was "not merely guided and directed in movement, but actually drawn from the materials of Nature at human behest as to time and place and quantity, much like the fires used now in the arts and crafts."[103] In this poetic and portentous sweep, control of the elements would not be complete until water was turned into a source of power. Herein, hydroelectricity became part of dynamic geography's mix, with improvement in long-distance transmission allowing these renewable sources of energy to be tapped for the benefit of rural areas alongside water for irrigation and flood control. In this multiple-use and utilitarian application, waterways development would be the most revolutionary change in the history of the world, McGee concluded, and he would be alongside Pinchot and Roosevelt at the center of it. McGee could stare out of his Washington bureaucrat's window and see it all so clearly now.[104] Earth, fire, wind, and water—Progressive conservation would complete the global conquest of the elements.

This complete conquest of nature had especially strong connotations in the United States, because McGee, like many others, assumed that American abundance required only efficient manipulation to make the nation the dominant world power. Water again was the key, since in other respects the nation was still relatively rich in resources, despite waste in timber and coal that he explicitly condemned. McGee measured water not in absolute terms but against the population and area of the nation. Contrasted with other major countries, the figures showed how the United States could continue to grow demographically while sustaining a much larger industrial capacity, if only water were better distributed. "Our growth in population and industries is seriously retarded by dearth and misuse of water," he claimed. One million square miles of the West—"Fully a third of our territory"—was equal to the area of Great Britain, Germany, France, Spain, Portugal, Italy, Austria, and Denmark combined, and yet this "wasteland" remained "practically unoccupied and nearly unproductive by reason of aridity." The reduced rate of economic development threatened by prodigious waste elsewhere in the Republic could be balanced if a national hydraulic revolution were undertaken: "With half our land area and the same water, our capacity for population and industries would be as great as now; with twice our water equally distributed over our present land, our capacity would be more than doubled."[105] Reflecting his faith in experts (such as he believed himself to be), McGee anticipated a happy outcome resulting from the growth of knowledge and "mental capacity," which would "guide material progress" in such a "nation of science."[106]

Publicity, publicity, publicity—those three essentials were not to be forgotten either. The work of the Inland Waterways Commission burst into the national spotlight with the decision to have its members take a Mississippi

6.2. William John (W J) McGee, architect of water reforms. LCPP.

River cruise in early October 1907 to inspect key sites. There, Pinchot, other members of the commission, and Roosevelt joined local advocates for waterway reform. Aboard the aptly named steamer MV *Mississippi*, Roosevelt went from river port to river port, extolling the case for national action. The cruise revealed Pinchot's consummate knack for media manipulation. As a man who understood that the press had learned to "follow" Roosevelt, and believing that "action" was "the best advertisement" for government policies,[107] Pinchot spent nearly two weeks on the river with the president. Roosevelt spoke at river towns from Cairo, Illinois, to Saint Louis, Memphis, and Vicksburg. At Memphis he addressed the Deep Waterways Convention, announcing that rivers were to be part of a comprehensive national inventory of resources. This was soon revealed as his most significant ploy. In Vicksburg he proclaimed the need to build levees, using the example of "little Holland," and again drew a parallel with the vigorous but careful administration of the Panama project. Roosevelt also speculated that the South as much as the West might partake of the bounty of the waters for irrigation. That way the beneficence of government would be more widely spread—a most politic position—yet justified in terms of efficiency.[108] This was "probably the most successful presidential publicity event" that Pinchot and Roosevelt created, one press historian claims. Certainly the chief forester rejoiced at the "front-page news," not only in the United States but "in many other parts of the world also."[109] Only a financial panic on Wall Street threatened to divert political attention as 1907 waned.[110]

The *Preliminary Report of the Inland Waterways Commission* presented on 3 February 1908 called for coordination of local, state, and federal governments in the development of canals, rivers, lakes, and ditches. It supported multiple use (with flood prevention, navigation, and forest watershed protection as essential) and included irrigation as part of the plan. In his accompanying message, Roosevelt highlighted the commission's call for a permanent National Waterways Commission to coordinate water planning and implement the recommendations. As historian Donald Pisani has pointed out, Roosevelt "hoped to use the Panama Canal Commission, a largely autonomous, seven-member panel . . . as the model for a permanent body."[111]

At this move Congress baulked. There was some dissent within the commission to give the nation's legislators pause. The Army Corps of Engineers representative Mackenzie objected to the majority report. He opposed a permanent commission, because it would have put his own agency, which was responsible for national rivers and harbor improvement, under civilian control and unnecessarily usurped existing technical capacity. Some railroads

6.3. Pinchot and Roosevelt aboard the MV *Mississippi*, October 1907. LCPP.

likely to be adversely affected also opposed the plan as not economically viable.[112] Yet the idea of a national public works program was highly political in any case. Roosevelt's lame-duck status after the election of 1906 contributed to a stalemate, so too did Roosevelt's increasing resort to such sweeping executive unorthodoxy, a position made clearer from the time of the eleventh-hour withdrawal of forest reserves from the public lands in

March 1907. The commission did not become permanent, but a different five-year National Waterways Commission under (now) Senator Burton's chairmanship was created in 1909.[113] That Roosevelt did not get his way was merely a temporary setback. He had already decided that the waterways plan was only a stage in the development of his even broader program, and the report led directly to the appointment of the National Conservation Commission by executive fiat, which was tasked with compiling an inventory of the nation's natural resources. One section of the commission, under McGee's leadership, concerned itself with water.[114]

Even as Congress refused to support the work, and the Army Corps of Engineers fought against the recommendation that a separate and permanent waterways body be established, federal officials and influential civil engineers continued to advocate a coordinated public works program under federal control covering all existing government agencies. For several years, the Panama Canal remained a model for a new exercise in nation building. As John Hays Hammond argued, the Panama Canal experience should be applied to "the active work of the government in connection with the natural resources of the country." This would include centralized administration within a single bureau or department covering forests, the draining of swamps, irrigation projects, and "power sites on public waterways."[115] Pinchot and other Progressives took up this multiple-use philosophy now clearly articulated in the wake of McGee's work. They would fight for watershed protection and federal control of hydroelectric power over the next decade. Successes and failures followed, mostly failures, and the plan for comprehensive nationwide resource management was scotched when Congress passed the separate Federal Water Power Act of 1920. Not until the 1930s and the formation of the Tennessee Valley Authority did such all-encompassing watershed development take place. In the meantime, the water reforms were sidelined, and having been blocked on the national level, Roosevelt and Pinchot turned to still grander plans of international action.

SEVEN

The Problem of the Soils and the Problem of the Toilers

The frontier had "closed" in 1890, the pioneer days were over, and the United States would have to face that fact. So concluded the historian Frederick Jackson Turner in his famous address of 1893 at the Chicago meeting of the American Historical Association. Subsequent research showed that Turner's thesis needed many qualifications; for one thing, a good deal more "frontier" land was opened up to farming after 1890.[1] Yet, if we reinterpret Turner's judgment to include the use of cheap, plentiful, easily accessible resources of the soil that had no prior exposure to intensive agriculture, there is more than a kernel of truth in Turner's insight that the nation faced social, economic, and cultural readjustment. At the turn of the twentieth century, momentous changes were under way in agriculture and in the relation of country to city. The rise of the cities was drawing people off the land. Poor market opportunities for farmers in the 1890s depression gave way to fears of declining agricultural yields in the boom years that followed. The railroads' discriminatory freight rates squeezed farmers tightly and made cities seem more attractive. In many observers' eyes, the windfall fertility of "virgin" lands across the West had almost disappeared.[2]

Those who toiled upon the land responded to the crisis at the local and state levels. In the aftermath of the failed Populist movement of the 1890s, these toilers turned to organized reform. They urged relief from mortgage debt, spread the idea of farmer cooperatives, and sought to improve the conditions of rural life. Europeans had pioneered the cooperative movement, but even before 1900, American farmers began to form their own. Established federally in 1905, the National Farmers Union was one such organization aimed at the reform of agriculture. Though less overtly radical than the rural protest movements of the 1880s and 1890s, and "less representative of tenants" than of landowners, the Farmers Union championed a range of other Progressive interests, including natural resource conservation, farm credits, and postal savings banks. By 1910 there were 1,600 Farmers Union cooperative warehouses in the cotton states. Because farmers relied on the rural press and sympathetic urban reformers to spread their message, the media was important in stirring a public consciousness about rural problems. The increasing publicity for and political lobbying by farmers brought rural disquiet to the notice of intellectually and politically influential people. This internal anxiety coincided with geopolitical fears that the nation's international comparative advantage in agriculture was being squandered.[3]

No one more strongly articulated concern over farm conditions than James Jerome Hill. At first blush, the Canadian-born American railroad baron was an unlikely choice for one of Pinchot's handpicked speakers at the Conference of Governors in 1908. He had fallen out with Roosevelt when the administration prosecuted the railroad trust as a restraint of trade. Hill bristled at the president for this act of treachery against business in the Northern Securities case.[4] But Roosevelt and Pinchot could not ignore Hill, who by 1906 was an outspoken campaigner for conservation. Across the Midwest he told audiences that the United States was indeed a wasteful nation, and his comments were reported nationally, and approvingly.[5] Foreshadowing Roosevelt and Pinchot's onslaught of 1908, Hill drew attention to all kinds of natural resource waste, but especially that centered on agriculture.

For Hill, "American wastefulness" was most obvious in rural areas precisely because the frontier and its "free" land had so recently been "closed." At the Conference of Governors, Hill voiced a common perception that soil had been "mined" the way gold and silver had. "We are only beginning to feel the pressure upon the land," he warned. "The whole interior of this continent, aggregating more than 500,000,000 acres, has been occupied by settlers within the last 50 years. What is there left for the next 50 years?" Soil erosion in the South produced a lamentation fit for Jeremiah: "Millions of acres, in places to the extent of one-tenth of the entire arable area, have been so injured that no industry and no care can restore them."[6]

The railroad magnate hoped to counter the crisis with diversified farming. Livestock should supplement grain production in the most arable sections, he urged. With wheat farms declining in productivity, livestock would add to wealth creation and, through manuring, return the fertility stripped from the soil by wheat's monoculture. Hill put his money where his mouth was by investing in agricultural technology and offering prizes for high-quality farming in the tracts along his railroad line in Minnesota and the

Dakotas.[7] At his North Oaks Farm near Saint Paul, Minnesota, he assumed the role of innovative farmer and, championing modern commercial butter making, installed a Swedish De Laval separator and barrel churns.[8]

What turns a businessman (an unlikeable "robber baron" in some eyes) into a conservationist? In Hill's case, the Great Northern Railway, extending from Saint Paul, Minnesota, to the Pacific coast, had a hand in it. The line carried more agricultural freight than any other in the nation, giving Hill good commercial reasons to seek continued farm prosperity that would entice people onto the land and keep them there. But Hill also believed fervently in the values of rural society, those traditions that held yeoman farming as the "most dignified and independent occupation" and the chief source of republican virtue, in contrast to vice-ridden cities.[9]

Rather than stress economic matters alone, Hill emphasized the same fear of class upheaval that motivated Progressives in their campaign for more efficient energy. Hill thundered: "Not only the economic but the political future is involved. No people ever felt the want of work or the pinch of poverty for a long time without reaching out violent hands against their political institutions, believing that they might find in a change some relief from their distress." Though Americans had not yet taken such drastic action, Hill believed that "the unnecessary destruction of our land will bring new conditions of danger."[10] The recent organization of the Farmers Union was a straw in the wind. This social restiveness necessitated prudent steps to reverse the environmental damage that lay at the heart of the political and economic challenges.

In addition to concern for internal social stability, Hill's perceptions as a transportation tsar were strongly fashioned by direct external threats. He foresaw the rise of Japan and rebirth of China, arguing that, given low labor costs in East Asia, those countries would eventually outcompete Western nations in manufactures. The true superiority of the United States lay in its vast and fertile landmass. However, this greatest national asset in the peaceful war for commercial supremacy faced the slow death of rural decline because many farmers were forsaking the land, while the cities were bulging at the seams through mass immigration. Failing to predict the future trends in commercial farming in the twentieth century, Hill nevertheless correctly perceived the specter of increasing national and global population pressure. Farmers could not keep up with the growing demand for food. Soon the nation would be unable to feed itself, let alone provide the exports upon which the American economy depended for prosperity. Hill raised the stakes on this topic in his landmark October 1906 speech at the Minnesota State Fair.[11] The Saint Paul resident told listeners to "GO BACK TO THE LAND," as

7.1. James Jerome Hill, railroad baron and model farmer. LCPP.

the headline screamed. The future was "Menaced" by the "Rush from Farms to Cities."[12] This address was widely received and applauded.[13]

The geopolitical consequences were not lost upon bureaucrats and politicians. James Wilson, the long-serving secretary of agriculture, worried that international competitiveness would slide because the nation's trade surplus still required strong agricultural exports, now imperiled. Government officials were "fearful of the international consequences" of a "static agricultural productivity." More efficient agriculture was a necessity for the future, since the bounty of the land remained "closely related to the maintaining of national power."[14] As Cyril Hopkins of the University of Illinois put it, the United States was "one of the great agricultural nations," and "without agriculture America is nothing."[15]

Hill correctly identified soil as the leading resource in immediate danger. Soil erosion caused by inadequate attention to flooding and drainage was well known. At the governors' conference, Andrew Carnegie concurred with the diagnosis, telling delegates in alarming tones that echoed W J McGee's views: "More than a thousand millions of tons of our richest soil are swept into the sea every year, clogging the rivers on its way and filling our harbors. Less soil, less crops; less crops, less commerce, less wealth."[16] Hill agreed with the experts that this erosion was a serious problem but went further: "Far more ruinous, because universal and continuing in its effects, is the process of soil exhaustion. It is creeping over the land from East to West. The abandoned farms that are now the playthings of the city's rich or the game preserves of patrons of sport, bear witness to the melancholy change."[17] Soil exhaustion led to the declining yields that many feared, resulting in rising costs, which meant more expensive food and, eventually, an inability to feed the nation's people.

In theory, the calamity of ruined farms could be averted easily enough, since soil could be restored. Yet while nitrogen- and phosphate-based fertilizers could help, these were either in short supply or wastefully used. To be sure, nitrogen was abundant in the atmosphere, but enormous amounts of energy were required to convert it to a usable form,[18] and fossil fuel energy itself was, as we have seen, subject to projected future shortages. American farmers and the USDA were well aware that major supplies of the naturally occurring phosphates in guano were far away in the Pacific islands and Peru, and they were dwindling. Nations had already squabbled over these prizes, and as early as the 1850s the United States had acquired its own "guano islands" to shore up the losses of soil productivity attributed to cotton and tobacco overproduction, as soils became spent in the eastern states. These distant supplies had been quickly scoured. Mined phosphate could be an alternative, and mining operations had recently commenced in the United States, but phosphate itself was something that Roosevelt's advisers regarded as threatened. Charles Van Hise affirmed this in the discussions on the NCC's draft report at the Joint Conservation Conference of December 1908. Far too much phosphate went down the drain as sewerage and was dumped into rivers, a loss that did not occur in Asian societies. The "fertility of the soil in phosphorous may be perpetually maintained," but American efforts in this regard fell sadly short of Japan's.[19]

The other source of critique over fertilizer concerned exports and revealed the growing importance of international resource competition in perceptions of scarcity. A form of quasi mercantilism was on the verge of developing. Experts alleged that valuable mined phosphates were being sent overseas despite the fact that, as Cyril Hopkins contended, phosphorus was a relatively rare mineral in the United States.[20] Though almost as important in corn production as abundant potassium, Americans sold this golden cargo abroad with gay abandon, and the nation's reserves were, F. B. Van Horn of the USGS asserted, "drained for the benefit of the worn-out farm lands of foreign countries."[21] Fully 46 percent of phosphate rock left the country. Van Hise warned the Joint Conservation Conference that these exports should be forbidden, because to do otherwise would be "nothing short of agricultural suicide."[22]

Not new, the widespread concern over lost farm productivity went back to the 1890s, along with the perception of a dwindling supply of free land. In 1894 a Division of Agricultural Soils within the Weather Bureau was established, but not until 1899 did a National Soil Survey begin, and only in 1901 did the soils division gain greater stature as a separate bureau in the USDA, due to increasing demands for the rehabilitation and preservation of this precious resource. Recall that W J McGee of water reform fame worked in this bureau in 1907 and dealt with hydraulic erosion of soil at the time he was chosen to head the Water Section for the NCC in 1908.

The facts of soil exhaustion were hotly contested, however. The Bureau of Soils became a source of obstruction for Roosevelt's plans when its leadership proceeded to refute the jeremiad of waste and destruction. This outcome differed from that for other resources. In the case of forests, Pinchot and his allies exercised close control while, under the influence of the USGS, oil and coal experts favored Roosevelt's position too. In contrast, an entrenched bureaucracy in the USDA had been dealing with threatened agricultural yields for two decades, and it required more convincing that the new alarm was justified. Adding to the resistance was the idiosyncratic stance of Milton Whitney. A chemist and geologist who had lectured at the

University of Maryland, Whitney was poorly qualified in academic achievement and temperament for the position, but he nevertheless became head of the soils division in 1894. There, Whitney pushed ideas that verged toward crackpot science. He sometimes confused his own tendentious scientific positions with official policy and clashed with his field researchers.[23]

To be sure, Whitney started out from a similar premise to that of the alarmists, demonstrating the wider consensus on the need to act on conservation for efficiency's sake. He agreed that "fertility of the soil" was "a great national asset" and that "practically all available new lands in this country" had already "been taken up." Thus, improvement in the use of existing farmland would have to "be more rapid than in the past to maintain the integrity of the nation and its commercial supremacy among the nations of the world." Whitney agreed that the national government's role in averting the crisis of soils was extremely important, as did Roosevelt, McGee, and Pinchot: "It must be a national effort participated in by all the people, as upon it depends the very foundation of the continued prosperity of the nation."[24]

That said, Whitney used international evidence to refute the alarmist claims of the Conference of Governors. From comparative studies of soil fertility trends in many countries over the course of the previous century, Whitney drew conclusions sharply at odds with those of Hill. Whitney concluded that management of the world's soils "appears to be improving," not declining, "at least in modern times." This improvement was "greater in Europe" than in the United States, due to the denser population and therefore "more pressing" need to farm intensively and efficiently.[25] That is, a more closely cultivated society would divine ways to counter the deficiencies of pioneer agriculture. There was nothing to fear. Further, Whitney disputed the idea discussed at the governors' conference that American yields were declining, claiming that they were actually increasing on a decadal average.[26]

A USDA colleague, E. C. Chilcott, supported Whitney's skepticism, though from a different direction. In charge of dryland farming in the Bureau of Plant Industry, Chilcott rejected the opinions of "the superficial observer," such as those who studied the subject "from the viewpoint of the great transportation companies." This was a transparent jibe at Hill. The perceived falling off in yields was due to other causes, Chilcott advised. "Changed economic or sociological conditions" or more conservative farming were likely to be responsible for any apparent decline. He argued, "Agricultural production can best be increased by adopting vigorous measures to remedy faulty economic conditions," not environmental ones. At bottom, it was necessary to insure to farmers "a fair share of the price paid by the consumer," without "exorbitant profits" going to "middlemen." Provision of agricultural machinery at affordable prices also needed encouragement. This approach was highly sympathetic to the idea of agricultural cooperatives.[27]

Yet Whitney went further than Chilcott. He announced the extreme position that soil fertility was inherently "indestructible, immutable," and perpetual in character. The policy implications were plain: the alarmist call for fertilizer conservation was wrong, and the energy implications of increased fertilizer use were therefore irrelevant. Whitney maintained that soil texture was the most important factor. Correct tillage and crop rotation, already being implemented under the tutelage of the USDA, would stem any crop yield slide attributable to the frontier's end.[28]

For all these reasons, the chapters on agriculture in the report of the NCC in 1909 did not endorse the anxieties of Hill or Carnegie or of scientists such as Van Hise. Inadvertently, however, the USDA encouraged questionable schemes of agricultural rejuvenation and inexhaustibility on the coattails of Whitney's advice. Whitney's concern with soil texture and correct tillage (greater cultivation to increase the health of the soil) encouraged both the vernacular contention that rain followed the plow and certain arguments of dryland-farming advocates that deep tillage could conserve moisture. Though not officially propagated by the USDA, dryland techniques became widely advocated for marginal farming areas. Ultimately, Whitney's viewpoints on soil depletion and inexhaustibility would be discredited. From experience in the county-based soil survey project that commenced under the Bureau of Soils would come the genesis of soil conservation initiatives of the New Deal era. Hugh Bennett, the chief of the Soil Conservation Service (1935), first worked on soil erosion under Whitney in 1903.[29] But the short-term result of USDA policies in 1908–9 was to turn the issue of agricultural reform toward the sociology of agrarian life and to broader questions of social and environmental sustainability. Nature would no longer be seen apart from its cultural, social, and political context. Humans would be part of what must be conserved.

The widespread calls for greater agricultural efficiency were always closely associated with systemic cultural anxieties over the future greatness of America. The closing of the frontier and the shrinking of the nation's remaining "wild" territory spurred efforts to preserve the rugged pioneer ethic of the frontier, symbolized in Roosevelt's penchant for the strenuous life and the cowboy culture. But whatever his personal sentiments, Roosevelt's abiding political and cultural concern was farm life rather than wilderness. His 1903 message to Congress referred to "the unwholesome tendency" for people to move to the cities. The "more active and restless young men and women" had been driven from farms, rebelling "at loneliness and lack of mental companionship."[30]

These views were commonplace. A vocal defender of rural towns and New England villages was the *Independent*, a liberal, originally Congregational magazine of social, political, and religious affairs edited by Hamilton Holt. A peace and Prohibition advocate with a long family pedigree of reform, Holt had studied at Columbia University and held Progressive views shaped by anxieties over the rise of industrial society. He highlighted dire changes going on "with astonishing rapidity in American country towns" caused by "the centralization of industrial capital" and a concentration of power at the state level rather than the local community. "Country life," he announced in 1901, was "unquestionably endangered" and with its demise would go "the most cherished traditions of American republicanism." This cultural concern was tied to gender anxieties. The *Independent* wrote in Rooseveltian terms of the "deplorable" decline of rural neighborhoods that were "once abounding in vigorous manhood."[31]

In a manner that revealed the geopolitical dimension of these fears, the *Independent* carried the debate one step further and pointedly linked the problem of rural society to the anxieties of empire. Following Brooks Adams's line of argument, Holt's magazine editorialized on the "internal decay" that Adams identified in fin-de-siècle America and called it the underlying reason for "the overthrow of once mighty empires." The "ruin of villages, and even of entire agricultural provinces, by the flow of wealth and power to the great cities" was in fact responsible for the "disintegration of imperial Rome." Most likely, in the centuries to come the same would be said of the United States, for what was the nation after 1898 if not a latter-day Rome?[32]

Roosevelt shared such concerns over the impact upon farms of internal industrialization and urbanization, on the one hand, and external pressures, on the other. He made it clear that economic efficiency alone was not the aim of Progressive reform. After all, those farmers who stayed on the land were already benefiting from rising prices in the years before World War I. Agriculture was "not the whole of country life," the president concluded. "The great rural interests are human interests, and good crops are of little value to the farmer unless they open the door to a good kind of life on the farm."[33] More than better roads, schools, and postal services, true rural reform meant enrichment of social life. It was imperative to develop the "great ideals of community life" and "personal character." Roosevelt favored strengthening rural churches through provision of YMCA facilities, as in cities. More fundamentally, he sought to bolster the rural family because it was the source of the next generation of citizens and workers.[34] The "especially important" duty of the state was to "prepare country children for life on the farm," in order to stem the exodus to urban areas.[35]

Partly for this reason Roosevelt had long thought about gender roles and their implications for country life. The "condition of women on the small farms, and on the frontier, the hardship of their lives as compared with those of the men," he found both impressive and worrying. Their welfare must be foremost in "whatever was done for the improvement of life on the land."[36] This notion was consistent with his views on racial "vitality," views that made the procreation and raising of healthy, moral children the most important duty a woman owed to the "life of the nation." This was a highly gendered ideal since, he asserted, "neither man nor woman is really happy or really useful" unless doing "his or her duty." A woman who shirked her duty "as housekeeper, as homemaker, as the mother whose primary function" was "to bear and rear a sufficient number of healthy children" was "not entitled to our regard." There was "no more important person" in national life than "the farmer's wife." She must provide for farm labor through this endless stream of procreation. Thereby, she could hold up more than half the world. Without healthy, procreating families, "the farms of America could not sustain the cities," which in turn could not feed "the hungry nations" abroad nor remain the "stay and strength of the nation" in peace or war.[37] In this way, the political and intellectual critique of the flight from the farms was much more than economic and reflected a wider social concern that Roosevelt sought to address. He saw a need to satisfy "the higher social and intellectual aspirations" of country people and to build the countryside "upon its social as well as upon its productive side." At the same time, the proposed rural agenda placed a heavy moral burden on women and implicitly censured those who wanted something different from babies and life on the farm.[38]

For Roosevelt, rescuing farmers was fundamental to social sustainability on a national, as well as an international, level. Not an ecological sustainability, to be sure, this social and natural resource–based sustainability became the core of Pinchot and Roosevelt's conservation mantra. It became manifest when, in mid-1908, Roosevelt appointed yet another commission in addition to the recently established NCC. The Commission on Country Life was particularly important, not in its immediate results, but as a revelation of Roosevelt's personal vision. He closely supervised its proceedings.[39] As Scott J. Peters and Paul A. Morgan state, the commission was "one of the first high-profile, comprehensive attempts to sketch out a broad-gauge vision of sustainability in American agriculture and the technical, political, cultural, and moral demands its pursuit would make on individuals, professions, associations, institutions of civil society, and all levels of government." Few historians have dealt with this quest for a sustainable rural life as part of the conservation movement. They have misunderstood wise-use conservation in purely economic terms or dealt with conservation as separate from the reform of rural life. But Rooseveltian Progressives did not make these distinctions.[40]

Rather than seeing preservation of nature and conservation of resources as separate agendas, Progressive conservationists most often incorporated both in a larger ethic of social efficiency. The term "sustainability" was not used, but the kernel of the idea was present. Many who supported the national park movement illustrated the mix in societies, such as the ASHPS, that urged protection of scenic beauty. These groups supported rural reform by means of country life as well as the development of not only a national park system but also regional and local parks.[41] A submission paper from the ASHPS to the Joint Conservation Conference in December 1908 expanded on the theme: "The beautifying of the country village or of the farm; the utilizing of the land by the roadside, in place of allowing it to degenerate into a weed bed; the building of proper roads, and the fencing in of attractive points, will do much to keep the young man of [*sic*] the farm where he belongs, where he will be most happy, and where his labor will increase the annual sum, of our agricultural production, the greatest asset of the nation."[42]

They were allied in this work with Horace McFarland. Best known as a champion of national and city parks, McFarland saw the role of farms as an important complement to them in the promotion of scenic beauty and a sustainable society. At the 1908 Conference of Governors, he put it thus: "The smiling farm, the glowing orchard, the waving wheatfield, the rustle of the corn[,] all these spell peaceful beauty as well as national wealth, which we can definitely continue and increase." McFarland identified rural landscapes as aesthetically pleasing compensation for industrial ills. Drawing on the ideas of English socialist and art critic John Ruskin, he attempted to insert beauty and well-being into the utilitarian conservation debate over coal and iron: "Can we not see to it that the further use of our unrenewable resources of minerals . . . is no longer attended with a sad change of beautiful, restful, and truly valuable scenery into the blasted hillside and the painful ore-dump, ugly, disturbing, valueless?"[43]

The same blend of utility and beauty applied, country life advocates believed, to a resolution of the threatened split within conservation over

7.2. Ugly culm (waste coal) dumps ironically came to be seen as a tourist feature or mark of regional distinctiveness, as in Scranton, Pennsylvania. LCPP.

national parks. Liberty Hyde Bailey, a professor at Cornell University's School of Agriculture, would best articulate the point in his statement to the National Conservation Congress held at Saint Paul in 1910. He believed that national parks and national forests were complementary, not projections of alternative visions of preservation, on the one hand, and of efficient economic conservation for maximum sustained forestry yield, on the other. They should have "mutual regard for each other's claims."[44] In this scheme of things forests were, as much as national parks, both aesthetic and utilitarian. Landowners would gain, stated the ASHPS, from the "beautiful prospect" adjacent to a national forest.[45] Bailey believed that nature and culture were not separate. He criticized the overly mechanized world as being at the heart of the modern crisis and advocated restoring the balance of nature as the solution. "Soon or later, every person feels this desire to plant something. It is the return to Eden, the return to ourselves after the long estrangement of our artificial lives."[46] In this view, the real world of nature was not the national park but the farm. Humans needed to work with this "nature"; they could justifiably change it only within physical limits that respected the earth's life-support systems.[47]

Not only have historians failed to address the role of Country Life and its supporters as a project in attempted social sustainability, but they have also failed to see the international connections that pushed Roosevelt himself to see the world in this way. Roosevelt's ideas about the sources of the nation's rural problems were formed by many experiences, but the inspiration for action came from an Irish land reformer more than anyone else. Roosevelt's *Autobiography* recalled: "One man from whose advice I especially profited was not an American, but an Anglo-Irishman, Sir Horace Plunkett. In various conversations he described . . . the reconstruction of farm life which had been accomplished by the Agricultural Organization Society of Ireland, of which he was the founder and the controlling force."[48]

Plunkett's life was a transatlantic reform story. Born in Gloucestershire in 1854, the third son of a landed aristocrat, Plunkett graduated from Oxford University in 1877 and returned to manage his father's hereditary estate in County Meath. He did not stay long. A brother had died of tuberculosis, and fearing development of the same complaint, Horace sought "a drier climate." For ten years from 1879, he became a rancher in the foothills of the Rocky Mountains in Wyoming. Coincidentally, Roosevelt also went west during the same decade, spending much of the period from 1883 to 1887 raising cattle in the Dakota Territory. The two did not meet on the frontier, but Plunkett did develop a lifelong connection with and love of the American West.[49]

Forced in 1889 to take control of his late father's property, he soon became deeply involved in reforms designed to stem the decline of Irish agriculture. Not to the United States did he look principally for answers but to the agricultural cooperation movement pioneered in Denmark. Plunkett worked to alleviate the poverty of Irish farmers by introducing state-of-the-art Danish dairy technology and the cooperative methods of marketing. He wished also to improve the farmers' social environment. Deeply committed to his homeland as he was, Plunkett nevertheless retained farming and other business interests in the American West. Indeed, he found there certain clues to the Irish agricultural dilemma. It was the comparative productiveness of American agriculture and its need for better organization that stirred him. As historian W. G. S. Adams has stated, "He saw the rising tide of agricultural production" in North America and its consequences for the rural economy of Europe[50]—namely, that Europe must introduce better technology and that farmers must combine if they were to match their American competitors' economies of scale.[51]

Though Plunkett met Roosevelt briefly in 1895 in New York,[52] his close association with the president did not begin for another decade. From

1905 he worked through Pinchot, lobbying Roosevelt to create within the USDA a "Bureau of Rural Social Economy." The latter would spread the benefits of better social organization of farm life, including social advances available in the city (such as electricity and comprehensive schooling). Involvement in Roosevelt's work was accentuated by the president's embrace of efficiency in conservation. Plunkett understood Roosevelt as few did, shrewdly noting that "Roosevelt's chief test of national efficiency seems to be the manner in which the mental and physical energies of the people are applied to the resources of the country."[53] After a Washington meeting in January 1906, Roosevelt asked him to put on paper his views about American agriculture, and the two continued to meet on a regular basis through the rest of president's second term. Roosevelt was very happy with these meetings, telling Plunkett, "I wish you were an American and either in the Senate or my cabinet!" Plunkett stressed that if the president acted on rural reform, his actions would be welcomed "throughout the world."[54]

In 1907 Plunkett presented TR with "a memorandum" on the "entire social economy of the rural population of the United States." He noted that Roosevelt had already signaled through public speeches his in-principle support for such a program. Plunkett believed that other government agencies must supplement the worthy record of the USDA, bringing to farmers the "achievements of sciences, arts and inventions" that had already been "conferred upon the dwellers in cities and towns." The proposed Bureau of Rural Social Economy would coordinate such an agenda and "investigate the facts" on rural improvement by studying "analogous conditions abroad." Plunkett impressed upon Roosevelt the importance of agricultural cooperation along lines shown by "farmers of every progressive European country."[55]

At Roosevelt's request, Plunkett conferred on this position paper with Secretary of the Interior Garfield and Pinchot in the spring of 1908, and Pinchot suggested the appointment of a Commission on Country Life "as a means for directing the attention of the Nation to the problems of the farm, and for securing the necessary knowledge of the actual conditions of life in the open country."[56] There was delay in getting the Country Life idea up and running, partly due to the increasingly fierce battles with Congress in 1906–7 over presidential executive power. This, Pinchot told Plunkett, "drove me entirely away from agricultural matters for a time."[57] But in August 1908 Roosevelt finally announced the creation of the commission.

The original nominations for membership were Liberty Bailey as chairman, Kenyon Butterfield (president of the Massachusetts College of Agriculture), Walter Hines Page (editor of *World's Work*), Pinchot, and Henry

Wallace (editor of *Wallace's Farmer*). Charles S. Barrett of Georgia[58] (president of the National Farmers Union) and William A. Beard of Sacramento, California (a commercial farmer and editor of the *Sacramento Bee*) soon added regional expertise and grassroots knowledge. Thirty public hearings followed, and the commission undertook a pioneering sociological survey by distributing to farmers and other citizens over half a million questionnaires. Some 120,000 were returned. Newspapers were contacted and they, too, spread the message. Fifty-eight percent of the respondents were farmers. The information presented was overwhelming in volume, and there is some indication that Bailey paid little attention to the actual questionnaire answers, though their detail was at least partly reflected in the final report. The answers emphasized the absence of effective organization for cooperative marketing as the critical weakness in the farming economy.[59]

Through Pinchot, Plunkett wrote the draft of the Message to Congress of 9 February 1909 presenting the *Report of the Commission on Country Life*, but this influence was theatrically concealed. Pinchot and Plunkett exchanged telegrams in code in a clandestine effort to hide the Irishman's influence, apparently because of fears that congressmen might see him as an interfering foreigner. Due to these circumstances, Pinchot had to apologize that the report transmitted to Congress was "imperfect" in reflecting or acknowledging Plunkett's views.[60] In his accompanying letter, however, the president made his allegiance plain. Admitting defensively that it was not the business of government to force farmers to reform but to provide information, Roosevelt followed Plunkett further than the final report itself did in telling farmers what to do, stating that the cooperative ideal was indeed the most important conclusion to draw from the study, because it "develops individual responsibility and has a moral as well as a financial value over any other plan."[61]

The commission responded to the widespread anxieties over soil exhaustion but set the environmental problem in the context of social relations. Soil depletion had social causes and social consequences, the study found. These included absentee landlords and capitalist farming ignorant of agricultural and climate conditions. Through soil depletion, farmers became a dependent class or "tenants in name, but laborers in fact and working for an uncertain wage."[62] Because the causes were not purely economic and reflected inadequate social capital, the report treated education reform as vital. In one of its central recommendations, the commission called for "a system of extension work in rural communities through all the land-grant colleges with the people at their homes and on their farms." Such practices could diversify farming and allow the tillers of the soil to "redeem themselves" from the "bondage of an hereditary system." Other recommendations included specific calls for rural transport through an expert-led "highway engineering service" that would cooperate with the states on road improvements, as well as legislation for a parcel-post system and rural savings banks.[63]

Though the commission is rarely considered as part of Roosevelt's conservation program, Pinchot's presence ensured that soil and forest depletion would be emphasized as causes of the farmers' economic problems. In unmistakably Progressive rhetoric, the report condemned "wastage" in the "control of forests," where "private gain" had prevailed over the national interest.[64] Streams had been "ruined for navigation, power, irrigation and common water supplies," and whole regions were "exposed to floods and disastrous soil erosion." To combat these evils, the commission seconded the "strong demands" of farmers for forest reservations in the White Mountains and the southern Appalachian region, measures pushed by Pinchot's supporters.[65] It also called for an inquiry into the control of streams in order to protect their public ownership for agricultural uses.[66] Stern strictures against monopolization of water rights further reflected Pinchot's influence and the conservation values that he and Roosevelt worked to promote at every opportunity. In an important concession to the scenic-beauty lobby, which had the ear of Bailey, the commission urged that all policy be based on "a realization on the part of all the people of the obligation to protect and develop the natural scenery and attractiveness of the open country."[67] This would include "preserving the natural features and developing the latent beauty in such a way that the whole country becomes part of one continuing landscape treatment." Estimates of "natural resources" must now include the "value of scenery," the commission recommended. Without conserving and appreciating "all the beauty of landscape," it would be "impossible to develop a satisfactory country life."[68] Given its sweeping agenda that united farms, forests, and parks under its proposals, the Commission on Country Life was not a sideshow in the conservation story but central to Roosevelt's plans for perpetuating his legacy. It was another weapon with which to deploy an increasingly complex and sophisticated program that he and his aides were devising in his second term for the reform of American society along sustainable lines.

What the commission failed to recommend was the creation of Plunkett's proposed bureau in the federal government.[69] Bailey recoiled instinctively from external or nonfarm interference in the renaissance of American agriculture; he believed rejuvenation had to come from within the farm community itself. Besides, a new bureaucracy was not likely to win support in

Congress, let alone the USDA. Nor was there any immediate political backing for Plunkett's idea of systematic foreign studies, again partly due to the potential expense and also because of Bailey's pragmatic belief that, in the United States, European collectivist ideas would turn out to be instruments of capitalism, not community. "Many of the so-called 'cooperative' " organizations are really not such," the report concluded, "for they are likely to be controlled in the interest of a few persons rather than for all and with no thought of the good of the community at large. Some of the societies that are cooperative in name are really strong centralized corporations or stock companies that have no greater interest in the welfare of the patrons than other corporations have."[70]

This is not to say that the commission rejected the cooperative idea entirely. Reflecting the influence of the National Farmers Union, the report recommended an expert investigation of the role of middlemen in the handling of farm products and revealed such economic concerns as transportation rates and rural credit.[71] The commission also accepted that existing cooperatives might have useful social spin-offs by training "workmen in habits of thrift, if the men were encouraged to join them."[72] The ill ease with which the commission regarded cooperatives was manifest, however, in the linking of any benefits to the homily that "the primary cooperation" was "social and should arise in the home, between all members of the family."[73]

Despite the reputed neglect of the social survey of citizens, the commission's findings largely complied with the drift of the replies, and the legislation forthcoming from Congress followed the main lines of interest revealed in that survey. The widespread dissatisfaction manifest among farmers in their responses to "economic concerns" and "training offered by schools for life on the farm" was eventually heeded. Yet other, more socially oriented complaints concerning the "opportunity to get together for mutual improvement, entertainment and social intercourse" could never result in legislation.[74]

Plunkett's direct impact was limited further by the animus of Congress toward the whole commission idea and by the refusal to pay for the publication of the report and its evidence. Though Roosevelt asked for a budget of twenty-five thousand dollars to cover publishing the replies to the survey, legislators flatly refused, and the report languished as a Senate Document, with only a few copies in existence—but not for long, since Roosevelt and Pinchot had played a clever card in authorizing the social survey. It stirred the newspapers and created a mass constituency. Publicity for the cause of rural reform was deliberately generated in this way, and a connection with Rooseveltian conservation goals established.[75] Ultimately, the attempt of

Congress to stifle the report backfired by actually increasing demand. The Spokane Chamber of Commerce published and distributed it for free, before it was republished commercially in 1911.[76]

Despite the obstacles put in his path, Plunkett had forged a strong alliance not only with Roosevelt, an alliance that persisted after the latter left the presidency, but also with Pinchot. Just as Pinchot was a close adviser to TR, Plunkett remained an adviser to Pinchot.[77] Pinchot was attracted to Plunkett as a practical man but also because he offered a way of linking the struggle against waste with the broader social and political objectives of the former president. Pinchot saw Anglo-American agriculture as particularly wasteful, and he looked to European examples of sustainability. Among "the English-speaking peoples," the farmer "had been one of the chief wasters," Pinchot told Roosevelt, and "must in the end become the chief conserver of our resources and our civilization. So the Country Life and the Conservation policies are really the two great parts of the supreme whole." Herein lay the interconnectedness of rural problems and conservation of the "natural" environment.[78] To further these objectives Plunkett revisited the United States several times and supported the Progressive conservation causes that Pinchot continued to espouse. In the following years, Plunkett also wrote indefatigably on the "rural life problem," endorsing conservation as rural life's twin.[79]

Yet the impact of the commission was not merely intellectual or theoretical. The technical and specific recommendations were eventually carried through despite congressional hostility to Roosevelt's high-handed executive style of government. Taft adopted the postal savings bank proposals (1911) and domestic parcel-post system (1913). Then the major recommendations for extension education were legislated in 1914 and for farm credits in 1916.[80] Finally, after World War I came national action on agricultural cooperatives. Progressivism did not die when Roosevelt left the White House, nor did conservation. The ideas that he, along with Plunkett, Bailey, and Pinchot, pioneered would be taken up, albeit in narrowed form. The story of this process involves two more presidents, but Taft and Wilson would have to deal with the consequences of the ideas that Roosevelt's presidency set loose, not only on Country Life but also on the preservation of the nation's wild and scenic heritage and on the entire spectrum of conservation issues.

EIGHT

Conservation, Scenery, and the Sustainability of Nature

The Americans who worried about coal shortages, farm fertility, and forest depletion espoused an ostensibly utilitarian version of conservation. Yet many of the same people were also concerned about the fossil fuel revolution's effects on aesthetics, social life, health, and the future of the Republic. The anomalies and paradoxes could be stark. At the 1908 Conference of Governors, ACA president Horace McFarland upbraided millionaire industrialists for foisting "ugliness" upon the American landscape while seeking to escape from the grime of eastern cities across the Great Lakes to holiday homes in Canada. The audience, which included a good few millionaire magnates and hard-nosed politicians, applauded.[1] The contradictions of the president himself were palpable. The same Roosevelt administration that created national parks and wildlife refuges also took the first steps to designate the Hetch Hetchy Valley in Yosemite National Park as the site of a dam and hydroelectric facility for the earthquake-ravaged people of San Francisco. Those who see Roosevelt as a nature-loving president have not properly reconciled this tension. Perhaps Roosevelt "was too many-sided and paradoxical to be pigeonholed," perhaps the tension was just an untidy fact of practical politics and its compromises, and certainly one about which he admitted twinges of regret, but he never changed his mind over Hetch Hetchy. Instead of examining this contradiction, historians have swooned over Roosevelt's preservationist legacy.[2]

Long before the Hetch Hetchy dispute, Roosevelt had gained a reputation as a great advocate of saving nature for its own sake. During his presidency, the key move in this regard was the setting aside from the public lands certain areas for protection, for forests and wildlife or for scenic or historical preservation. By the time he left office, the president had signed acts creating five national parks and proclaimed fifty-one wildlife reserves and eighteen national monuments.[3] These achievements are signally important in posterity's perception of the twenty-sixth president. Since Roosevelt was an avid student of natural history, the public and historians have concluded that preserving natural habitat was at the core of his values. He was the "wilderness warrior" and the enthusiastic naturalist. An aficionado of game hunting, Roosevelt stood against the senseless slaughter of wild creatures and yet favored the rugged pioneer values of the West that included hunting.[4]

This chapter argues that preservationist sentiment was not completely separate from utilitarian conservation but actively engaged with it and was part of a more general sensibility concerning the disruption of nature and need for repair. Roosevelt's activist support for nature protection grew in tandem with the more utilitarian conservation that he also fostered. Many of his forest reserves, national monuments, and wildlife preserves were created in the last two or three years of his presidency, when the drive for efficiency-oriented conservation was ascendant. Actors in the debates over conservation did not see a hard and fast choice between conservation for future use, on the one hand, and preservation, on the other. Some of Roosevelt's supporters insisted that wild game would flourish in the forest reserves he created and that distinctive American species would thereby be preserved. Other ardent nature lovers saw national forests as themselves impressively scenic, even when destined for distant future use.[5] Looked at from the perspective of 1908 rather than in the wake of modern ideas concerning wilderness, it seems more correct to interpret preservation and utilitarian conservation as part of a larger shift in attitudes. Moreover, this sensibility was not American alone so much as international, and it involved transnational influences upon American conservation.

Just as utilitarian conservation emerged full blown in the last few years of Roosevelt's presidency, so too did the preservationist cause quicken and mature quite suddenly. European-inspired preservationist impulses were filtered through the American experience of the rise of urbanization, industrialism, and imperial expansion. Exposed to an international drive for nature's conservation, especially on issues of wildlife destruction and the rise of a civilization of steam, steel, and electric power, American preservationists broadened their campaigns from the built environment of historic homes, landscape gardens, and city parks to the protection of national monuments, areas of scenic beauty, and national parks. Scenic preservation ideas and practices that regarded nature as part of cultivated landscapes developed more and more into interest in the wild as part of a preservationist ethic. The imminent desecration of Niagara Falls due to faulty regulation of the newfangled hydroelectric power sites in both Canadian and American jurisdictions became a particular international cause célèbre and was an important yet now "overlooked" galvanizing agent, but there were many episodes and forces that built momentum toward preserving American nature.[6] Elite preservationists joined hunters, scientists, philanthropists, humane society advocates, and a broader middle class of Americans in seeking to stop the erosion of American heritage and the nation's natural patrimony, at a time when other countries were doing likewise and were creating their own national parks and game reserves.[7] A member of the old elite who shared much of its value system, Roosevelt was at the center of turmoil over this bundle of issues during his presidency.

One of Roosevelt's earliest acts, bird conservation, seems unequivocally an example of his love of nature. When in 1903 he proclaimed Florida's Pelican Island as the nation's first wildlife refuge, this was a sign that TR was going to pursue an unorthodox and vigorous agenda for nature protection. He declared the Florida bird refuge without explicit legislation, basing his executive action on the general oversight of public lands and the power to withdraw land from sale.[8] Of course, Roosevelt was not acting alone. His move followed lobbying from the American Ornithologists' Union and the amateur naturalists of the Florida Audubon Society.[9] Grassroots support among an assortment of amateurs and professional groups helped Roosevelt on his way.

Preservation of wildlife had many roots. Some lay in the attitudes of scientists, particularly zoologists and ornithologists, and others among hunters seeking to preserve stable supplies of game for their own sport. The history of bison protection is familiar because that iconic species came so close to extinction. Allies of TR, especially Bronx Zoo director William T. Hornaday, took leadership roles in the bison rescue mission at the same time that conservationists in the European empires were questioning the indiscriminate slaughter of elephants in Africa.[10] A less spectacular yet equally important part of American conservation anxiety was bird protection. With birds used prodigiously for feathers, eggs, and meat in both national and international trade, the need for national action on species vulnerability was pressing. Already the passenger pigeon was thought unlikely to survive. Yet bird protection had ramifications for other aspects of conservation too, notably retention of forest habitat and water tables.[11] Nor was bird protection a matter for hunters and scientists alone. The issue resonated in the eastern cities and in middle-class homes, where products derived from birds were used, especially for fashion accessories. There, preservationists tapped the conscience of middle-class women in the club movement,[12] who drew upon the tradition of the humane societies for the prevention of cruelty to animals and gave the campaign against bird slaughter "the flavor of a moral crusade."[13] While bird protection laws began to be passed in many states in the 1890s under the lobbying of the state-based Audubon societies, there was no uniform national law, nor was there action in every state.[14] The Lacey Bird and Game Act of 1900 laid the groundwork for federal preservation of wildlife, even though that measure was a modest one that merely stopped the trade in birds and feathers in interstate commerce to and from states that had introduced their own laws.

"Firsts" implies a bold break with the past, but Roosevelt's Pelican Island edict of 1903 was a limited one. Pelican Island was a tiny mangrove island off the east coast of Florida. Not useful for development, the land was therefore more easily protected, and the action was noncontroversial at the national level. Roosevelt could make any executive order to preserve public lands that he liked, but without congressional funding and a bureaucracy to police the ruling, the position of the pelicans was problematic unless private aid was forthcoming. For this reason the Pelican Island reserve developed as a public/private partnership, under USDA control, with the Florida Audubon Society providing and, initially, paying for a warden. The warden could not stop the clubbing to death of hundreds of pelican chicks in later years, but the principle of a wildlife refuge had been established, and the president created fifty others before his second term ended.[15]

Such preservationist actions required a strongly supportive political network. Audubon societies and the Boone and Crockett Club of gentlemanly hunters that Roosevelt himself had helped form in 1887 were examples of public mobilization necessary to sway legislators and useful for presidents. Yet strategically placed power brokers were best of all, Roosevelt discovered. No one was more aware that, domestically, the presidency was mostly a "bully pulpit," since, after all, he coined the phrase. He needed allies and had a strong and, indeed, almost indispensable agent in John F. Lacey. An eight-term congressman from Iowa, Lacey's roots in the nation's agricultural heartland were important. For several years Roosevelt was closely reliant on Lacey in the House, since the Iowa Republican finished his congressional career with twelve years' seniority on the Public Lands Committee, a body with considerable influence over development issues. From the passage of the act creating Yosemite National Park in 1894 onward, Lacey had been a supporter of conservation and aligned with Roosevelt from his time as a city reformer, through his vice presidency, and to the Oval Office.[16] The Iowan also sponsored the bill for Wind Cave National Park in 1903 and for Mesa Verde National Park in 1906.[17] An ally of the forestry chief as well, Congressman Lacey publicly called Pinchot "my dear young friend" and professed to "look to him for the future of the forests" and for the preservation of wild birds that depended on the forests for habitat. As Lacey stated, "While preserving the forests you will preserve the animals that roam therein; while preserving the forests you will give shelter to the birds of the air that make their nests therein."[18]

Drilling down into the passage of the Lacey Act of 1900 reveals the complex machinery of Progressive Era conservation: the mixture of utility and preservation; the equally complex overlapping relationship between the local, the national, and the international; and the legislative compromises that were necessary. At first glance Lacey might seem an odd example of the conservation ideal. As he frequently remarked, he came from a farm state not covered with forests, but Iowa had a flourishing movement for state parks, a movement that revealed attitudes toward preservation and conservation that were similar to those of Roosevelt.[19] Lacey's politics were far from orthodox Progressivism, since he supported the railroad interests within Iowa. A member of the conservative, stand-pat faction in his state's political brawling, he nevertheless pursued matters of "conviction" at the federal level, and his conservation efforts led to his being held in "greater respect" in Congress than back home.[20]

In part, Lacey's motives for bird protection involved deeply held ideals that he shared with the humane societies, ideals that went back to Henry Bergh's American Society for the Prevention of Cruelty to Animals, founded in 1868.[21] Yet the alliance with Roosevelt also assumed practical importance to him, for sheer political survival. As a stand-pat, he was a target for insurgent Progressive Republican forces within his state. Fearing enemies encircling him at home, Lacey wrote to Roosevelt in 1906 for a testimonial. The president complied, though the endorsement did not prevent Lacey's defeat in the midterm congressional election in November, when the Republicans suffered substantial national losses. Nevertheless, since the 1890s both Roosevelt and Lacey had gained political and personal dividends from their collective work for conservation.[22]

Lacey's speeches reflected a larger imperial and global worldview strikingly similar to Roosevelt's. The Iowan was proud of all the nation's far-flung "possessions."[23] He repeated the old cliché that the "star of empire" had "taken its way, ever to the west" from "the Euphrates, the Tigris, and the Nile."[24] Rather than singling out wild areas or animals for protection, he advocated conservation of all resources and pointed to the failure to conserve in ancient civilizations. He accepted the conventional late nineteenth-century jeremiad of the American Republic that gained shrill support in the wake of 1898. "Without these forests in our western mountains," he argued before Congress in 1902, "the desolation of the mountains of Palestine and southern Italy will be soon duplicated in the United States. We must learn from the mistakes of others."[25] He saw the Louisiana Purchase as the foundation of the nation's newly acquired global status, a greatness that the robbery of resources threatened, and therefore backed Roosevelt's utilitarian conservation plans. He was also influenced by the passage of bird and animal protection laws overseas, especially the 1900 Convention on the Protection of the Wild Fauna of Africa. In the *World Review* (an illustrated "journal of progress and civilization" produced in Chicago),[26] Lacey wrote: "Mankind are becoming aroused at last to the importance of protecting what has been spared of the birds and game once so plentiful. Even in darkest Africa the great powers of Europe, which have partitioned the wilderness among them, have recently made rules and regulations to prevent the indiscriminate slaughter of the remaining creatures of the forest." In contrast, the "struggle for wealth for the individual" had long prevented "adequate attention" to preservation in the United States. In Lacey's indictment, the rape of all resources threatened both wise use and wilderness values. "Our coal, gas, oil, forests, fishes, birds, and game have been wasted and destroyed with a recklessness utterly unworthy of so intelligent and progressive a people. It is high time to call a halt," he stated in 1901.[27]

Neither preservationist nor pure utilitarian, Lacey argued rather for sustainability based on the perpetual restoration of nature, reminding Americans: "the earth is our mother. She must be clothed." Echoing George Perkins Marsh's ideas in *Man and Nature*, he claimed that a quarter of the earth must always be kept forested. For this reason, Americans had to "give up some of our land to nature, in order to keep the remainder for ourselves." Basic life-support systems on the planet could not be violated forever. This sustainability did not mean at just one moment in time but for all time. Lacey imbibed the growing concern about intergenerational equity, warning, "Posterity cannot escape punishment for the sins of their ancestors," and for this reason, Americans "should not do to our children what our fathers have done to us."[28]

Underpinning the drive for the bird laws were yearnings for a balance in nature. To Lacey and his allies, that balance was upset by European land settlement. The answer was not to abandon European farming methods but to improve them, setting them on a more viable path. Farmers in the fledgling Republic had removed tree cover and encouraged, through monoculture, rampant insect populations while depleting the insect-eating birds. This utilitarian appeal to the needs of the farming interest was vital to getting bird protection laws, but it was also a deeply felt belief reiterated in the debates. The role of birds as biological control agents against pest insect species added support from agricultural experience to the moral impulses of others.[29] Huntsmen were similarly alarmed at wildlife depletion rates, with strong advocacy of bird conservation coming from the journalist George O. Shields, president of the League of American Sportsmen, a group that self-policed game laws and advocated limits to shooters' kills. In meeting the needs of hunters, the 1900 act provided for the importation of game birds for breeding, and Lacey was a guest of honor at the League of American Sportsmen's annual meeting in the wake of the passage of his law.[30]

The state and federal bird acts provided one contributory stream to the growing concern from the 1890s onward. Lacey engineered the Refuge Trespass Act of 1906, which provided the first federal protection for wildlife refuges and excluded hunting and fishing from those places.[31] Ultimately, agitation coalesced in the cross-national protectionist objective of the 1913 Weeks-McLean Act and the Migratory Bird Treaty of 1918 with Canada. Adequate numbers of birds for hunting and insect control, together with ethical concerns about the survival of species, were all issues that Lacey and his allies took into account in this long struggle.[32] Yet there were other problems where conservationists considered nature protection equally essential, among them preservation of areas deemed significant as national monuments.

It was Lacey who framed and steered through the House of Representatives the Antiquities Act of 1906, an extraordinarily important piece of legislation.[33] The act gave the president power to declare national monuments within the US public lands without congressional approval. As historian Douglas Brinkley has stated, the act "was to Roosevelt a contraption with which he could dictate land policy in the West."[34] A succession of US presidents have used this power extensively, moves pioneered by Roosevelt, who deployed it to declare monuments when the prospects for creating a national park through congressional action were dismal. The law was framed to placate Congress, to lull any sense of alarm by its restriction to sites of "scientific" or "historic" interest. According to the act and to assurances that Lacey gave during the congressional debates, the parcels of land set aside would be no larger than necessary to preserve the monuments. The act was, however, "loosely written," a characteristic that facilitated passage at the time.[35] Only in 1908 did the implications become clear, when the president went well beyond the ostensibly narrow framework of the act and exploited its potential to the full on behalf of the Grand Canyon, protecting a gigantic "monument" of more than 800,000 acres. Whether by design or accident,

Lacey had colluded with the president to create the single most powerful preservationist weapon Roosevelt or any successor could have.[36]

Equally important for the development of conservation were campaigns to preserve scenic beauty and natural wonders from the Adirondacks to the California sequoia groves.[37] Among the natural spectacles in danger was Niagara Falls. Though historians interested in national park creation stories have treated Niagara as a lost cause and an object lesson in what not to do, the Niagara agitation itself had more significance. Niagara's message was not in the failed attempt to preserve an unspoiled area. Niagara led to more than cooperation between eastern and western preservationists to prevent the "loss" of the falls to development from being duplicated in the American West. From the struggle for Niagara's preservation came important grassroots tactics, strategies, and aesthetic principles employed in the effort for national park creation. Through their Niagara campaign, preservationists also provided an important opening for international cooperation on conservation, since saving Niagara inevitably meant negotiations with Canada. Niagara was a training ground for the tactics and strategies of applying a preservationist ethic and for the development of an international sensibility on the preservation of nature in North America.[38]

A significant but largely forgotten step in raising eastern American consciousness of such scenic "wonders" as Niagara came from the ASHPS. Formed in 1895 and originally called the "Society for the Preservation of Scenic and Historic Places and Objects," the society began to extend its activities beyond New York to other states in 1900. The ASHPS reflected the anxieties of New York society's upper echelons (of which Roosevelt was, by birth, part), which felt threatened by immigration and rapid industrialization. They organized to preserve buildings and natural monuments; they developed an aesthetic of landscape and garden preservation to promote the continuity of tradition and culture that they saw in their own personal achievements as members of the elite.

Though the ASHPS was one of several important local and state preservationist and antiquities societies to emerge at the end of the nineteenth century, it was less inward looking and conservative than those that clung to local issues and preservation of old buildings. The ASHPS lacked the states' rights inhibitions espoused by the Virginia Society for the Preservation of Antiquities against national government intervention to create national spaces of "sacred interest." It was the ASHPS that first called in 1902 for the creation of a national park at the Jamestown, Virginia, settlement that was (to its southern supporters and many other Americans) the cradle of American development.[39] The ASHPS and the Virginia society had some shared membership; but the former was studded with many of the movers and shakers of society: leading industrialists, financiers, and other socially prominent people. New York Democrat Andrew H. Green, a city comptroller and Central Park commissioner, was the first president and had collaborated with Roosevelt as early as 1895 when the future chief executive was an energetic civic reformer in New York City.[40] Dismayed by the popular redefinition of Central Park recreation to include the Park Menagerie, with its "mangy and uncultivated beasts," Green had campaigned with Roosevelt in that year to use municipal funds to establish the Bronx Zoo as a private "zoological park" in which scientific interests prevailed.[41] After Green's untimely and tragic murder in 1903 in a case of mistaken identity by a jealous husband, others equally committed assumed leadership of the ASHPS. Beginning in 1907, George Kunz, a self-trained mineralogist and onetime special agent of the USGS, was the driving force behind the ASHPS. President of the New York Academy of Sciences, Kunz created mineralogical exhibits for US participation at several world's fairs and expositions from 1889 to 1904 and, as Tiffany's jeweler, was an associate of J. P. Morgan.[42] Among influential members of the ASHPS were Morgan himself, who served as a vice president (and later honorary president) before his 1913 death, and US Steel's George W. Perkins. ASHPS projects also received Rockefeller family funding.

Kunz and William B. Howland, the ASHPS official and New York publisher of the *Outlook*, were closely allied to (or members of the same social class as) the president.[43] Roosevelt wished that he could have been the founder of the ASHPS and, like its members, opposed "vulgarizing charming landscapes with hideous advertisements."[44] As they developed from 1901 to 1909, ASHPS campaigns expressed the themes of combined utility and beauty in national park preservation that Roosevelt advocated in his annual messages to Congress. Roosevelt called for "free camping grounds for the ever-increasing numbers of men and women who have learned to find rest, health, and recreation in the splendid forests and flower-clad meadows of our mountains."[45] The scenic preservation movement produced similar statements reflecting a shared class position and values.

Little remembered, the ASHPS was nevertheless influential as a conduit for European ideas of park development into the United States—adding to the momentum for national parks coming from indigenous American sources. These country landscape preservation and restoration campaigns came to prominence at exactly the same time that similar organizations were appearing in Europe, groups with which they were closely connected by active correspondence and reciprocal emulation. That itself is an interesting

8.1. Three ASHPS members—William B. Howland of the *Outlook*; George W. Burleigh, a director of the Lackawanna Steel Company; and John Aikman Stewart, a wealthy banker and onetime confidante of President Grover Cleveland—with Henry Vivien, second from left, member of Parliament, Great Britain, in 1913. LCPP.

parallel rarely remarked upon by historians on either continent. American scenic preservation work was very similar to that of the societies for the protection of the countryside in Europe, discussed in the pages of ASHPS reports. For instance, bicycle clubs in France and tourist interests in Belgium advocated for improvement of industrial and rural landscapes.[46] "The problems with which they [the Belgians] have to contend," wrote the ASHPS president, "are the same as those which confront the advocates of scenic and historic preservation in America."[47] Not only did American scenic-beauty people think like Europeans, but they also worked with them. In 1900 the society joined with Britain's National Trust in a campaign to stop American millionaires from alienating British heritage sites, by raising money to buy properties for the trust. The American organization saw itself as the trust's American equivalent and its partner.[48]

Moving from historical to scenic preservation, the society soon worked for protection of natural environmental "wonders" as well. Two key causes were the protection of Niagara Falls from the diversion of waters for hydroelectric power and the restoration of the scenery of the Palisades Cliffs on the Hudson River in New Jersey against the sound and sight intrusions on the view of the Hudson from New York brought about by stone quarrying.

Characteristically, the ASHPS compared its work to nature conservation undertaken on the Rhine in the same period and called the Hudson the "American Rhine."[49] When Roosevelt was governor of New York, the ASHPS collaborated successfully with him to protect the Palisades through the creation of the Interstate Palisades Park Commission, which stopped the quarrying.[50] In many ways the Palisades Park established in 1900 was a measure of the group's legacy; it was essentially a labor of what in Britain came to be called "nature conservancy."[51] Partly privately funded, the commission was headed by financier and industrial magnate Perkins, with land donated by Edward H. Harriman and money from J. P. Morgan and others. From its formation in 1900, it undertook much restoration work. Its founders clearly did not set aside untouched wilderness, but Roosevelt admired its achievements. He sought to use the ASHPS to help preserve great national monuments and "thought that Kunz was among the most cogent voices for protection of American scenery."[52]

The ASHPS's other high-profile campaign concerned Niagara Falls. Like the Palisades project, the protests against excessive development began in New York. Under Green's leadership as a commissioner for New York's Niagara Falls State Park and as ASHPS president, the fight over the diversion of Niagara River waters went back to 1895, the same year that the first successful hydroelectric power plant project at the falls opened. The ASHPS thereafter denounced all "schemes on a larger scale that would destroy this grandest of the world's natural phenomena."[53] Because the ASHPS had among its members a pair of prominent investors in the existing Niagara Falls Power Company hydroelectric scheme, the society's work on this project could be deemed hypocritical.[54] Most prominently, J. P. Morgan stood to gain by stopping rival firms from entering the market, but the ASHPS strongly opposed any further expansion of the site and went on to campaign with equal force against other capitalist projects elsewhere while Morgan remained a high-ranking member. The ASHPS accepted the magnate's money because he was always "ready to save an object of beauty, whether the product of nature, of art, a painting, a mineral collection." Though pro-hydroelectric supporters also accused the Niagara campaign of favoring coal interests, a further business with which Morgan, Harriman, and some other ASHPS backers had connections, this did not stop the ASHPS and its allies from denouncing the disfiguring effects of coal upon the environment.[55]

The Niagara defense that the ASHPS undertook anticipated the opposition to the flooding of Yosemite and became closely linked with the latter through Horace McFarland's newly founded ACA in 1904.[56] A Harrisburg, Pennsylvania, horticulturalist and journalist, McFarland first came to prominence championing rose gardens, urban parks, and city beautification.[57] His aesthetics matched that of the scenic-beauty group, and he opposed billboard pollution and the Niagara River's desecration.[58] After 1904, the ASHPS continued with letters, petitions, and political lobbying to support the latecomer McFarland's campaigns against any hydroelectric expansion at the falls.[59] With an overlapping membership, the two organizations worked in tandem.

The Niagara threat provoked a rallying cry against ruining the sublime spectacle of the falls, a point that McFarland vividly made in 1906. Using his monthly Beautiful America column in the *Ladies' Home Journal*, he asked readers, "Shall We Make a Coal-pile of Niagara?" and called the great cataract "a wonder of God," just as Muir regarded Hetch Hetchy. Without conservation controls, water would be ripped from the falls, McFarland warned, leaving a mass of factories and bare cliffs that exposed a maze of "wires, tunnels, wheels and generators."[60] McFarland used the common muckraking language of the Progressives to condemn hydroelectric developers as greedy speculators taking something for nothing through a free gift of nature. The language was strikingly similar to the jeremiad against all resource depletion offered by the journalist Rudolf Cronau.

What was most conspicuous about the ACA campaign was not the tone of its indictment but its distinctive tactics. McFarland took the aesthetic and preservationist critique nurtured in the ASHPS and repackaged its elitist message in a form more accessible to the middle and lower middle class. Using his access to the *Ladies' Home Journal*, a magazine that reached millions, McFarland developed mass media propaganda, solicited a cadre of workers, and implemented an effective means of delivering the wrath of the middle class to Congress. Starting in 1905 he relied on petitioning targeted at congressmen and the administration, though not to oppose Roosevelt but to support him. To accomplish this McFarland needed the ASHPS. The latter reportedly had money equaling that of "a foundation"[61] and access, through McFarland's friend, the "genial" William B. Howland, to one of the most prestigious magazines of its time, and from there to Roosevelt. Howland's "acquaintance with the great of the earth was pleasant for his friends," McFarland recalled. "Always welcomed by presidents, cabinet officers and other responsible officials, he was glad to take with him some of us who might otherwise have had difficulty to obtain [an] audience." There "our friend was 'in action,' and at his genial best. He was a hard man to say 'No' to."[62] In the Progressive Era, mass mobilization for conserving nature needed special channels or networks of influence at the level of national politics, and this the ASHPS supplied through Howland.

8.2. "Save Niagara Falls—from This": *Puck* cartoon of 18 April 1906, showing the falls run dry due to diversions of water for hydroelectric power stations. LCPP.

All this was uncannily similar to the contemporary campaigns undertaken from 1903 to 1906 by missionary support societies and by moral reformers led by the Reverend Wilbur Crafts, the superintendent of the International Reform Bureau. The conservation crusade borrowed heavily from missionaries via the evangelical churches and their mass petitioning to stop the evils of drugs and prostitution in the new American colonial possessions. Form letters were sent out, seeking groups to sign. Just as Secretaries of State John Hay and Elihu Root used Crafts to direct a political effort that combined government and voluntary organizations to change public opinion on legalized prostitution and opium in the American-controlled Philippines, McFarland met with Roosevelt in late 1905[63] and developed an almost-identical understanding on Niagara's preservation: the administration would act on conservation if backed by a public opinion and lobbying campaign, Roosevelt told him. He urged McFarland to get as many people as possible to write letters and send telegrams to Congress to strengthen the administration's hand.[64] As McFarland boasted, senators and congressmen were soon "overwhelmed" with *Ladies' Home Journal*–inspired drives of "thousands of letters and petitions."[65] Boosted in this way, in 1906 the administration achieved passage of an act that halted further hydroelectric development, a measure that McFarland and Roosevelt sought for the "preservation of Niagara Falls." Backed by knowledge of the many earnest letters "promptly sent" in the latter part of 1905 to provide Roosevelt with ammunition, the "vigorous president" highlighted the necessity for Niagara protection in his annual message to Congress: "There are certain mighty natural features of our land which should be preserved in perpetuity for our children and our children's children. . . . It is greatly to be wished that the State of New York should copy as regards Niagara what the State of California has done as regards the Yosemite. Nothing should be allowed to interfere with the preservation of Niagara Falls in all their beauty and majesty."[66]

McFarland was prominent in his local church, the Grace Methodist Episcopal Church in Harrisburg, which had an active missionary support circle that included his daughter, Helen.[67] His reform campaigns in that city mobilized the local churches for the "gospel of civic decency."[68] Combining in his work what he called "real politics and real religion" to "create better conditions on earth as a preparation for the world to come,"[69] McFarland applied evangelical mass mobilization tactics to his own city's environmental problems and to Niagara. This missionary borrowing was yet another of the ways in which the greater transnational reach of American culture to the far corners of the earth in the first decade of the twentieth century influenced events and social movements back in the United States.[70] Rather than popular grassroots conservation being merely a vehicle tacked on opportunistically to spread the message of a scientific conservationist elite from Washington, civil society was an active force in the shaping of conservationist policies.[71]

More than that, conservationists absorbed the rhetoric and the moral fervor of the international missionary movement. It was Pinchot who stated that the conservation cause was "going to win" because it was "a moral movement."[72] At a Laymen's Missionary Movement dinner in Washington in 1909, he argued that conservation was a Christian crusade to "preserve our civilization," and one that required Americans to "use every effort to spread the kingdom of Christ to every corner of the earth." Pinchot advocated "the conservation of the conscience of mankind" through missionary work. The movement indeed began with forests but had mutated to "conserve for the good of ourselves and future generations the moral conscience of humanity." In this light, conservation of resources aided the movement of "us laymen in promoting Christian religion . . . over all the world."[73] Although Pinchot used the missionary channels in a pragmatic bid to gain foot soldiers for a coming war of Armageddon against the anticonservationist forces, he was also an established financial supporter of missions and a member of the Domestic and Foreign Missionary Society of the Protestant Episcopal Church.[74]

McFarland's Niagara campaign also had support from missionaries anxious to prevent the United States from displaying lower aesthetic standards than those that applied in "heathen" lands. In Japan, American Christians noted a high regard for nature. "May we not learn a lesson from this beauty-loving people?" asked Emma E. Dickinson, a Methodist missionary and a graduate of Mount Holyoke College. A returning colleague, Georgiana Baucus of Hinsdale, Illinois, was "shocked to find" how much the "practical and utilitarian" approach had gained the upper hand at home during her nine-year absence. As she applied this "Eastern" aesthetic alternative to judge American action, her reference point was not the American West but the cultivated Japanese Zen gardens that she had come to love.[75]

Not only did the ACA campaign dovetail with Pinchot and Roosevelt's plans by bringing a new morality to conservation, but it also highlighted "the international importance" of the scenic preservation issue, because hydroelectric development was threatened on the Canadian, as well as the American, side.[76] To McFarland Niagara Falls was "a possession of America, of the world"—it was a heritage asset beyond nation, even as lobbying necessarily had to target the national seats of power.[77] It was the ASHPS that alerted McFarland to the full international implications of the Niagara

8.3. Horace McFarland, president of the American Civic Association.
Courtesy Pennsylvania State Archives.

issue, and it urged a dual-nation lobbying campaign. In this way, McFarland came to understand that a transnational approach by non-state actors could shoehorn local efforts to preserve Niagara. No other man than Howland, McFarland averred, "had more to do" with this "internationalizing of the great scenic wonder which was formerly only a state and a provincial prey."[78]

To defend this scenic wonder internationally, McFarland and his supporters in the ACA and ASHPS did more than lobby US congressmen and Roosevelt. They wrote directly to the Canadian governor-general Earl Albert Grey, so that North American leaders would "find a way of international action" concerning a "great international danger." McFarland urged supporters to get every club, association, and church to write, mobilizing "that great American public of the United States and Canada." Thereby, "true-seeing and right thinking people" would, through "extra-legal and international action," make the heads of government of the two countries listen.[79] McFarland's approach was an exercise in, not exceptionalism, but transnational shaming, because it emphasized the dishonor that would be "divided between the richest two nations [on] the face of the earth."[80] Simultaneously, scenic-beauty supporters also lobbied against the introduction of Canadian hydroelectric power to the United States, in order to limit the Canadian interest in diverting more water from the Canadian side of the falls.[81] Without a US market for Canadian power, the expansion of Niagara facilities north of the border would be uneconomical.

Yet McFarland's case against Niagara expansion could not be aimed at preservation of pristine nature. Rather, it was an aesthetic balance that McFarland promoted. All American waterways came under his scrutiny for an aesthetic quality that might add "an increment" of "beauty travel" to the nation's patrimony. Noting the Roosevelt administration's plans for the rejuvenation of major rivers through engineering works, he argued that the nation was "going to build planned waterways," but nothing in the plans indicated any commitment to canals that "are beautiful." This unaesthetic form of utility was quite unlike European canals, to which American tourists flocked for holidays and recuperation from the stresses of industrial life, McFarland opined. The beautiful waterways of Sweden and Holland had "been planned to be so," while American rivers seemed condemned to permanent ugliness.[82]

This aesthetic campaign's broadened implications became clear when McFarland appeared at the 1908 governors' conference to denounce an "ugly" America.[83] There he developed a full-blown critique of utilitarian conservation while attempting to transform from within the character of the Roosevelt-Pinchot project into a more holistic scheme. Separately published in 1909 in a revised version as "Shall We Have Ugly Conservation?," his paper indicted the United States not just for the desecration of nature but for the huge human costs of an environment disfigured by industrial overdevelopment: "The horrors of the pit-mouth and of the mining village" must have been clear to very few, he surmised, "else they could not continue in a Christian civilization." Even as Americans knew in theory that the waste of human life was wrong, McFarland charged, "the way we house our defenseless foreign labor which digs our coal and works our metals" was "no credit to our sociological intelligence." The moral lacunae explained the "crime, death, and bad citizenship" that followed as night followed day.[84] McFarland's urban and industrial reform efforts in Pennsylvania can be seen as part of the City Beautiful movement. McFarland had joined this reform impulse with the founding of the Harrisburg League for Civic Improvements in his hometown in 1901.[85] Even as municipal agitation had opened his mind to the travail of those who manned the factories and mines, deeper down he believed that the impact of industrial society was toxic to all, not just the working class. Though utilitarian themes did preoccupy most delegates at the 1908 conference, McFarland was not entirely alone, receiving endorsement from, among others, the governor of New York, Charles Evans Hughes, and from the ASHPS's Kunz and Liberty Hyde Bailey as well.[86]

As an alternative to "ugly" conservation, McFarland and the ASHPS posed a nascent form of sustainability, using the term "habitability," to be achieved by putting physical resource conservation together with human health and "beauty."[87] The ASHPS used the expression "a more satisfactory society in which to live,"[88] one in which scenic value was taken into account as contributing to human well-being. At the governors' conference, Thomas C. Chamberlain, the well-known University of Chicago geologist and president of the American Association for the Advancement of Science, agreed. He spoke of "renewed habitability" and the need to "cooperate with nature" and its own regulatory system.[89] Though it was a struggle, scenic-beauty people collectively succeeded in having the underlying concepts added to the Conference of Governors Declaration of Principles: "that the beauty, healthfulness, and habitability of our country should be preserved and increased."[90] McFarland elaborated on that value in March 1909, just as Roosevelt left the White House. He urged that the NCC be changed to create a subgroup within each of its four sections to deal with scenic impacts and noted how other countries preserved scenery with tracts set aside for "appearance and for health." These four subsections should meet to consider these matters, thus constituting collectively a fifth section of the NCC.[91]

This approach would have involved de facto environmental impact statements for developmental policies where public lands were involved. McFarland believed the idea to be inherent in the conservation program of the president, but it had not been institutionalized by the time he left office. Roosevelt's actions certainly showed that he was sympathetic. The president indicated as much in establishing the Commission on Country Life, on which Pinchot served and which, we have seen, was aimed at obtaining a balance between development and the "latent beauty" of the countryside by strengthening rural America, especially farming and the ability of small towns to survive. He and Pinchot had the same concept written into the Declaration of Principles of the North American Conservation Conference of February 1909, which also proposed, as part of its ongoing agenda, measures to prevent pollution of streams that crossed national boundaries, in another nod to the Niagara agitation.[92]

Yet, in the wake of Roosevelt's departure, and with the NCC deprived of congressional funding, the chance to create a comprehensive and sustainable approach to economic development was lost. Though the continuation committee of governors and federal officials established under the Joint Conservation Conference existed on paper after 4 March 1909 and even had, at Pinchot's direction, assigned office space in Washington, it lacked funds and presidential support. With no secure institutional base to implement the governors' Declaration of Principles concerning "beauty," "habitability," and "healthfulness," preservationists had to seek instead a rearguard action to shore up the integrity of the national parks as separate aesthetic spaces in a larger world of material and utilitarian development.[93]

The struggle to "save" Niagara Falls made two vital contributions to the conservation movement. One led inward to the campaign for a national park *system* after 1909; the other, outward toward an international agreement over the Canadian-American boundary waters of which the Niagara River was part. The Niagara dispute contributed in both ways to the strengthening of the American nation-state. As Gail Evans has explained, "The Niagara debate had larger significance . . . for the future participation of the federal government in nature preservation." It "effectively galvanized and politicized public sentiment for protecting America's unique natural spectacles."[94] But the episode also shaped nascent environmental diplomacy and contributed new international institutions and frameworks for environmental regulation. National development and international development of conservation not only coexisted but also became intertwined. Within the United States, agitation coalesced around the struggle to save the Hetch Hetchy Valley from inundation. Scenic preservationists quickly applied the experience gained in organizing for Niagara to the equally vexing problem of Hetch Hetchy's proposed dam. They employed lawyers to lobby for the protection of natural spaces deemed unique in value and to disseminate their views in the press.[95]

During a long contest from 1906 to 1913, the Hetch Hetchy issue reputedly pitted utilitarian-minded people in the US Forest Service led by Pinchot against preservationists led by John Muir. The dam's proponents had much on their side. Not only was San Francisco in need of a new water system because of the 1906 earthquake, a factor Roosevelt was loath to ignore, but the city authorities also proposed a municipally owned system. This accorded with Pinchot's growing concern about the monopolization of public power sites by private interests and introduced complex ethical considerations of resource allocation that overshadowed the simple division of scenic beauty versus utility.[96] In practice, there was no such clear-cut division in the first place. Muir himself, the best-known advocate of preservation, made compromises with economic development and so did McFarland.[97] Despite his suspicions that Pinchot was playing a devious political game over the issue, McFarland was a pragmatist. He told one correspondent, "one cannot obtain things in this world always by direct methods," and denied he was a "goody-goody" idealist.[98] Throughout 1909 McFarland tried to work with Pinchot to convince him that scenic beauty could be reconciled with utilitarianism. Pinchot, too, avoided conflict and, without Roosevelt's executive power behind him after 4 March 1909, cultivated McFarland as his "friend." Pinchot tried, as Roosevelt had done, to straddle the issue and supported Secretary of the Interior James Garfield's prior decision that the Hetch Hetchy dam could proceed under restricted conditions, but only after a separate San Francisco water project for Lake Eleanor had been developed. Pinchot reassured McFarland that the valley would not be flooded for a considerable time, if ever. Pinchot went so far as to concede "the esthetic [*sic*] side of conservation" as important, yet he did not understand (or want to understand) that McFarland's critique was based on incorporating aesthetic value into every issue, not balancing separate issues on a case-by-case basis.[99]

Even when it became clear that Pinchot was for all intents and purposes favoring San Francisco, McFarland did not declare war on utilitarian conservation; in fact, he cooperated later with a new organization, the National Institute of Efficiency, to minimize waste in every walk of life.[100] Instead, scenic-beauty advocates turned their attention to a new compromise, one that tacitly accepted the political impossibility of a comprehensive conservation policy that gave effect to the governors' Declaration of Principles of 1909. In response to pressure for compromise, but also as part of their deeply held landscape aesthetic of balance, park advocates addressed the question of how to measure the worth of parks and assign tangible value to nature's protection. They tried to raise public awareness of national parks, reasoning that only a low level of public understanding had allowed the threat of national park alienation to materialize in the first place. This action spurred calls for a federal park bureaucracy and encouragement of tourism within the parks. Though initially unsuccessful, the ACA joined with the ASHPS, the Appalachian Club, the Sierra Club, and other groups to demand uniform national park legislation to replace the scattered and incoherently administered inventory of parks.[101]

The ASHPS was a persistent supporter of this move, and colleges, universities, museums, the press, and members of the public sought its views.[102] Legal counsel represented the ASHPS at congressional hearings on Hetch Hetchy, and the society appeared alongside the ACA at the final testimony in 1916 on establishment of the national park system.[103] In the latter case, these two organizations were the only lobbyists present apart from the Burlington Railroad (which had a Yellowstone concession). The ACA took the leading role, but H. K. Bush-Brown, a sculptor and a trustee of the ASHPS "since its inception,"[104] represented the latter organization. He had spent two summers in Switzerland and drew attention to the pioneering role of the Swiss in creating parks in Europe. He regarded the Swiss achievement as seamlessly connected to his own aspirations for an American equivalent centered on recreational and aesthetic values. For Bush-Brown, nature and nurture came together in the creation of a nationality. "The foundation of the physical, mental, spiritual, and moral forces that are latent among our people" would be manifest through national parks. Though America lagged behind Switzerland in recognizing the educational value of its parks, "the time [was] coming," Bush-Brown asserted, when Americans would "go into the open and study nature at first hand instead of devoting [themselves] to the study of nature through books." He linked this aesthetic to the creation of physical spaces for recreation because the United States had not yet "made very much headway in the physical development of our people." In order to achieve that, Americans would have to "live out of doors in the open." Conceiving of parks in this way would make them a practical "resource" for the nation just as more obviously utilitarian resources were.[105]

This activism in the service of a cultural and "physical" nationalism used international examples to draw attention to the tourist potential of parks and the benefits of centralized park management. From 1906 to 1910, railroads with national park concessions inaugurated sporadic "See America First" campaigns, but from 1915, they worked in tandem with park advocates.[106] At a memorable park conference held in Berkeley in 1915, Mark Daniels, the general superintendent and landscape engineer of national parks, who was based in San Francisco, claimed that four to six hundred million dollars was annually "spent by American tourists in Europe for the purpose of seeing scenery," money that should remain in the United States.[107] Yet it was not simply economic utilitarianism that prevailed. Rather, the discussion linked the aesthetics of recreation with nation building in a competitive international system. As Daniels conceded, "Economics and esthetics [*sic*] really go hand in hand."[108] The gain from creating parks was educational but needed to be cast in a national framework to develop love of country.

Advocates of this approach held up Canada's Dominion Park Branch, established in 1911, as their model. The ASHPS publicized the need to compete internationally for the tourist trade, to copy international innovations, and to reframe aesthetic values to fit the new political environment. It tried to enlarge the concept of "value," praising the success of the Canadian park system for combining an economic contribution with an intrinsic value to support Canadian nation building.[109] "The value of the parks as an asset of the Dominion of Canada cannot be measured by immediate results in dollars and cents; but they have been a means of spreading the fame of the beauty of Canada to all parts of the world where it otherwise would have been unheard of. Hundreds of visitors have said that they were induced to visit the Rocky Mountains through reports of its beauties given by tourists who had been there in former years." Using the information provided by the parks commissioner in Canada, the ASHPS drew attention to the success of the tourist trade in Banff compared with Yellowstone, whose park superintendent rued that "more than three times as many people visit Canada's National Park as visit the famous tourist resorts of [the] United States."[110]

When Congress established the National Park Service in 1916, this lobbying came to fruition. The guiding mission was to "conserve the scenery and the natural and historic objects and the wildlife therein and to provide for the enjoyment of the same in such manner and by such means as will leave them unimpaired for the enjoyment of future generations."[111] This omnibus statement reflected the scenic-beauty influence. The previously dispersed control of the national parks and the diversity of conditions under which they functioned were to be replaced with national consolidation. In this way, the international influence in the conservation movement that the ASHPS and the ACA imported served to strengthen nation building by fostering hallowed national spaces and by promoting national institutions to administer and champion them.

The other major influence of the Niagara campaign was directly cross-national and flowed from the lobbying that McFarland and the ASHPS undertook. Internal agitation put pressure on the United States to seek international agreement over water diversions to preserve Niagara Falls from the utilitarian dreams of hydroelectric power operators. Thereby, the Niagara issue became inextricably linked to the wider Boundary Waters Treaty, which determined water allocation in the shared catchment area of the Great Lakes system and in the North American West. The conclusion of an agreement over the boundary waters remains one of the understudied and undervalued aspects of the Roosevelt administration's conservation work. Three other (better-remembered) treaty negotiations were under way in Roosevelt's second term, with mixed results. These were early exercises in what today would be termed environmental diplomacy. The Fur Seal Treaty of 1911, on which work began in 1905, was the most successful. The groundwork was also laid for the Migratory Bird Treaty, ratified in 1918, but the signing in 1908 of an Inland Fisheries Treaty with Canada ultimately failed ratification during the Taft administration. Congressional objections representing the "short-term self-interest of a small band of constituents" prevailed.[112]

The origins of boundary-waters diplomacy did not lie in the Niagara dispute but became linked to it in 1902. As the 1890s debates over irrigation policy showed, an agreement on a Canada–United States commission on boundary waters came first from water politics on the southern border of the United States, not the northern. Piggybacking on Mexican discontent over Rio Grande diversions, Canadians had sided with Mexico at the National Irrigation Congress of 1895 to call for both boundaries to be considered. Canada had in mind diversions on the Milk River, which ran through Montana and Alberta to Hudson Bay. Though Canada sought US government action in 1896, nothing was done until after Roosevelt took office.[113]

When the change came in 1902, a different configuration of motives was involved. From a geopolitical standpoint, both Britain and the United States wanted to settle a raft of minor Canadian-American disputes that threatened the growing Anglo-American diplomatic rapprochement of the time. This coincided with Roosevelt's (and Secretary of State Root's) desire for greater emphasis on international arbitration rather than a continuation of bellicose diplomacy.[114] Equally important, grassroots campaigning over the Niagara question presented a prize opportunity to link conservation to geopolitical maneuverings. Andrew Green, the president of the ASHPS, had seen the possibility of using ASHPS contacts with the freshly installed Roosevelt administration to achieve his society's Niagara objectives through the wider framework of the Great Lakes Deep Waterways navigation issue, which had been the subject of an earlier US report in 1896. Green lobbied Roosevelt in 1901–2, putting forward a plan for an International Waterways Commission to deal with the diversion of all shared waters.[115] For Roosevelt, this pressure for action on Niagara was the key concern in the subsequent boundary-waters diplomacy.[116] Roosevelt was able to use Congressman Theodore Burton (R-Ohio), the chair of the Rivers and Harbors Committee (1899–1909), to push through the legislation that would halt further development while the nation negotiated a deal with Canada. "I am very fond of Burton," Roosevelt told James Garfield, "and regard him as a peculiarly high-minded and able leader." From Roosevelt, this was first-class praise indeed. To the *New York Times* he was "Busy, Brainy Burton the Administration's Utility Man."[117] McFarland's ACA rewarded him with its second lifetime honorary membership. (Roosevelt received the first.)[118] Burton did not disappoint. The 1902 Rivers and Harbors Act that he crafted gave the president the power to work with the Canadian government to create a joint commission of three US and three Canadian members. It would "investigate and report upon the conditions and uses of the waters adjacent to the boundary lines between the United States and Canada."[119] The remit included "the diversion of these waters from or change in their natural flow" and "measures to regulate such diversion." As was considered necessary for constitutionally valid legislation on rivers and harbors, the act stressed the effects on navigable streams.[120]

Initial progress was slow. By 1903 the United States had agreed with Canada upon the need for an International Waterways Commission, but appointment of the Canadian members was delayed until 1905 by internal Canadian resistance and "red tape" in the protracted trilateral negotiations with Britain. Canada had been disappointed by the separate Alaska Boundary Treaty negotiations breakdown in 1899, with the subsequent settlement in 1903 favoring the United States. The northern neighbor had to be coaxed back into supporting arbitration of another border issue.[121] Reporting on 25 April 1906, the commission recommended that a permanent organization be established for the Great Lakes system that would have authority to divert to each country "equal quantities of water from non-navigable boundary streams."[122] For Niagara, the report documented the extent of the existing hydroelectric development and examined the scientific evidence of its impact on the falls. Only 18,500 cubic feet of water per second on the American side of the falls and 36,000 on the Canadian side could be taken without affecting the scenic beauty of the river, the report advised. With the ACA lobbying campaign already under way, Roosevelt asked Congress to pass a law to "preserve the Falls without waiting for a treaty with Canada." This the able Ohio congressman accomplished, and the Burton Act of 1906 provided for a unilateral moratorium on US diversions to last three years and called for a treaty, to which it was considered a preparatory measure.[123] The Canadians passed their own legislation that same year to regulate hydroelectric power, but it did not resolve the major bones of international contention.[124]

Advancing the final negotiations from 1906 to 1909 was difficult. Because the Canadian falls were much larger and less in danger of diminution from hydroelectric diversion, the Canadian concern lay more in power development for industry in the province of Ontario than scenic beauty. For Root and Roosevelt, however, the demands of scenery took precedence because of their links with the ASHPS and the ACA. The secretary of state lobbied Governor-General Grey, with Grey reporting to Prime Minister Wilfred Laurier that Root "and the President have that subject [Niagara] very near at heart."[125] There were other differences, too. The Canadians wanted a treaty that would cover diversion of all waters "adjacent to the boundary" as part of what was, in effect, a total catchment approach, while the Americans wanted individual treatment of issues because this would benefit their conflicting objectives, allowing the United States to distinguish between its preservationist interests in regard to Niagara and utilitarian interests in Chicago on the diversion of water from Lake Michigan for the Chicago River Canal project. American and Canadian negotiators were appointed, with James Bryce, the British ambassador in Washington from 1907 to 1913, retaining oversight because Canada lacked formal control of its own foreign relations. As a friend of Roosevelt's, he proved an essential go-between for the three governments and steered the Canadians toward an amicable pro-American settlement at the partial expense of Ontario's economic interests.[126]

In the draft treaty of August 1908, Canada won more water to use for electric power than the United States, though limits were placed on both parties to preserve for scenic purposes the flow over the falls. Americans thought the Canadians got the better deal, but the latter pointed out that part of the extra water granted to Canada under the diversion agreement would produce power available for export to the United States. The able Canadian chief representative, Sir George Gibbons, argued that the practical effect was, therefore, an equal division of the water. While the framework for future administration of the waterways was broad, the provisions on the

Niagara Falls and River were specific, and became article V of the overall Boundary Waters Treaty. This article limited Niagara River hydroelectric diversions only above the falls, thus underlining the importance of the scenic issue to the agreement. Meanwhile, the Chicago River diversion that lowered lake levels was deleted from the calculations on the river flows out of the Great Lakes. This made the treaty easier to pass in the Senate, since this concession benefited the United States. Yet even within this modified agreement, a rider had to be added to guarantee private US power companies' existing water rights at Sault Ste. Marie on the Canadian border, to which Canada acquiesced at Bryce's urging to prevent failure of the treaty.[127]

At its most basic level, the treaty signed in January 1909 and ratified by all parties in 1910 established a permanent Joint International Commission to settle water disputes over lake and border river waters and secured US navigation rights on the border rivers, which had previously relied upon Canadian discretion where those waters passed through Canadian territory at the Welland Canal.[128] Of long-term importance, article IV included a clause that reflected the educational efforts and political lobbying of McFarland and his allies. The signatories "agreed that the waters herein defined as boundary waters and waters flowing across the boundary shall not be polluted on either side to the injury of health or property on the other." Inserted at the behest of the United States, this was a small step toward "habitability" through transboundary regulation of pollution in North America, though it was to be many years before the permanent international commission achieved adequate supervision of the pollution issue.[129] Aided by McFarland's lobbying, the prospect of an international agreement simultaneously spurred the extension of the Burton Act,[130] with a joint resolution of the US Congress early in 1909 preventing additional power diversions on the American side for a further three years.[131] Throughout the treaty-making process, Root had strongly supported preservation of Niagara Falls from further depletion of water, reflecting agreement with the ideas of the ASHPS and the ACA on scenic beauty. In fact, Root was a member of the former organization.[132]

At the national level in the United States, this policy making contributed, one authority concludes, to the growth "of federal power over natural resource management" concerning not only preservation of scenery but "power generation and distribution" as well as "multiple-purpose river basin development." In effect, the outcome furthered Roosevelt's broader objectives of national consolidation.[133] Almost unnoticed south of the Canadian border was the external dynamic in American policy, however. The United States had successfully defended its economic interests in the larger

Great Lakes system, including the entitlement of American companies to existing hydroelectric power rights, while asserting preservationist objectives that limited the utilitarian agenda of the Canadian provincial power operators, though without affecting the existing power stations on that side of the border at Niagara Falls. Preservation of nature had become tied to the quasi-imperial hierarchy of a rising world power, as much as to internationalist ones. While the agreement was an international conservation victory for the Roosevelt administration, its implications were only partly those of the cooperative internationalism that Roosevelt espoused toward the end of his second term. The needs of American economic interests abroad still took precedence, due to the difficulties of getting Senate ratification.[134]

Even as Roosevelt and his allies used the Niagara dispute to forge a pathbreaking international agreement on the falls, alarm over future threats to the falls did not abate. McFarland continued to campaign against further power development as the congressional moratorium drew to a close. In 1911 he argued that Niagara should be declared a national park. The falls were "never more in danger," he warned, and must "come under the federal mantle as a national reservation." Only in this way could the United States "be saved from the lasting disgrace that now threatens our most notable natural wonder." As he so often did, McFarland pointed to the inconsistency between the nation's larger global ambitions and its imperfect efforts for beauty, health, and permanence at home: "A nation that can afford a Panama Canal cannot afford a dry Niagara!"[135] But neither McFarland nor the ASHPS could turn back existing hydroelectric developments or create a national park.[136] While the Root-Roosevelt diplomacy had prevented the complete destruction of Niagara, the power companies and their diversions remained. More efficient generators were added to the landscape in future years, and the International Joint Commission would negotiate further increases in water diversion (in 1929 and 1950) for hydroelectric power.[137] The falls themselves would never return to their pre-1895 state nor regain their original level of sound. Roosevelt had tacitly accepted this compromise, just as he favored a similar one for Hetch Hetchy. But by the time these compromises were made manifest, the president's last term in office was well and truly over, and he had turned his attention to the world stage in different and much more spectacular ways.

NINE

Lessons for Living: Irving Fisher, National Vitality, and Human Conservation

The term "conservation" usually conjures up an image of the natural world, meaning either preservation or wise use of "nature" understood as the non-human world. Yet conservation in the United States came to connote something more than that by 1909. It incorporated humans as objects of concern through the concept of "national vitality." A sizable portion of the NCC's report of that year was devoted to this concept.[1] An academic of wide interests and influence promoting the idea was Irving Fisher, a mathematical economist based at Yale University. Later to be president of the Econometric Society and the American Economic Association, Fisher was not simply a man at ease with numbers and equations. Born in 1867, he survived tuberculosis in the late 1890s and thereafter developed a passion for health and hygiene. In the *NCC Report*, his long chapter on "national vitality" occupied a prominent place. The chapter was vetted by Pinchot and given the imprimatur of an official government report, serving as a manifesto for the Roosevelt administration's considered outlook on race, population, and health policies. Though neglected by historians, this work was extremely important at the time as an effort to deal with the otherwise-intractable problems of the social geography of class in industrial society—rooted in pollution, industrial accidents, and ill health. Subsequently, Fisher's chapter was issued independently under the same title and was widely reprinted and discussed. The US Senate released "several large editions" at the suggestion of Senator Robert L. Owen. Copies of the report and its findings "were spread throughout the country" with "the direct co-operation" of *McClure's Magazine*, *World's Work*, *Survey*, and *Good Health*.[2] The Yale professor became a prolific speaker on these issues for several years and wrote a national best seller entitled *How to Live: Rules for Healthful Living Based on Modern Science* (1915).[3]

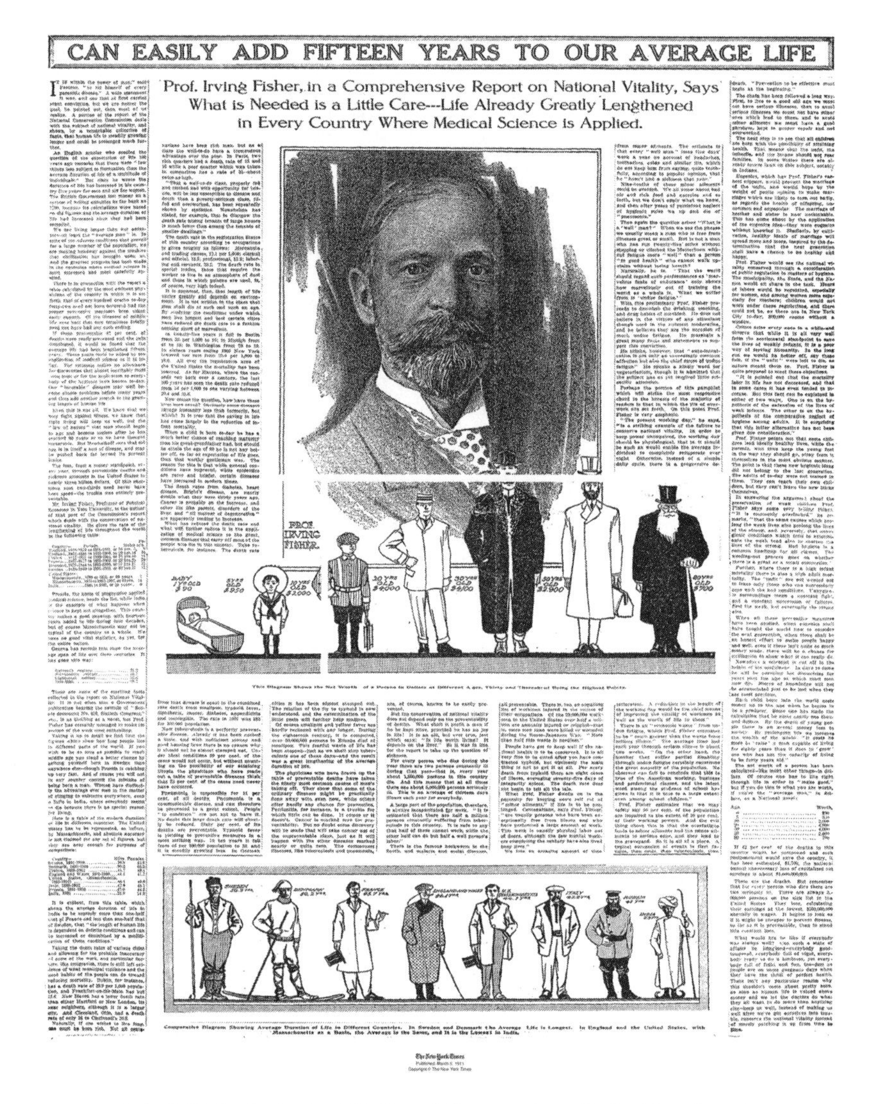

CAN EASILY ADD FIFTEEN YEARS TO OUR AVERAGE LIFE

Prof. Irving Fisher, in a Comprehensive Report on National Vitality, Says What is Needed is a Little Care---Life Already Greatly Lengthened in Every Country Where Medical Science is Applied.

9.1. Publicity for *National Vitality*: a full-page review of Irving Fisher's work from the *New York Times*, 5 March 1911.

Using Fisher's *National Vitality*, Progressive reformers stretched the concept of conservation to treat humans as subject to the pressures of medical, physiological, and biological constraints and opportunities. It now included the efficient use of human life in the interests of the nation and denounced unhealthy practices as "waste." Though manifesting internal

American anxieties over the growth of cities, slums, and disease, the new interest was also part of a transnational movement, whose origins lay in the question of empires at the turn of the twentieth century. The Boer War, which lasted from 1899 to 1902, was a catalyst for this new way of thinking. With Britain bogged down in protracted military conflict in South Africa, many thousands of soldiers were needed, but British troops proved to be deficient in health. Recruits were too short, too sick, too often infected with rickets and other maladies.[4] Forty percent from the large industrial towns were rejected, and commissions of inquiry investigated the causes and consequences.[5] A Royal Commission on Physical Training in Scotland as well as an Inter-departmental Committee on Physical Deterioration in 1903 raised British alarm. The latter was the most important investigation, recommending "social reform to ensure healthy national stock" and endorsing military drills "to improve physical fitness."[6] Racial decline blamed on the falling birthrate of educated women also became a transnational concern. The Royal Commission of 1903 on the Birth Rate in Australia, cited by Fisher, was a bellwether. Anxious Antipodeans blamed white women for the failure of the white race to reproduce rapidly, and the faltering birthrate among Anglo-Saxons was lambasted in both Britain and the United States.[7]

The perils of ill health also troubled Americans in the far corners of their newly won Pacific empire. US Army generals in the Philippines worried about the health toll of war and colonialism, especially in relation to venereal disease. The creation of a large army to fight the Philippine insurgency of 1899–1902 brought increased prostitution in its wake and a reactive military interest in sexual hygiene. Venereal disease rates doubled in the army from 1897 (84.59 per thousand) to 1910 (197) and prompted preventive health measures through segregation of soldiers from prostitutes, instruction of troops in "clean living," and, eventually, the issuing of "prophylactic packages."[8] Fisher's *National Vitality* cited these alarming figures and found the naval record equally disturbing. He also drew parallels with the British military in India, where the authorities had condoned regulated prostitution. Of the troops returning home to Britain after completing their time of service on the subcontinent, "25 per cent were found to be infected with syphilis," Fisher disclosed.[9]

In public health, the late nineteenth-century epidemics of smallpox and yellow fever that swept across Asia and the Atlantic world and the last great wave of bubonic plague that affected the Pacific Rim from the 1890s to 1908 provoked international agitation.[10] Outbreaks of these and other contagious diseases increased because international travel and trade had accelerated. The Philippines became a site of government experimentation with compulsory vaccination. Colonial officials used this evidence to show "the efficiency of compulsion" in vaccination procedures that doctors wished to implement back in the continental United States, even as resistance to vaccination became vocal.[11] Imperial experiences and a series of epidemics at home, notably a smallpox outbreak from 1898 to 1903, spurred efforts to reform health services. Doctors, social welfare reformers, and social scientists sought to raise human health levels through an improved environment, with better control of pollution and positive promotion of healthy life habits at the state level. The spread of tuberculosis, a disease that flourished in the poorly ventilated tenements of industrial cities, was a particularly important spur. This disease touched many; not only did Fisher suffer directly himself, but his father had died of the "white plague," as it was called. But there was much that governments could do. European improvements in school hygiene, including physical education programs and medical inspections of students, impressed Fisher.[12]

In 1906 Fisher became head of the Committee of One Hundred on National Health. Formed by the American Association for the Advancement of Science, it included medical experts and public figures such as Thomas Edison, Booker T. Washington, and Franklin Giddings. Giddings was vice president of the American Association for the Advancement of Science and a strong advocate of American imperialism as economically helpful to the project of exploiting tropical resources and morally important as a character-building exercise in global responsibility.[13] The Committee of One Hundred advocated greater intervention in health issues by the national government, and many Progressives saw public health as essential to the projection of American power through a strong nation-state. Patriotism became a motive for improved public health in this way, a dynamic that Roosevelt himself promoted.[14]

Professionalization of medicine motivated the core leadership for reform. Instead of qualified medical intervention in health, Fisher lamented, quacks and such groups as the Christian Science Church catered to a growing consumer demand for better health. To him it seemed that questionable methods flourished at the expense of the serious and competent medicine of scientific experts.[15] Nevertheless, Fisher was as aware as Pinchot that any conservation measure depended on effective advocacy and democratic opinion. It could be led from above but not mandated from above, and Fisher envisaged an enhanced role for the federal government in winning the struggle for public consent by spreading authoritative health information. Fisher therefore became a ubiquitous publicist for the health efficiency movement, and he saw Roosevelt and Pinchot as allies in this cause.[16] Along with Pinchot, Fisher was a member of the Yale Skull and Bones Club in the 1880s, graduating from Yale University with a bachelor of arts degree in 1886 and a doctorate in 1891. With this pedigree and a professorship in political economy at his alma mater from 1898, it is not surprising that Fisher became a contributor to the NCC investigations that Pinchot organized. Fisher referred to his fellow Yale Skull and Bones alumnus as "our leader" and a man whom Fisher "foresaw" in the early 1890s as likely to render "great service to the nation."[17]

At the Sixth International Congress on Tuberculosis in Washington (September–October 1908), the Yale professor advocated the creation of a National Department of Health. He pointed to the work that the federal government had already done successfully in investigating through the USDA the issue of tuberculosis in dairy products, in promoting a vaccine program for animals, and in providing scientific information to the public. Because this program was currently executed unevenly and inadequately in some states, his national plans "met with general approval" among eminent medical authorities attending the congress.[18] Roosevelt's second-term initiatives to strengthen national institutions attracted Fisher, since he believed, like the president, that a strong national public health campaign alone could save the United States from externally generated epidemics.

Government should fulfill three key public health functions, Fisher insisted. The first was research, namely the investigation of "preventive medicine and public hygiene"; the second, the dissemination of knowledge concerning that research, warning the medical profession and the public about dangers and providing evidence for pathways to better health. The third was "administration," by which he meant extended federal regulations concerning public health and their strict enforcement—for example, higher standards of hygiene on interstate transport. Underpinning his approach was the call for "better and more universal vital statistics." Without assembling such statistics and observing health trends and their correlation with socioeconomic circumstances, it would be "impossible to know the exact conditions in an epidemic, or, in general, the sanitary or insanitary conditions in any part of the country."[19]

This idea of using the state to strengthen public health paralleled Roosevelt's vision of a strong government. In his Seventh Annual Message to Congress in 1907, Roosevelt acknowledged the role of the Committee of One Hundred. "There is a constantly growing interest in this country in the question of the public health," Roosevelt began. "At last the public mind is awake to the fact that many diseases, notably tuberculosis, are

National scourges." How could he help as president? Characteristically, Roosevelt referred here not to a "federal" government but to a "national" one with a capital *N*: "The work of the State and city boards of health" should be supplemented by "a constantly increasing interest" on the part of that "National Government."[20] In 1908, after the NCC had been established and after the international congress on tuberculosis had met, Roosevelt went further to stress international benchmarks for reform. He called on Congress to meet the challenge from European nations where health agencies and efficient conservation had, he argued, already come into alignment. Further, he noted that USDA successes in protecting animals against disease ought to be duplicated in the protection of people's health. The international congress had made Americans "painfully aware" of the inadequacy of their public health legislation. As he so often did, Roosevelt's international reference point was intended to goad Americans to action: "This Nation can not afford to lag behind in the world-wide battle now being waged by all civilized people with the microscopic foes of mankind, nor ought we longer to ignore the reproach that this Government takes more pains to protect the lives of hogs and of cattle than of human beings."[21]

In line with Roosevelt's nationalist approach, Fisher's recommendations in the NCC's 1909 report also matched other themes of Progressive conservation, notably efficiency. The professor targeted waste that went beyond the obvious loss through human mortality, believing that "economic waste from undue fatigue" was probably more significant in economic terms than "serious illness."[22] Even "relatively slight impairment of efficiency due to overfatigue" could have striking economic effects on labor, increasing the risk of industrial accidents. Fisher's support of workers' compensation and industrial safety laws contributed to the contemporary transatlantic conversation on these matters.[23] American states in the North and Midwest were introducing such laws in this period. Nevertheless, through public health improvements, the horrific rate of industrial accidents in the United States could be indirectly addressed with federal help. The government could not interfere under the realities of the US Constitution, which favored the primacy of capital over the health of workers, but it could advocate positive health policies that would mitigate the worst effects. Quite apart from the impact of accidents, Fisher emphasized the insidious morbidity of fatigue. "A typical succession of events" was "first, fatigue, then 'colds,' then tuberculosis, then death."[24] Prevention of such a syndrome was a key aim of the new human-centered program of conservation. Aside from the tragedy of unnecessary deaths, morbidity undermined both economic efficiency and social equity long before that end occurred.

Nation was paramount in these efficiency calculations. Economic prosperity depended on a safe, sane, and healthy nation, and these national considerations outweighed private interest in short-term profits. Fisher factored into economic efficiency the externalities of industrial pollution and accidents to come to these conclusions. For Fisher, the economic advantages of lunch breaks and reduced hours needed to combat worker fatigue were not so much for the employee's or the employer's benefit as for the "race" and therefore the nation. As in other aspects of conservation, longer-term sustainability was the goal. "Continual fatigue" was "inimical to national vitality," and however much it affected the "commercial profits of the individual," in the end morbidity would deplete "the vital resources" on which "national efficiency" depended.[25] These and other examples showed, Fisher argued, "that the return on investments in health are often several thousand per cent per annum." Nowhere else could such rapid and rich opportunities for improved productivity occur.[26]

In making his arguments Fisher drew upon European examples, but the transatlantic exchange of ideas on the reform of industrial society was part of a larger imperial circulation of information concerning public health.[27] Fisher mentioned the Australian birth control and postnatal health evidence and cited the record of Great Britain in reducing child mortality, as well as the American public health campaigns undertaken in Cuba by military governor Leonard Wood. The latter work, health campaigners proudly announced, supplied world leadership on the control of malaria and other diseases in the tropics. Why, Fisher asked, could such success not be applied within the United States? "If the same thoroughgoing measures used in Habana and at Panama were employed among our own people, the resultant blessings would be almost equally striking." After all, General Wood had declared that the "discovery of the means of preventing yellow fever" saved each year "more lives than were lost in the Cuban war."[28]

Fisher also emphasized the military advantages of health improvements, pointing both to the Boer War's sorry record in terms of preventable illness and disease and to the greater success of the Japanese in the Russo-Japanese War. He considered army hygiene in time of war very important, and the absence of such hygiene had "grave consequences." As was widely known, the British army in South Africa "lost more men from typhoid fever than from wounds received in battle." In sharp contrast was the efficiency of Japanese hygiene, which was manifested in startling facts: "General Oku's army of 75,000 . . . had but 187 typhoid fever cases in a seven months' active campaign." In the Russo-Japanese War, the Imperial Army had worked hard to improve the health of the troops, as shown in the reduction of illness rates from the Sino-Japanese War of a decade earlier.[29] The strength of empires evidently depended on national efficiency in health. Japan was showing itself to be a leading member of what Roosevelt and Fisher called the civilized world by simultaneously defeating a European power in battle and demonstrating the value of public health reform that other imperial powers must adopt.[30]

As with many pro-imperialists, Fisher's anxieties over the nation's future came with reflections on history that were meant to prove his case, and the history he chose concerned the rise and fall of an ancient empire: "When in Rome foresight was lost, care for future generations practically ceased. Physical degeneracy brought with it moral and intellectual degeneracy." Instead of conserving resources, "spendthrift Romans, from the emperor down, began to feed on their colonies and to eat up their capital. Instead of building new structures they used their old Coliseum as a quarry and a metal mine."[31] As Amy Kaplan has argued, fear of becoming another Rome was commonplace in the imperial discourse of Rooseveltian America. Health and disease were powerful metaphors for this larger process of anticipated imperial decay.[32]

Fisher's Progressive conservation credentials extended to intergenerational equity as well as the international benchmarks that Pinchot and other Progressives referenced. When emphasizing the "foresight" lost by the Romans, Fisher joined Roosevelt in seeing this concept as a social, moral, and cultural value. Yet Fisher gave foresight a mutable, evolutionary explanation, arguing that "the conservation of health will promote the conservation of other resources by keeping and strengthening the faculty of foresight." The need for such a precaution Fisher attributed to racial decay rooted in an inefficient public health system and poor personal habits. "Lack of forethought" was "one of the first symptoms of racial degeneracy." Just as healthy men cared for and provided for their own descendants, Fisher opined, "a normal, healthy race of men, and such alone, will enact the laws or develop the public sentiment needed to conserve natural resources for generations yet unborn."[33]

This was a two-way process. In large part, Fisher allowed, the physical betterment of Americans necessarily depended directly on the conservation of nature. An "important method of maintaining vital efficiency [was] to conserve our . . . land, our raw materials, our forests, and our water." Only in this way could citizens "obtain food, clothing, shelter, and the other means of maintaining life." While efficiency and progress in the use of such raw materials of nature should produce healthy populations, the reverse was also true. A healthier population would "tend in several ways to the conservation of wealth." As the race became "more vigorous and long-lived" it could make "better utilization . . . of its natural resources." Through the prism of Progressive conservation, Fisher believed he understood the interrelated nature of the economic and social ills that American society faced and that conservation of natural resources must address. This was "not a series of independent problems, but a coherent all-embracing whole." If the nation cared to make "provision for its grandchildren and its grandchildren's grandchildren," it "must include conservation in all its branches."[34]

Yet Fisher went further to posit humans as the nation's most fundamental resource. He accepted that raw materials would soon be in short supply. Precisely for this reason, the intellectual and physical quality of the nation's population was vital to manage external nature better. "The development of our natural resources in the future will be more dependent on technical invention than upon the mere abundance of materials," he predicted. Fisher supported the idea that innovative technology would be the best way to stave off the impending crisis, arguing that the labor power of the "race" that practiced preventive health would be "more intense, more intelligent, and more inventive."[35] Fisher thus sided with those who felt that the challenges Americans faced over future raw-material shortages could be met by improvement in human capital rather than by purely regulatory restraint. In the long run, this approach was potentially at odds with the tendency of Roosevelt and Pinchot simply to withdraw resources such as land from the market and pointed to deeper contradictions within conservation policy, contradictions that will be taken up in the final chapter.

One very controversial development flowed from many of Fisher's judgments: the proposed "improvement" of "racial" stock. Behind these concepts lay what seems in retrospect to be conservation's dirty little secret.[36] Efficiency arguments had begun to meld with the nascent eugenics movement. A science concerned initially with plant breeding (British scientist Francis Galton, a relative of Charles Darwin, coined the term), eugenics meant, when applied to humans, taking action to maximize the population's strength. Eugenicist ideas flourished in the wake of Social Darwinism's rise and in tandem with the mounting international discourse over the fate of empires in the Euro-American world. Within the United States, eugenicists welded the earlier racist concern over the demographic decline of the white population to analysis of the American population's supposed degeneration by the too prolific breeding of the poor or defective. Supporters of eugenics advocated preventive measures for improving health through natal care for women and improved diets for the working class to stop physical deterioration. In extreme versions, eugenicists tried to influence directly the genetic makeup of society. They sometimes advocated compulsory sterilization to prevent undesirable breeding of allegedly inferior racial stock. Some advocates adopted the neo-Lamarckian idea that acquired characteristics could be transmitted across generations, and they therefore condemned bad habits. Under the influence of the genetic theories of the late German scientist and Catholic priest Gregor Mendel, these neo-Lamarckian ideas were largely eclipsed by the study of genetic characteristics.[37] The dividing line between the two was, however, blurred due to moralists' attempts to adapt the new genetic approach to their neo-Lamarckian one through speculation about the effect of habits on the alteration of genes. Soon temperance reformers began to talk about drunkenness as a hereditary characteristic and called alcohol a "poison" that robbed infants "of their racial rights" (as Prohibitionist congressman Richmond Hobson would later put it).[38]

Eugenics began to gain political traction at the same time that Progressive conservation peaked. In the years leading up to World War I, a number of states passed compulsory sterilization laws to limit undesirable births. Starting in Indiana in 1907, draconian legislation introduced involuntary sterilization of certain groups in state custody "to prevent procreation of confirmed criminals, idiots, imbeciles and rapists."[39] Connecticut (1909), California (1909, 1913, 1917), Washington (1909), Iowa (1911, 1915), Nevada (1911), and New York (1912) were among the states affected as the eugenicist tide surged. Even then, eugenics was a hotly contested and controversial topic because of its association with attempts to exclude from the nation (or otherwise control the influence of) Jews, blacks, Slavs, and Asians.[40] It seems that the brush of pseudoscientific racism tarnished conservation insofar as conservationists accepted eugenics. In German historiography, this trend has fed into a debate on "how green were the Nazis?"[41] Though the alignment of race and eugenics with conservation was by no means exact in the United States, the trajectory of conservation reform was toward coalitions of reformers and reforms. For this reason, eugenics was bound to touch upon many conservation themes, and vice versa.[42]

Eugenic ideas were undoubtedly part of the intellectual mix that went into the conservation movement, but mostly after 1910. Established by Seventh-Day Adventist and health food promoter John Harvey Kellogg, the Race Betterment Foundation (1911) took a leading role in the early growth of popular eugenics in the United States and influenced certain conservationists.[43] Fisher had been a patient at Kellogg's Battle Creek Sanitarium. There he befriended Kellogg, who advocated a healthy lifestyle for all Americans. Pinchot, too, met Kellogg at the sanitarium after checking himself in at a time of physical exhaustion in 1911, and his (sometimes) association with

9.2. Irving Fisher, bottom right, with Pinchot, Horace Plunkett, John Harvey Kellogg, and muckraking publisher S. S. McClure at the National Race Betterment Conference, 1914. LCPP.

eugenics followed.[44] Despite disagreements with hard-line Mendelians, Kellogg embraced some key eugenics strategies. He supported a breeding register of suitable people as a strategy for strengthening the white race. "We have wonderful new races of horses, cows and pigs," stated Kellogg. "Why should we not have a new and improved race of men?"[45]

The case for a eugenicist link with the broader movement for conservation in the late Progressive Era after 1910 rests on both advocacy and association. Pinchot attended the first International Congress of Eugenics in 1912, alongside "conservatives" like Charles Davenport, the director of the Cold Spring Harbor Laboratory, later a notorious source of eugenicist practice.[46] In 1922 Fisher became a secretary of the American Eugenics Society, a body founded by Henry Fairfield Osborn and the openly racist Madison Grant. Grant was the author of the widely read *The Passing of the Great Race* (1916). Osborn wrote the foreword, which called for "the conservation and multiplication for our country of the best spiritual, moral, intellectual and physical forces of heredity."[47] Grant contributed to the nativist intellectual climate that finally won, in 1924, restrictive immigration based on racial quotas favoring the "Nordic races" in the United States. He also worked to "save the redwoods" in California in the 1910s and 1920s. For Grant, the redwoods campaign reflected an identification of these ancient trees with the anti-immigrant action that followed American entry into World War I. Redwoods not only were sublime American trees in danger from economic development but also were very old and hence symbols of old-stock Americans threatened by the immigration of "hordes" of eastern and southern Europeans.[48] Within the conservation movement, newsworthy figures opposed the tide of those immigrants said to "swarm" over the countryside killing every wild thing in sight. William Hornaday put it most famously in 1913: "Let every state and province in America look out sharply for the bird-killing foreigner; for sooner or later, he will surely attack your wild life. The Italians are spreading, spreading, spreading. If you are without them to-day, to-morrow they will be around you. Meet them at the threshold with drastic laws, thoroughly enforced; for no half way measures will answer."[49]

Faced with the growing fashion for racial explanations of social behavior, it is not surprising that the final recommendation Fisher made in his original *National Vitality* report of 1909 gave some sanction to the nascent eugenics movement. Fisher urged authorities to "gradually put into practice" the "prohibition of flagrant cases of marriages of the unfit, such as syphilitics, the insane, feeble-minded, epileptics, paupers, or criminals, etc." Indiana's example, Fisher added, "should be considered and followed by other States, as also in regard to the unsexing of rapists, criminals, idiots, and degenerates generally."[50] Four years later, Roosevelt admitted this stance to be rational and progressive. In a much-cited letter, the ex-president stated that the time would come when "society has no business to permit degenerates to reproduce their kind."[51] Supporting the reform position that had adapted Mendelian genetics, Fisher came to argue some years after the 1909 report that certain moral behaviors could alter the genes. In 1913 he claimed "that the alcoholic taint can and does affect the germ plasm of which future generations are made," but apart from the "evils" of alcohol and sex, he believed it "unreasonable to infer that any [other] unhygienic habit" could be inherited through that means.[52]

Yet the acceptance of eugenics within the conservation movement, even in this modified form, was far from straightforward or uniform. There are strong reasons for doubting the affiliation of Roosevelt, Pinchot, and, in 1909, Fisher with the extreme eugenicist position.[53] Fisher's support for eugenics was tentative in the *NCC Report*. As he put it, "Until more results have been obtained, it would be premature to make great claims for the possible future usefulness of applied eugenics." Eugenics was just one approach that should be studied further, the report recommended. Very little space was devoted to compulsory sterilization. In line with Pinchot's views on public opinion, much more attention was given to the development of "habits" that would encourage in the longer term the voluntary separation of the fit from the less fit than to authoritarian state-imposed laws.[54] Those who were healthy and vigorous would naturally choose partners equal in vitality. The state did not need to intervene except to provide the environment where natural selection could take its normal course. "A public opinion should be aroused which will not only encourage healthy and discountenance degenerate marriages, but will become so embedded in the minds of the rising generation as will unconsciously, but powerfully, affect their marriage choices." Nor was racism straightforwardly implicated, for Fisher held up the Japanese people as a prime advertisement for the benefits of the physical-culture regime they had embraced for their schools. Civilization, race advancement, and physical health seemed embodied in the rise of Japan to high esteem among the nations.[55]

Roosevelt also believed that, while certain races and classes were "backward," uncivilized, or underdeveloped, they could be lifted up through positive economic and social reforms. This was consistent with his stress upon moral character and his attitude toward Japan after the Russo-Japanese War: "What wonderful people the Japanese are," Theodore Roosevelt told his English confidante, Cecil Spring Rice. In his annual message for 1906, Roosevelt stated that the Japanese people had "won on their own merits" the right to be treated with respect.[56] Ultimately, for Roosevelt, strenuous and manly achievement on the field of battle and evidence of modernization trumped race. As successful imperialists, the Japanese were on their way to becoming honorary Europeans in his opinion. (This is what made the domestic, political necessity of undertaking the "Gentlemen's Agreement" of 1907–8 to restrict Japanese immigration to the United States so difficult for him.)[57]

In any event, Roosevelt did not propose to stop the so-called inferior from breeding, and he later denounced "twisted eugenics."[58] Indeed, the ex-president advocated tax relief for *all* the bearers of children and financial punishment of those parents who did not breed.[59] Even as he conceded that, ideally, "criminals should not have children," nor the "shiftless and worthless," Roosevelt did not dwell upon the negative because the "positive is always more important." The greatest danger for civilization, he argued, was the "failure to have enough children of the marriages that ought to take place. What we most need is insistence upon the duty of decent people to have enough children."[60] He did not rule out the possibility that the poor and the non-Anglo-Saxon could be lifted by positive improvement in their environments, and neither did Fisher (in fact, Roosevelt criticized the second-generation children of immigrants as much as old-stock Yankees for falling birthrates). This environmentalism was the sine qua non of the Progressive approach and of the idea of national vitality. Environment could alter the physical condition of human beings in a positive way through conservation reforms.[61]

What Roosevelt did retain was an older, late Victorian concept: race suicide. Roosevelt's views on this subject were shaped by ideas that had roots in the 1890s, influenced by the Anglo-Australian racial theorist and radical democrat Charles H. Pearson. Unless the white race reversed its relative failure to reproduce, dilution of the racial stock of the British Empire through depopulation must follow. The Anglo-Saxon race risked succumbing to competitors on a global level. Pearson's work, *National Life and Character,* was avidly read by Roosevelt and became part of the discourse over the survival of empires.[62] In 1901 prominent sociologist E. A. Ross applied the term "race suicide" to the birthrate's decline, and Roosevelt embraced the concept as an article of faith, with Fisher merely following.[63] The president raised the specter again in his annual message in 1906 and in his celebrated Sorbonne speech of 1910, given during his European tour. There he went further to label race suicide the greatest crime of civilization.[64] As in so many other ways, Roosevelt's moral views remained characteristically late Victorian.

Recognizing the impossibility of any scheme that could "prevent all undesirable people from breeding,"[65] as Roosevelt later put it, the *NCC Report* focused on campaigns to change personal habits. This objective accorded well with the rise of the national Prohibition movement, but it also incorporated tobacco restriction, a cause that the Woman's Christian Temperance Union and other reform groups of the Progressive Era championed. Fisher called for a "quiet revolution in habits of living, a more intelligent utilization of one's environment, especially in regard to the condition of the air in our houses, the character of the clothes we wear, of the site and architecture of the dwelling with respect to sunlight, soil, ventilation, and sanitation, the character of food, its cooking, the use of alcohol, tobacco," and "other drugs." Last and not least came improved "sex hygiene."[66] This was an all-encompassing call to social reform, not simply a eugenicist claim, even though certain conservationist ideas that spread to the alcohol prohibition movement in the decade after 1910 were consistent with eugenic sentiments.[67]

If Progressive conservation under the Roosevelt administration avoided the more virulent implications of eugenics, the flirtation left a mixed legacy for the conservation movement. On the one side were the dark and splenetic fulminations of Madison Grant and his ilk, whose work could be admired by Adolf Hitler. On the other was the growing concern for a comprehensive national public health agenda. Fisher's campaign certainly had an impact on the redefinition of what conservation was. In Roosevelt's landmark Osawatomie speech of 1910, as he prepared for a possible further tilt at the presidency, the ex-president emphasized "that the health and vitality of our people are at least as well worth conserving as their forests, waters, lands, and minerals, and in this great work the national government must bear a most important part." Conservation here was not a matter of genetic inheritance, Roosevelt repeated. Its ultimate goals included "*training* them [the people] into a better race to inhabit the land and pass it on." Conservation was not a scientific but a "great moral issue" because it involved "the patriotic duty of insuring the safety and continuance of the nation."[68]

In the Progressive Party platform of 1912, that moral issue was further elevated. There, justice took precedence. "The supreme duty of the Nation is the conservation of human resources through an enlightened measure of social and industrial justice." Justice included the extension of government power to guard the citizenry against ill health. The 1912 platform capped the burgeoning health campaign by calling for a national department to implement the objective of national vitality. This department must, Progressives believed, have "such additional powers" to protect the public "from preventable diseases."[69] The US Congress in that same year gave a down payment, in effect, on the implementation of Fisher's goals and Roosevelt's. The Public Health Act of 1912 included key provisions recommended in the *National Vitality* report. The newly named US Public Health Service was tasked with collecting and disseminating statistics on mortality and morbidity and given authority to "investigate the causes of disease and the factors at work in their propagation, including the pollution of the nation's lakes and streams."[70] In this way, the agitation of the Roosevelt years actually reached a modest sort of fruition.

This redefinition of Progressive conservation to include human health came partly as a result of lobbying from eugenics sympathizers. Charles W. Eliot, the former president of Harvard University and member of the Race Betterment Foundation, was a prominent supporter of the National Conservation Association (NCA). He urged the introduction of health objectives into Progressive and conservationist platforms. Eliot wrote Pinchot in 1915 that "[e]ver since the admirable treatment of national vitality by Dr. Irving Fisher in the National Conservation Commission Report, I have felt that the many problems involved in the human conservation are more important than those regarding the natural resources."[71] But Eliot did not believe the NCA under Pinchot's leadership had done enough. In reply, Pinchot conceded the importance of health reform but explained that the efforts to stop economic monopoly had become "more imminent" in the years since Roosevelt's presidency. In 1915 Pinchot still favored coordination of the nation's public health bodies, as in the 1912 Progressive Party platform, but he excused himself from agitation over the issue because political exigencies required lobbying Congress on waterpower as a first priority.[72]

Whatever the legacy, the idea of conservation had already crystallized by the time Roosevelt left office. The journey from the National Reclamation Act of 1902 to a comprehensive conservation policy was complete, or almost so. Conservation had taken shape, not as dealing with resources of water, wood, fossil fuels, and soils alone, but as involving human welfare as well. This movement toward a comprehensive approach was quite deliberate on Pinchot's part, as he and Roosevelt sought an ever-widening circle of reform and as many allies as possible. Yet something is missing from this itemization of reform, something that motivated the slew of reforms in the first place. Global anxieties over resource availability and fears of external competition had affected human aspects of conservation as much as any other. The resultant agenda was both national and international, but the publication of *National Vitality* in 1909 came when Roosevelt was leaving the presidency. How could the momentum of the conservation movement and its all-encompassing platform be advanced internationally and nationally when the power of the presidential bully pulpit and the arms of the bureaucracy were no longer Roosevelt's? A trip to Africa might be more than a distraction. It might be a way of sustaining the conservation project by means of the publicity that it would generate, bringing this global issue and Roosevelt's postpresidential future together in an exhilarating adventure. Roosevelt could thereby draw attention to species extinction, the burdens of empire, and himself.

PART THREE

The Global Vision of Theodore Roosevelt and Its Fate

TEN

To the Halls of Europe: The African Safari and Roosevelt's Campaign to Conserve Nature (While Killing It)

Exactly what should one do at the end of a term as American president? Some of Theodore Roosevelt's predecessors retired exhausted, sulked, wrote memoirs, traveled around the world, or died almost immediately. For Roosevelt, it was a strenuous hunting expedition to Africa that was on his mind. On 23 March 1909 TR and son Kermit boarded a steamer in New York, bound via Naples for Mombasa, British East Africa, arriving there 21 April. In the ten months that followed, father and son mixed with Maasai tribesmen and pious missionaries, haughty British officials and hunters full of derring-do. Roosevelt's party traipsed from Kenya into Uganda, then to the Lado, a small exclave of the Belgian Congo, and on to southern Sudan. They reentered a world of curious onlookers and feverishly scrambling newspaper reporters via the mystical Nile, arriving in Khartoum on 14 March 1910. Once they had emerged from what Roosevelt called "savagery" and "wilderness" into "civilization," it was time to take stock. TR's party had collected eleven thousand specimens great and small, including elephants, hippos, rhinos, leopards, and lions. Father and son alone shot 512 birds, mammals, and snakes.[1] These specimens were skinned and sent back with their skeletons to museums in the United States for stuffing, mounting, and exhibiting, though Roosevelt kept some animal parts to display as trophies at his Sagamore Hill residence. The total take from British East Africa alone included over four thousand mammals.[2]

In spectacular ways, the visit highlighted certain themes in the ex-president's complicated relationship to conservation and global power. Waste and efficiency, love of the wilds of nature and its challenges, the role of settler societies as agents of human evolution and natural resource conservation, and the building of a strong nation-state—all these figured in the story. Also present was Roosevelt's seemingly insatiable desire to maintain the close attachment that he believed he had forged with the American people, as well as the continuing urge to identify Progressive reform with himself. The visit had its own interesting bag of stories, but at its heart, it also advanced the ex-president's ambitious global agenda of permanence and power for the American Republic within the imperial state system. What better way was there to advance all these objectives than a newsworthy event that brought the personal and the political so close together that they could not be separated?

From the perspective of the early twenty-first century, one glaring contradiction in the African venture immediately stands out. How could a man who slaughtered animals en masse be a conservationist? In *African Game Trails*, Roosevelt's engrossing account of the trip, "conservation" is not mentioned, but "natural history" is, and Roosevelt was known as a great lover of nature, of the field called natural history, and of wildlife. Roosevelt and his party went to Africa in the name of scientific discovery and the documentation of natural history. He believed that greater scientific knowledge of this type would allow governments (and hunters) to know what species were in danger of extinction. Implicitly, "wise legislation" might follow. Roosevelt had cause to think that way. He gave strong support to the rulers of colonial Africa who had pioneered the idea of game reserves. Such legislation informed his own ideas about preserving game for hunters internationally and nationally.[3] Though he slaughtered many animals, he did what other Victorian and Edwardian hunters and naturalists in Africa had readily done before him. He was in that respect a man of his times. More distinctive was the fact that he wanted not only to study nature but also to be seen studying nature—as Roosevelt, the intrepid naturalist-explorer. His book and magazine articles, for which Scribner's offered him a handsome advance,[4] reflected upon the issue of adaptation to natural surroundings and intimated that his bloody campaign served the study of the natural world and of biological evolution. This theme squared well with the lecture, "Biological Analogies in History," that Roosevelt delivered in Oxford during his post-safari visit to Europe.[5] In Africa, he was thinking through the science of natural selection, the world-historical significance of evolution, and the place of the United States and himself in it as a force of both nature and civilization.

Despite the enormous numbers of animals slaughtered, there was relatively little criticism at home. Certainly, humane societies protested, cartoonists had a field day, and old enemies carped. General Nelson Miles, whom Roosevelt had "driven" from the army in 1903, remarked that a man who shot elephants, zebras, and wildebeest "must have a depraved mind."

Miles pointed out that "England recently passed a law setting apart a large tract of land in Africa for reservation purposes and for the protection of just such animals as elephants, zebras, antelopes and ostriches, which live in that region." What a pity that a man should "find enjoyment in shooting" such creatures as these.[6] But most American commentators treated the ex-president's exploits with adulation, though the occasional charges of bloodlust and exhibitionism raised hackles among Roosevelt's many adoring supporters, such as John Burroughs, a prominent naturalist, and William Hornaday, director of the Bronx Zoo.[7] When the latter jumped to TR's defense, the justification was largely scientific. Hornaday claimed that the ethics of this trip were not in any way "assailable." To "condemn the Colonel's work in Africa [was] to condemn the museum idea so far as it relates to zoological forms." The African venture was all done "in the interests of zoology and devoted wholly to science."[8]

Hornaday also linked the episode to a vigorous nationalism, an assertion of the American nation on the world stage. The trip was nationalistic in the way it highlighted the retirement of a president already admired in Europe for putting the United States on the geopolitical map. But it was more than that. In Hornaday's opinion, the slaughter was a price that had to be paid for national self-esteem. These specimens were needed for "this great progressive, and wealthy Nation" that the United States had become and because America's *national* museum had been scandalously underfunded and handicapped. Its collection was unrepresentative of so many species, some of them endangered. Roosevelt's vigorous campaign would fill the void.[9]

Although the trip was undertaken, or so it was claimed, in the name of science and nation, it also had to be seen to reflect, in Progressive reform style, calculation and restraint, as Roosevelt's own account put it in the case of the white rhinoceros: "Too little is known of these northern square-mouthed rhino for us to be sure that they are not lingering slowly toward extinction; and, lest this should be the case, we were not willing to kill any merely for trophies; while, on the other hand, we deemed it really important to get good groups for the National Museum in Washington and the American Museum in New York, and a head for the National Collection of Heads and Horns which was started by Mr. Hornaday."[10] The assumption here remained survival of the fittest. Killing for science, stated Hornaday, was "no more cruel or wasteful of life than the forces of nature herself." After all, just one lion killed 104 zebras, gazelle, and other mammals a year, on average. This was the way of nature. Species would die off anyway—humans could justifiably take for science because of this cycle of life and death. In a sense, killing species conserved them forever, whereas neither the

10.1. Roosevelt hunted rhinoceroses despite (or perhaps because of) concerns over the possible extinction of the northern white rhinoceros (*Ceratotherium simum cottoni*). LCPP.

animal world nor "savage" peoples had any such foresight. Hunting would help preserve the memory of and knowledge about the northern square-mouthed rhino even in the event of its demise. So Hornaday opined, and Roosevelt agreed.[11]

In hunting for science, Roosevelt justified the process in terms of the Progressive Era's efficient conservation practice. The jaunt in Africa was morally upright use rather than hapless waste. "We only shot for meat, or for Museum specimens—all the Museum specimens being used for food too." At its most elemental level, father and son found themselves "in a wild, uninhabited country, and for meat we depended entirely on our rifles; nor was there any difficulty in obtaining all we needed." Mostly, the abundance of game rendered the take innocuous: "The naturalists were as busy as they well could be" because "except when we were after rhinoceros, it was not necessary to hunt for more than half a day or thereabouts."[12] There was simply too much game to worry about extinctions in most cases, yet the desire to prevent waste remained a consideration in Roosevelt's retelling. We know that Chicago's meatpacking industry was said to leave nothing of the pig but the squeal. A similar point could be made of the hunt. Sagamore Hill received a wastepaper basket made from the hollowed-out foot of an elephant and an inkwell crafted from a rhino part.[13]

But the event was neither a scientific expedition alone nor an episode in efficiency conservation. It represented on one level Roosevelt's masculinist strivings for a "full-blooded picnic," as one author puts it, or an encounter with the wilderness.[14] He wished to experience the wild once more—to be close to elemental nature in all its bloodthirsty detail. There is evidence that Roosevelt coveted a hunt in the Belgian Congo because there were no limits on the take in that "greatest of all elephant countries."[15] That said, Roosevelt entered Belgian territory only marginally, mostly skirting it and traveling north from Lake Victoria via the more ordered, British-controlled Uganda and Sudan. Manliness evidently had its limits and was not in any case incompatible with a leadership and policy orientation. For this reason, Roosevelt's "Pleistocene vacation"[16] should be put in context. It was preceded by a concerted campaign for conservation and succeeded by three months of touring and lecturing the European powers on their duties and responsibilities for the uplift of the less fortunate people of the world. The African visit spoke directly to both of these themes—and to a third, his media manipulation, which underlay all Roosevelt's achievements. With the driving, scheming force of Gifford Pinchot behind him, Roosevelt had inaugurated the modern media's relationship with the presidency.[17] The African trip was a prime exhibit for this point and was calculated to enhance Roosevelt's reputation and to draw attention to his interlocking causes—conservation and imperial power.[18]

At first glance, Roosevelt seemed to be shunning the media whirlwind that had followed him during his presidency. The deal with Scribner's prompted a ban on press photographers (but son Kermit substituted in this role). Nor were reporters officially to accompany him into the interior. This was a masterstroke that did not escape his press secretary William Loeb, who calculated for Roosevelt exactly how effective the impact of denying press access was.[19] Back in the United States, the press covered the tour as best it could. Though Roosevelt ostensibly spurned publicity, intrepid reporters hung about, sneaking into meetings with local British officials; letters leaked out via Maasai runners; and information flowed across to Europe and the Atlantic world. In truth, the event had become a media circus even before Roosevelt landed in Mombasa, with reporters sailing on the same ship.[20] So much was he pestered that he relented for a while, allowing an

10.2. Roosevelt told Harvard classmate Ramon Guiteras that he wanted to hunt "as many good tuskers" as possible. LCPP.

old family friend, Warrington Dawson, the United Press International correspondent, special access as far as Nairobi to counteract spurious accounts of the adventure that Dawson called "nature fakery." Robert Foran of the Associated Press gained a similar, though less familiar, access. Roosevelt dictated drafts of *African Game Trails* to Dawson and later openly tried through Dawson to shape how the press remembered him—as taking "care of animals" during his trip.[21]

The extent of American engagement with the wider world is suggested by the ex-president's encounters with fellow countrymen. He met Americans everywhere, and they obliged newsmen with pictures and stories for the home audience. Upon completion of his own hunt, an old Harvard classmate, Ramon Guiteras, met the colonel in Mombasa and took back information for readers of the *San Francisco Call* and the *New York Herald*, providing a rundown on what TR *might* be doing.[22] Potboilers and travelogues appeared too.[23] Footage shot by the official film cameraman, Yorkshireman Cherry Kearton, provided almost an anthropological coverage.[24] Kearton strove for the unattainable goal of authenticity and refused to film hunting events that could take place only if they were staged for the camera, such as daytime shootings of lions. This inhibition did not hold back commercial interests. The Motion Picture Patents Company (commonly called the Edison Trust) released the film using Kearton's footage but announced it with the "sensational" description "The Far Famed American Hunter, Colonel Roosevelt, amid the man-eating monsters of the Wild African Jungle." It was a hit.[25] Resourceful entrepreneur William Selig filmed a successful fake, *Hunting Big Game in Africa* (1909), shot in a California back lot with an actor playing TR. In all, the trip was surely one of the best-covered hunts in history. But not until Roosevelt returned and published *African Game Trails* did the full extent of his blood-soaked activities become clear through the many photos that the expedition leaders themselves had taken.[26]

News stories often gave the impression of a backwoodsman combating the elements, with his firearms and unerring aim. Dawson told American readers that Roosevelt was a good shot who had the African tribesmen in "awe" over his marksmanship. This was quite untrue. Roosevelt's eyesight was defective, and on occasion, he and Kermit sprayed bullets indiscriminately. The acting governor of Kenya, Frederick Jackson, later called him "utterly reckless" in his thoughtless use of ammunition and senseless wounding of animals that had to be finished off by others.[27] But none of this detail made Dawson's reports. The only hint of comedy came in Dawson's observation that the natives called TR "Bwana Tumbo," which meant "portly master." Though Dawson claimed that the name was a term of endearment, Roosevelt took silent offense and tried to preserve a grander self-image. He told home audiences that "I was always called Bwana Makuba, the Chief or Great Master." Yet the "portly master" tag stuck in some reports.[28]

The Africa through which Roosevelt trekked was no "wild, uninhabited" space.[29] Wherever he went, Maasai or other tribesmen were all about, with dozens of porters to carry the ex-president across even the slightest of streams, in pith helmet, full Bwana style, however that term was interpreted. The total touring party resembled a small army. Porters (260 of them in all) carried every conceivable personal item, among them nine pairs of glasses, Roosevelt's books, champagne, chocolates, and ample amounts of dental floss (presumably to preserve his toothy grin).[30] Several other European and American hunting expeditions were going on in the same area of British East Africa where most of the killing took place.[31] One was that of the famed American taxidermist Carl Akeley, who orchestrated the collection of specimens for the Smithsonian Museum. This was the era in which, as historian John MacKenzie has argued, the safari as tourist experience was being created, facilitated by construction of the railway from Mombasa that, after

10.3. Roosevelt's 260 porters on the march, photographed by United Press International newsman and family friend Warrington Dawson. LCPP.

1899, made the sleepy crossroads settlement at Nairobi the launching pad for European adventures. Roosevelt reenacted that story, though with his own publicity-centered twists. His visit gave big-game hunting in East Africa "celebrity" status and, in effect, made it more attractive to rich tourists following in his footsteps.[32]

By no means was Roosevelt's visit only an effort to escape to the wilderness. Rather, it reflected a persistent ambivalence. His time in British East Africa was peppered with return visits to Nairobi to meet "society" and deliver speeches. At night by the campfire, Dawson reported, Roosevelt seemed less than satiated by wilderness alone. The ex-president counterbalanced the world of the wild with that of politics and intellectual life in an elaborate juggling act. The pull of the latter reflected his desire to use the trip to comment on Europe and its imperial outreach. Arguably, this became more important than the hunting itself.[33] The ambivalence was especially reflected in his attitude toward the land. While he stressed similarities with the Wyoming and Montana of his youth, it was the "ranches planted down among the hills" and homesteads surrounded by gardens of flowers that caught his eye. Even as he lamented the fading of the American West's wild phase with its teeming game long gone, he advised government-sponsored irrigation for the African uplands on the western American model specifically because it would turn the country from an untamed wilderness into a land of white farmers with families. He praised those settler farms where this irrigated transformation into American-style farmland was already under way.[34]

The official film of the visit followed the story of advancing modernization under British rule and of tribalism in the culture of the natives. It contrasted the progress of civilization through the railway journey to Nairobi with the ethnographic presentation of the tribal rituals of the Maasai and Kikuyu (wrongly labeled in the film as Zulu).[35] However, the indigenous people were essentially depicted as human beings capable of (varying degrees of) progress. The humor of the native people who performed dances and other feats on camera is clear. They partake in a ritual for European benefit, and crowds of Africans in tribal or Islamic dress show that they understood the nature of the performances as they watched.

Leaving British East Africa and the Uganda protectorate, Roosevelt's party journeyed through the Sudan for three weeks, ending up in Khartoum. There Roosevelt toured the scenes so recently contested between the forces of the millenarian Islamic leader Muhammad Ahmad (the self-proclaimed "Mahdi") and the British under Lord Charles George Gordon, who died at the siege of Khartoum in 1885. As Roosevelt reminds us in his book, Lord Kitchener had relieved Khartoum from Islamic control and avenged the death of Gordon only in 1896. The bloody conquest of Africa was close in time, and Roosevelt identified the extension of British rule with the recent American subjugation of the Philippines. In this and in many other ways, Roosevelt's visit was inseparable in strategy and impact from events back home and his subsequent visit to Europe to receive the approval of European empires.

The entire trip cast Roosevelt as a world leader doing a lap of honor. At last he was going to Europe to claim the Nobel Prize won in 1906. In this narrative, TR's African prelude made him an actor in an imperial drama, aligning him with the social ethic of colonial hunters. "Hunting represented the most perfect expression of global dominance," John MacKenzie has stated in noting how, in Africa and Asia, British hunting lay behind the creation of the first forest and game reserves. Roosevelt sympathized with this movement.[36] Kruger National Park had its origins in one such reserve proclaimed in 1898. After the Society for the Preservation of the Wild Fauna of the Empire was established in London five years later, Roosevelt commended (and joined) its conservationist efforts as part of a global movement.[37] The hunt drew attention to Roosevelt's identification with a conservation sentiment that preserved wild game for the selective use of wise white marksmen, at the expense of indigenous use.[38] Accompanying him on early parts of the tour was the noted English hunter Frederick Selous, who had, years before through his writings and correspondence, interested Roosevelt in the conservation of African animals. This was an expansion of Roosevelt's interest in saving the American bison, which had come so close to extinction.[39]

When in Africa Roosevelt could not help but reflect, through the Euro-American world's quest for national efficiency, upon civilization's onset in that "dark" continent. Taking up this question at the Guildhall address in London, he referred to "the spread of civilization over the world's waste spaces," thus linking conservation, empire, and efficiency in ways that occluded indigenous occupation. He repeatedly praised the work of British Empire officials. "British Rule in Africa" was the title of the Guildhall address—these Anglo-Saxon cousins were engaged in the task of "subduing the savagery of wild man and wild nature." The *New York Times* called it a "world-stirring address."[40]

But this was imperialism endorsed with a *settler* inflection in mind. In a telling after-dinner speech in Nairobi, he had already linked the British project in East Africa with the history of the "English-speaking peoples" and their "remarkable spread" over so many underutilized "places of the earth." This "expansion," he argued, must continue through a conscious settler colonial process, which required conservation. Roosevelt was careful in his talk not to offend the British audience by claiming, as he did for the readers back home in his book, the equality of Boer settlers in Africa with the English. He knew that the British thought that some of these "Afrikanders" (as he called them) had regressed into "savagery" under African conditions. Rather, the types he met in East Africa fulfilled his ideal of a composite American-style white identity on the Great Plains, where sturdy pioneers mixed and thrived under the frontier challenge. The editor of the *Nairobi Leader* rejoiced, stating that Roosevelt had justified white rule. The speech would "serve to correct some of the rather misleading descriptions" about the settlement process and show the area to be "perfectly fit for white colonization" rather than a region to be "devoted to Asiatic or other colored occupation." Whether deliberate or not, this speech was a highly political intervention, as South Asian laborers had begun to flock to the area in the 1890s.[41]

Roosevelt used the hunt as an opportunity to emphasize his support for settler expansion. The need to demarcate what might be called a "living space" he gave priority. The two topics of hunting and empire came together with the racial foundations of his thought when he noted the travail of the white settler under the threat of the wild. Game laws should favor the interests of these "settlers," meaning the white Afrikaners, British Kenyans, and

Australians with whom he hunted, not "well-meaning persons" who "apparently think" that "man could continue to exist" if "all wild animals were allowed" to roam "unchecked."[42] The latter stance was incompatible with "civilization." Roosevelt's account of the hunt was thus justified in terms of this settler-based game management, not preservation. Game animals left to run wild threatened agriculture. Instead, he held in high esteem "the happy mean which is healthy and rational" between the extremes of preservation and slaughter.[43] As with his domestic conservation policy, he did not adhere to a management strategy based on either scientific expertise or democratic principles; rather, he favored the amateur values of a social elite of manly hunters. In the interests of the same civilizing values, however, he supported ideals of gentlemanly fair play, since "game butchery" was "as objectionable as any other form of wanton cruelty or barbarity."[44] This was an extension of his attitude toward American game hunting expressed in the Boone and Crockett Club.

It was not only masculinist hunting that Roosevelt endorsed as a settler value. The very first photograph of his travels in the interior to reach the American public's newspapers was taken at Kijabe, British East Africa, on the edge of the Rift Valley, at the Africa Inland Mission.[45] Heading this outpost was the Presbyterian Charles Hurlburt of Pennsylvania, who appears in the photograph. TR had met with Hurlburt in 1908 while the missionary leader was in Washington, and Roosevelt promised to return the visit. The president sought from Hurlburt hunting advice and maps of East Africa before he departed, and he heaped praise upon the missionaries.[46] Like other American politicians of the time who went abroad, Roosevelt used missionaries to delineate American cultural endeavor around the world both spatially and morally. That missionary work, however, was to be nondoctrinaire and practical.[47] "I earnestly wish you well in your work; all missionaries who do honest, practical work, whatever their creed, are entitled to the heartiest sympathy and support and it will be a particular pleasure to me when I go back to my own country, to report what is being accomplished by this Interdenominational Mission."[48] Medical assistance he especially endorsed. At the Sobat Mission further north he praised "the faithful work" that Christian doctors were doing "under such great difficulties and with such cheerfulness and courage."[49] Missionaries thereby served God and the state by rendering the indigenous "useful in the development of the country's resources."[50]

Tied to the missionary advance was the settler theme. Roosevelt was emphatic that missionaries must serve the white farmers as equally deserving of spiritual attention as the indigenous Africans. Country churches very

10.4. Roosevelt, second from right in front, wearing pith helmet, at the dedication of the Rift Valley Academy established for the children of missionaries and settlers by the Africa Inland Mission, 4 August 1909. LCPP.

similar to those he advocated for the United States in his response to the Commission on Country Life should be introduced for these scattered settlements of Europeans in Africa. Such affinity with Hurlburt did Roosevelt have that he returned to lay the cornerstone of the Rift Valley Academy, a mission school that would cater to settler children.[51] Not all of British East Africa was suitable for the white man, to be sure, but the higher elevations should be reserved for this purpose, Roosevelt advised. Kenya provided "an excellent opening for small farmers, for the settlers, the actual home-makers, who, above all others, should be encouraged to come into a white man's country like this of the highlands of East Africa."[52] In effect, the wilderness would have carved out of it special areas as demonstrations of the settler society model to lead the uplift of British East Africa as a whole.

TR emerged from his safari to pronounce upon the relationship of the colonized and their conquerors and, in effect, the entire fate of empires. He proffered advice on how Europe could learn from the Philippines and the judicious economic development that Americans were undertaking there and in Panama. In Khartoum on 16 March 1910 he declared that the "reign of peace and justice" that the British had brought should continue and that the Sudanese should "uphold the present order of things."[53] In Cairo, he preached complete religious toleration between Christianity and Islam and chastised the "noisy, foolish and sometimes murderous agitators" among the "anti-foreign element." Condemning the assassination of Boutrus Pasha (Boutros Ghali), the Coptic Christian prime minister of Egypt, by radicals, he opined that independence was a work not of years "but of generations."[54] His comments were controversial in Egypt, but in the United States, these speeches won him acclaim. Eight hundred men attending a fund-raising banquet for the Laymen's Missionary Movement in San Francisco endorsed his Khartoum speech and cabled him at Aswan stating that they "count largely upon your co-operation in enlisting the church for the evangelization of Africa and of the world."[55] Thus, the African trip had religious and moral, as well as scientific and cultural, significance.

Then it was on to Europe. There he basked in the glory of kings, emperors, prime ministers, and, in France, the president of a sister republic. In Italy, a worried Pinchot had met him after hurrying incognito across the Atlantic to confer on conservation policy at home regarding President Taft, a development that would ultimately draw the hunter back into American politics. In the meantime, he carried to the halls of Europe the same messages of "progressive Conservative" reform, including conservation and enlightened imperialism, that Britain's *Daily Mail* had praised upon his passage from office.[56] At the Sorbonne, he spoke of the strenuous life, and *Le temps*, the Parisian daily, had 57,000 copies of the translated speech distributed to France's schoolteachers.[57] In Norway, he received the Nobel Peace Prize. On horseback in Berlin, he swapped stories with Kaiser Wilhelm II, as part of his unofficial peace mission on behalf of Andrew Carnegie against the arms race in Europe,[58] and delivered his lecture "The World Movement," in which he slotted conservation into the story of what today would be called globalization. For Roosevelt this entailed the interdependence of all civilization and its spread across the world with extraordinary rapidity. "Any considerable influence exerted at one point" of the globe was "certain to be felt with greater or less effect at almost every other point," he counseled.[59]

As a perfect instance of his larger theme, he drew the attention of those European rulers once more to the global need for conservation: in the "modern scientific development of natural resources," some part concerned nonrenewable assets. There, development entailed destruction, Roosevelt warned. Exploitation on a "grand scale" meant an "intense rapidity" of advance, "purchased at the cost of a speedy exhaustion." Across the Euro-American world, the crisis of fossil fuels seemed nigh. With the "enormous and constantly increasing output of coal and iron," the day was "necessarily" hastened when "our children's children, or their children's children, shall dwell in an ironless age—and, later on, in an age without coal—and will have to try to invent or develop new sources for the production of heat and use of energy."[60] This was uncannily reminiscent of Rudolf Cronau's rhetoric and told a story not just for American consumption but the world's.

Roosevelt also discussed renewable resources, because "scientific civilization teaches us how to preserve . . . through use." Years of conversation with Pinchot were registered here. The "best use of field and forest" would make these assets "decade by decade, century by century, more fruitful." Even waterpower, better known and proven as it was, had "barely begun" to be exploited. This "harnessed water" would be an "indestructible power."[61] This was his conservationist plea, following up on his call for a World Congress on Conservation.[62] But aside from the specifics of his pet causes, the message was peace, justice, righteousness, and order, an order that included the imposition of civilization on the "savage" peoples, an order in which the bearers of Western ways should not flinch even if peace had to be brought, as it had in the Sudan and the Philippines, with a sword. And so on to London, where his visit coincided with the death of Edward VII and the ascension of the new king. The republican Roosevelt became a "Special American Ambassador" for the funeral. He returned to the United States on 18 June.[63]

Whether intended or not, the trip enhanced Roosevelt's reputation in American politics precisely because it cast him as a world figure. This was how acolytes, journalists, many Europeans, and a sizable section of the American public responded to the adventure and conceived of its significance.[64] The case can be illustrated through Marshall Everett's *Roosevelt's Thrilling Experiences in the Wilds of Africa Hunting Big Game* (1910), an account already mined by scholars to expose the hunter-statesman's adherence to standards of masculinity and for a gendered reading of its presentation on African exotica.[65] Yet the book devoted a good deal of its copious detail to commending Roosevelt's public diplomacy in Europe and positioned him for a possible future presidential bid, just as academics had remarked at his Romanes Lecture given at the renowned Sheldonian Theatre, Oxford. Peter MacQueen wrote one chapter for Everett, "Return of Col. Roosevelt from the Jungle." In MacQueen's uncritical eyes, Roosevelt confirmed as a result of his trip a "Remarkable Reputation . . . as a Man, a Hunter and a Statesman." Indeed, "The Eyes of the Whole World" were constantly on "This Great American, His Speeches and Striking Personality." A traveler, explorer, and war correspondent, MacQueen had been born in Scotland in 1865. Migrating to the United States at the age of seventeen, he studied

10.5. Roosevelt in Paris at Les Invalides, visiting the Tomb of Napoleon, with friends Jean Jules Jusserand, the French ambassador to Washington (in top hat, fourth from left, front row), and Robert Bacon, Roosevelt's last secretary of state and US ambassador to France (to the right of Jusserand), and French general Jean-Baptiste Dalstein (second from left, front row, speaking to Roosevelt). LCPP.

at Princeton University and Union Theological Seminary. After serving as a minister of religion, he acquired wanderlust for exotic places before the Spanish-American War. Accompanying the future president and his Rough Riders at San Juan Hill, he acted as a war correspondent and later followed the heroic, ill-fated General Henry Ware Lawton to the Philippines.[66] Then he turned to lecturing. MacQueen's slogan held that "to travel is to realize the world's intelligibility." By 1910 he had been in "forty different civilized and uncivilized countries" and gave lectures on anything from the new empire in the Philippines and the "Panama Canal and the tropics" to "old" Europe and the wonders of ancient Egypt. On one occasion, he was able to lecture various public audiences for thirty days in succession in Boston on "Africa as Roosevelt Saw It."[67]

With popularizers like MacQueen touting his praise and the newspapers full of his African and European exploits, no wonder that Roosevelt returned to rapturous crowds and full-strength publicity in his homeland. The nation, or large parts of it, could not get enough of the man and his adventures. Americans looked forward to watching where his future led. As MacQueen surmised, Roosevelt would "very likely sustain President Taft as long as he reasonably can." But the Scot had "no doubt at all" that Roosevelt would go into the Midwest, "be popular with the [Progressive] insurgents" against Taft, and take "strong, advanced ground on conservation of our national resources." Soon after, Roosevelt would probably "be returned to the Senate or else elected speaker of the house." In the opinion of "many people" he would be the next president of the United States.[68]

Clearly, Roosevelt's European and African trip was more than a holiday, more than a journey back to nature, more than a striving to reincarnate frontier masculinity.[69] In addition to these things, it was a calculated move that drew attention to him and his causes through a highly newsworthy series of events that dragged on for more than a year. The causes that he advanced included the need for conservation, especially a white-settler form of conservation, and he asserted the trajectory of American history toward hegemony over the Euro-American imperial world. Before departing for Africa, Roosevelt had already made a crucial move to align the planets of settler conservation with an American sun by calling for a World Congress on Conservation. But how would the world respond to that?

ELEVEN

Something Big: Theodore Roosevelt and Global Conservation

Frank Munsey, the millionaire newspaper and magazine chain owner, was flabbergasted. Roosevelt had conjured up a stunning way to end his presidency. A World Congress on Conservation was to be held at The Hague under American leadership. A Republican Progressive, Munsey was gushingly pro-Roosevelt, and it is not surprising that the media magnate's *Washington Times* (which "seldom noticed subjects not approved by the owner") provided a very revealing insight into this decision. Munsey's paper reported the gossip: "I knew he'd do some big, striking, spectacular thing . . . which would set the whole world talking about him before he left; but who'd have thought of this!" The account was almost certainly written by the proconservation Judson Welliver, Munsey's right-hand man and friend of Harry Slattery, a Pinchot aide, who fed information to the press on Roosevelt's agenda.[1] The account was uncannily true to Roosevelt's psychology. The president was soon to leave for Africa to hunt and had invitations from many of the rulers of Europe to visit when his African jaunt ended. There, at the courts of old Europe, this New World dynamo would "receive such an ovation" that "the chieftaincy of the great movement he will have started will be but a short and obvious step. This is the career which suddenly opens its possibilities to the retiring President." For Munsey's paper, it was a foregone conclusion that Roosevelt would effectively be "president of the world" in this most important matter affecting the destiny of nations: the conservation of resources.[2] Other newspapers in the United States that carried the story agreed, and the media interest coincided with Roosevelt's need to develop a postpresidential career. The entire train of events pointed to an obvious conclusion. He wanted to be remembered, and a global initiative on conservation would keep the memory of his achievements alive.

One might therefore be excused for thinking that the news-savvy Roosevelt's call for a World Congress was a publicity stunt or the personal gratification of a monumental ego.[3] It was that, but more. The decision reflected political, as well as personal, agendas and was a logical culmination to his thinking on the future of the American empire and on the renovation of the international state system. The idea's germination went back to 1908 and was entangled with the prior decision to hold a North American Conservation Conference on 18–22 February 1909. Replete with a White House reception, an address from the president, and a farewell party at Pinchot's Washington residence, the North American conference exuded continental camaraderie. Freed from congressional restraint, Roosevelt and Pinchot revealed their bold international agenda. Though pitched at cooperation with Canada, Newfoundland, and Mexico, the Declaration of Principles, sent to Congress and repeated in publications across the land and in diplomatic correspondence to the governments of the world, was a sweeping program framed in "the interest of mankind." It made central the preservation of the world's life-support systems to ensure "the habitability of the earth," as the Conference of Governors' own Declaration of Principles had decreed and as Horace McFarland had urged. The attending nations accepted the centrality of water and forest protection to conservation, the "public ownership of water rights" for hydroelectric sites, and disposal of mineral lands only by lease under national supervision to prevent waste. All agreed on public health as tending "strongly to develop national efficiency in the highest possible degree in our respective countries."[4] Naturalist William T. Hornaday addressed the delegates at Roosevelt's personal request and bemoaned "the resistless onward march of development" that would render extinct the "large" and "magnificent" wild game across the whole of North America.[5] In response, Pinchot succeeded in inserting in the declaration the importance of "game protection under regulation," with "extensive game Preserves."[6] Finally, the meeting endorsed the World Congress idea.

Both the proposed World Congress and the North American conference revealed Roosevelt's second-term bid for international cooperation as the focus of American relations with the rest of the world. This move had been revealed in his administration's environmental diplomacy with Canada and other nations over Niagara and fur seals. Yet the World Congress had a further significance. Thereby, Roosevelt reflected and sought to channel the growing enthusiasm in Europe and in the settler world of the British Empire for conservation. Preliminary feelers for such a congress went out with a Department of State memorandum on 6 January 1909,[7] prior to the holding of the North American conference, and were intimately linked to the negotiations over that conference. The idea for both events was already in Pinchot's and Roosevelt's minds at the time of yet another, earlier meeting: the Joint Conservation Conference that convened in Washington on 8–10 December 1908. That event included the members of the NCC, state governors, and other invitees, who were called to debate the key findings of the NCC draft report.[8] The attendance of Canadian delegates there did not occur by chance. Pinchot had close contacts with Canadian lumbermen and Prime Minister Wilfred Laurier and had addressed the 1906 Canadian Forestry Conference.[9] Proconservation newspapers featured the speech of the Canadian senator William C. Edwards at the December meeting. Himself a lumber producer, Edwards "startle[d] delegates" with a "recital of his country's experiences." Edwards accepted Roosevelt's and Pinchot's depiction of an imminent continental crisis in which North American resources were assumed to be on the path to exhaustion.[10] The Canadians already seemed to be in agreement. Edwards returned to Canada lauding the American achievement, telling Laurier, "I cannot think of anything more important and interesting." Laurier, who professed to be an "old friend" of Pinchot, responded enthusiastically to the president.[11]

It is likely, however, that the decision to hold a North American conference stemmed from deeper processes than Pinchot's contact with the Canadian conservationists. The personal and political imperatives of the president contributed. To achieve his goals, TR needed to bypass an increasingly recalcitrant US Congress that viewed him as a lame-duck leader from 1907. He accomplished this by creating the executive commissions for waterways and country life, which drew attention to conservation issues and recommended legislation. But congressional leaders rejected many of the Republican president's proposals, refusing outright to fund them. No matter—the more Congress resisted, the bolder the ideas that Roosevelt produced. Convening a North American conference with Canada and Mexico was part of this battle of wills.[12] The invitations did not go out until Christmas Eve, a week after House Speaker "Uncle" Joe Cannon announced that Congress would not fund the printing costs or future activities of the NCC. With less than three months left in his presidency, and Congress in open revolt against his style of executive government, Roosevelt sought an alternative power base for his postpresidential career by creating institutional networks and loyalties that tied the US government to his conservation policies. Roosevelt openly vowed to "drive" Congress in that direction.[13]

International agreements seemed to be an invaluable way to fulfill that aim. By 6 January 1909, TR and Pinchot had played their World Congress card for all it was worth. Secretary of State Elihu Root's official memorandum cited "preliminary and informal discussions" with the North American conference participants. Since the Mexican authorities had not met Pinchot at that stage, that could only have meant conversations with Britain via the governor-general of Canada, Earl Albert Grey, with whom Pinchot had conferred in Ottawa on 29 December 1908.[14] Rumors circulated that the World Congress idea was a diplomatic maneuver to placate the Foreign Office, which had been concerned about Britain's exclusion from the North American conference talks. After all, the British still controlled Canada's foreign policy, yet Roosevelt proposed to deal directly with the Canadians on an issue close to the foreign policy prerogatives of the mother country. Holding a World Congress might allay suspicions that the United States wished to woo Canada as a truly independent country, to the detriment of British economic interests. An Australian paper reported these suspicions, but hard evidence to support them is lacking. The British made no public or private complaints and had been kept fully informed since 24 December.[15] Only the *Chicago Daily Tribune* backed the Australian claim that the North American Conservation Conference had "brought out the earnest wish" of certain Europeans "that they might be included in a general conservation program," and that rumblings from across the ocean "probably gave birth" to the World Congress idea.[16] To portray the plan as simply a diplomatic sleight of hand overlooks Roosevelt's (and Pinchot's) trajectory of thought. Their embrace of a concept that environment was borderless led logically from the national to the continental to the global level. They were already convinced of the need to hold a world meeting, and were determined to do so.

The benefit for TR himself was as obvious as the *Washington Times* outlined. It was widely assumed that, as president of a meeting held at The Hague, Roosevelt would duplicate for conservation what the first and second Hague peace conferences had done for accord between nations. The formal diplomacy of "internationalism" and conservation would become connected. In this telling, Roosevelt's initiative for the rational allocation of raw materials foreshadowed the future of the whole planet, not just the United States. "Peace and prosperity" that brought "the nations more and more to know and understand each other" would foster cooperation among them "around the council board at The Hague." Echoing Root's own directive of 6 January, the *Washington Times* saw internationalism as necessary because "the world after all is but a neighborhood, and . . . no nation or people is independent nowadays of the others."[17] These words aligned perfectly with the administration's press releases and its messages on the World

Congress to foreign embassies. In his statements to the North American Conservation Conference in February, Roosevelt was already articulating his sense of a "world movement" of civilization based on increased interdependence. He revealed the theme most fully in his Berlin address of that name in 1910, with Kaiser Wilhelm II present.[18] Undoubtedly, Roosevelt and Pinchot accepted even before the meeting of the Joint Conservation Conference that the damage humans did to their environments by waste and destruction was global, not national. Roosevelt's Annual Message to Congress of 8 December had hinted as much.[19] Nearly a decade of American efforts to engage with conservation globally now reached a peak.

Roosevelt's decision on the World Congress reflected a shift in his diplomacy away from the blunter forms of imperialism.[20] In the early years of his presidency, the big stick had taken precedence. From the would-be Rough Rider's push for war with Spain in 1898 through to the Panama issue of 1903, the flexing of a muscular expansionism prevailed. But by 1907, Roosevelt thought more about immortality than masculinity, and his experience regarding the Russo-Japanese War indicated the payoffs of peace and arbitration. These had come handsomely in personal kudos, since the Nobel Prize was his in 1906. Yet he had not been able to find time to go to Europe to accept it. Now, as his presidency ebbed, he was ready to receive the accolades of the world.

The conservation proposal fed into this stream of thought.[21] The State Department memorandum of 6 January articulated the hope that the World Congress would lead to "general understanding and appreciation of the . . . material elements which underlie the development of civilization."[22] The North American Conservation Conference endorsed this viewpoint in mid-February, where Roosevelt announced a post-imperialist agenda: he claimed it "normally to the interest of each nation to see the others elevated," not exploited.[23] TR was not the only one at that meeting to endorse international cooperation. Senator Newlands rejoiced that there would be an organized body for conservation, not only in the United States and Canada but also "perhaps later on in other countries," a move that would promote "the peace of the world."[24] Not surprisingly, the pro-Roosevelt *Outlook* dutifully took up this idea of "international co-operation" to dissolve the "purely artificial lines" between the United States and its neighbors.[25]

Of course, this "internationalist" initiative was still arguably a form of imperialism aimed at strengthening the United States and Western civilization. This would be a resource-based imperialism led by developed nations. On 20 February 1909 the *Washington Times* put it in terms strongly reminiscent of William Elliot Griffis and Benjamin Kidd: because of the leading

11.1. North American Conservation Conference, at the White House, opening day, 18 February 1909. The Newfoundland representatives had not yet arrived, due to poor weather. Canadian delegates (left front), with Sydney Fisher (third from left, next to Roosevelt). Mexican delegates (right front), with Miguel Quevedo (far right). Back row includes Secretary of State Robert Bacon (second from left), Gifford Pinchot (third from left), James Bryce (third from right), and Secretary of the Interior James Garfield (fourth from right). LCPP.

powers' growing need for resources, "There will come an epoch when the tropics with their vast productiveness must be developed" and when "the sea must be made to give up more to man." Yet, because of Roosevelt's specific conservation objective, this undertaking was cast as an imperialism of recycling: "there must be systematically put back into the soil an equivalent for all taken from it."[26]

This call harmonized with the idea of repairing damaged nature, a concept of protosustainability contained in Secretary of State Robert Bacon's formal call for a World Congress, issued the day before the *Washington Times* article: "that *reparatory* agencies should be invoked to aid the processes of beneficent nature, and that the means of *restoration* and increase should be sought whenever practicable."[27] A Wall street banker who gave up his career to join the administration as assistant secretary of state in 1905, the handsome Bacon exuded Ivy League values and far excelled his Harvard classmate Roosevelt as a sportsman. An ardent conservationist, he saw eye to eye with his "old friend" the president. Though not yet defined or clearly articulated as such, sustainability, understood as the recycling of basic resources within the space of a single generation, was now being asserted by the Roosevelt administration as its ultimate goal.[28] This recycling process would need to be global because the "peoples of to-day" held "the earth in trust for the peoples to come after them." The *Washington Times* highlighted the administration's idea of an international council that would, like the permanent state and national conservation commissions established during 1908–9, do research, compile a "world inventory," distribute information, and proffer advice on "conservation, development and replenishment." Such an international body would also "promote easy exchange," take down "tariff and customs barriers," and provide "the solution to the problem of commercial warfare" through the rational distribution of raw materials.[29]

This conservationist internationalism contained the outward projection of American anxieties but was forged in an experience of exchange with other peoples. Roosevelt did not simply present a novel message to the world in 1909. There was already another universe of resource management taking shape outside the United States. Some of this environmental reform originated in Europe, but a great deal was occurring on the periphery of the major metropolitan centers of empire. It crossed national boundaries easily. Roosevelt did not create these concerns, though he tried to take advantage of them.[30] The activities of Roosevelt and his associates were part and parcel of this changing global sensibility. Atlantic exchanges of ideas contributed, but the range of countries was broader still. A nascent world interest was revealed in the response to the 1909 congress call. Enthusiasm for Roosevelt's proposals came from such faraway places as Australia, India, and Brazil, as well as from the core European countries of Germany, France, Italy, and Britain.

Underpinning foreign support for a world congress were four strains of conservationist concern. In Europe, several nations were already seeking to protect their countryside from excessive industrial development and to preserve scenery. A second European anxiety concerned timber supplies, with Europe's empires, particularly the British, French, and German, beginning to work on forest conservation in their respective colonies. India's British-run government broadly fitted this category too.[31] From the periphery of Europe came two variations on these themes. Countries of the Caribbean basin and Central America uniformly supported Roosevelt's plan. These were areas of already large US investment and direct interference in internal Latin American politics, accentuated from the time of the Spanish-American War and the acquisition of the Panama Canal Zone. Supporting the World Congress idea expressed modernizing Latin American elites' desires to cooperate with American capitalist development and yet at the same time nationalist concerns over the social, economic, and environmental impacts of mining, forestry, and agricultural activities. Mexico and Brazil fitted this type.[32] Finally, there were the responses of the settler societies, in which the white European "race" had gained demographic or political domination and sought to duplicate its civilization not simply by ruling these lands but by colonizing them with white people. In Argentina and the self-governing former colonies of the British Empire then known as dominions, the conservation-minded "settlers" were attracted to Roosevelt's initiative.

Despite early moves to create national parks in Australia, Canada, New Zealand, and Argentina, the issue in these settler societies was not so much nature protection as achieving efficient and less wasteful development. In many places, support for this idea went back years. Conservation sentiment had been rising since the 1880s in Australia, New Zealand, South Africa, and Canada. Historian William Beinart notes that a "progressive" South African "landowning group or class" favored economic improvement, "regulated capitalism, and conservationism, in which the state had a role." They "absorbed some of the economic and political ideas" of the American Progressive movement, which was "at its height under Theodore Roosevelt's presidency."[33] By 1908 Argentina joined the list of those interested in this type of settler conservation.

Proconservation sentiment in settler societies took many forms. Under the impact of European expansion in areas newly "settled" by Europeans, dramatic environmental change was occurring. Wars and disease had reduced the numbers of the indigenous, and much forest destruction had occurred. In the midst of soil erosion and other problems that deforestation caused, scientists, government officials, reformers, farmers, and other colonists worried about how the balance of nature could be restored if not by concerted cross-national action. But that sentiment in settler societies coexisted with a potentially contradictory concern for more intensive development of the land through using resources, albeit without unnecessary waste. Roosevelt's dual approach of a broad-gauged utilitarian conservation and simultaneous advocacy of national parks and wildlife protection appealed to non-American conservationists in the settler societies as a way of resolving this contradiction.

In addition to this preexisting seedbed of interest abroad, the dissemination of Rooseveltian ideas was aided by the formal networks of diplomacy and by civil associations that breached national boundaries. The latter included important nongovernment organizations and professional groups, such as international forestry conferences and similar bodies associated with farming. The conservation interests of agricultural groups emerged in and were shaped through not only the national and international irrigation congresses already discussed[34] but also the less well-known agricultural cooperation groups. These received an important stimulus from the founding in 1905 of the International Institute of Agriculture in Rome, under the sponsorship of King Victor Emanuel II. The United States took part in this organization and designated an American living in Rome as a permanent delegate.[35]

Piggybacking on irrigation exchanges came a growing international interest in dryland farming as a way of conserving water and making marginal land productive where irrigation techniques were not possible. The United States provided tantalizing prospects in this respect. Dryland farmers believed that conservation of the little rain that fell upon the Great Plains could be achieved in a simple way, though the claims were far from uniformly scientific and expert-driven. "Maintaining a pulverized dust mulch" and fallowing half the land in any given plowing would miraculously double moisture, dryland messiahs proclaimed.[36] Fundamental to the dryland methods was the gospel-like work of Hardy Webster Campbell. *Campbell's 1907 Soil Culture Manual: A Complete Guide to Scientific Agriculture as Adapted to the Semi-arid Regions* laid out his theories. The self-taught soil expert pleased railroad corporations, which saw in dryland farming the perfect way to make their remaining US public land-grant tracts around the railroad rights-of-way attractive for sale to would-be yeoman farmers. Stimulated by his evangelism and backed by railroad boosters, annual dryland-farming conferences began, the first held in Denver in 1907.[37] These conferences were not just for Americans. A transnational network of enthusiasts emerged. Many Australian settlers faced landscapes that were similar to the low-rainfall areas of the American West, and the newly federated nation had recently experienced crippling drought. Yet the transnational circulation of ideas for soil and water conservation went much further than this obvious example. Russia, Brazil, Mexico, Transvaal, and Canada sent delegates to the 1909 dryland-farming meeting. Following the connections is key to transnational history, and strong personal ties and influences across national boundaries certainly developed from these dryland meetings. The Australian senator James H. McColl toured the western American states to gather dryland farming data in 1905, which he published back home, and he returned to serve as chairman of the Wyoming conference in 1909. From there he carried back news of agricultural conservation schemes, as did William MacDonald in the case of the Transvaal.[38]

Matching this grassroots circulation of ideas by non-state actors was formal diplomacy, though the two types of action overlapped. No one individual was more important in the role of transnational conduit than a diplomat who mixed personal contacts with nongovernmental organizations, on the one hand, and formal international diplomacy, on the other. The Irish-born British scholar and politician Lord Bryce was a key figure in creating transatlantic networks on conservation and stimulating interest in Roosevelt's plans in both Canada and Britain.[39] Best known as the author of *The American Commonwealth* (1888),[40] his detailed survey of American political and social institutions that has often been compared in scope if not insight with the work of Alexis de Tocqueville, Bryce met Roosevelt before he became Britain's Washington ambassador in 1907. A friendship developed, though Roosevelt had a low opinion of his companion's intellect, calling him a "worthy and dull old person."[41]

Notwithstanding Roosevelt's typically snap judgment, Bryce's abilities and his contacts with Roosevelt should not be underestimated. The two men had a good deal in common, which enabled Bryce to become an advocate for the president's conservation policies while acting as a go-between with the British and Canadians. This mediating role was an important one because Canada's official foreign relations remained solely with Whitehall until mid-1909, and Bryce spent much of his time managing these Canadian-American affairs, as he did on Niagara and the boundary-waters question. Like TR, Bryce loved the outdoors, and he became a firm advocate of national parks. He also spoke frequently to the American and Canadian publics. Toward the end of his term, the ambassador gave a widely reprinted address on national parks to the ACA and won the praise of then secretary of the interior Walter Fisher.[42] Bryce also joined the Save the Redwoods League and befriended John Muir. With their flowing white beards on show, the two made an impressive impact as sages of conservation when they kept a Sierra Club dinner meeting of 1912 spellbound.[43] Like Roosevelt, Bryce had a strong emotional attachment to nature, expressing scorn for westerners who could see only dollars in the land on which trees stood and sympathizing with those who "raise a voice for the preservation of natural beauties, where that preservation comes into conflict with 'development.' "[44] But the ambassador also enthused on the Roosevelt administration's plans to rationalize resource use to prevent waste and promote intergenerational equity. In 1908 he praised the Conference of Governors as a meeting "of vast importance" to those ends.[45]

Bryce adopted the Progressive reform outlook on the contest of "the people" versus "the interests," especially over the resources of the American West. He reported home to the British Foreign Office that Pinchot had "been working hard to prevent the reckless squandering of those resources and their exploitation by Syndicates and Companies, whose methods are frequently as unscrupulous as their policy has been disregardful of the public interest."[46] As impressed with the chief forester as with Roosevelt, Bryce called Pinchot "a man of much ability and energy."[47] And after the end of TR's presidency, Bryce argued that "Taft must keep up the conservation programme."[48]

Though to the ebullient and egotistical Roosevelt he was dull, Bryce was as tireless in his own way as TR demonstrably was. The ambassador made particularly strong public and private efforts to interest Canadians in continental conservation because, he stated, the United States and Canada had "certain aims in common," and "more notable" was the "protection of natural resources." Bryce distributed to the Canadian and provincial governments literature on the state of American forests and the effects of deforestation on watersheds, along with news on American agricultural innovations. The 1909 North American Conservation Conference, he stated, was the perfect vehicle to effect his plans for greater cooperation between the two countries.[49]

As the first result of Bryce's agitation, the ground was smoothed for Pinchot's whirlwind visit of December 1908 to Ottawa—to gain Canada's support for his continental scheme. The Canadian governor-general and the prime minister both warmly received the American. Grey told the British Foreign Office that the lanky forester "made a most impressive appeal to the members of the Canadian Club at a luncheon . . . , urging them to take such action as might be necessary for the purpose of protecting our national resources from unnecessary waste and preventable destruction."[50] The second achievement of Bryce's intercession was evident at the North American conference itself. According to Sydney Fisher, the Canadian minister of agriculture and conference delegate, it was Bryce who supplied the initial "encouragement . . . in our labors."[51] Fisher returned from Washington "thoroughly well pleased with the results." The meeting would "awaken public opinion in Canada." The Canadian government would endorse, Fisher predicted, "a Commission to take an inventory of our resources" as a very important step toward the "education and enlightenment of our people."[52]

Almost immediately Canada did establish its own permanent national conservation commission and acknowledged the American model in this venture.[53] The legislation passed in Parliament with both government and opposition party support. Liberal parliamentarian and former minister for the interior Clifford Sifton was appointed chair of the Canadian Conservation Commission (officially the Commission of Conservation, Canada). Among the members of its four constituent committees was the prominent University of Toronto professor Bernhard Fernow, the former US forestry chief and the "sometimes mentor, always competitor" of Pinchot.[54] The commission was said to reflect "the epitome" of Pinchot's wide-ranging approach and included human health under conservation issues.[55] It was primarily a research body, however, and implementation of its agenda was divided between four dominion government departments. Little could be quickly achieved in practice. Critics and even some members of the commission became frustrated over its lack of impact on public policy, and its "permanent" status lasted only till a reorganization of resource policy occurred in the wake of World War I.[56] Yet under Sifton, the minister who had been responsible for the development of the Canadian West before 1905, the commission functioned along the efficiency and intergenerational-equity lines that Pinchot wanted.[57] Despite his earlier support for railroads that "linked mining and forest resources into Canada's east-west rail network," Sifton did not favor development at any costs.[58] Arguing that Canadians must stem the loss of heritage to "rapacious Americans,"[59] Sifton attacked critics who claimed Canada was too sparsely settled and underdeveloped for such fears about the future to be worrisome: "the best and most highly economic development can only take place by having regard to the principles of conservation," he averred. Paralleling the American jeremiad, he drew upon the examples of Egypt, Assyria, Rome, and other fallen civilizations, warning that Britain faced the same fate of imperial catastrophe. "Decline must inevitably set in" when that nation's "iron and coal disappear."[60] All these sentiments paralleled those of Roosevelt.

For Canada, its southern neighbor became a bogeyman, where resources had been laid waste and made "totally unproductive" or monopolized by a few, but the United States also ironically provided for Sifton a model of how to escape the American trap.[61] Sifton used the Canadian Conservation Commission to stop the projected damming of the Saint Lawrence River at Sault Long Rapids near Cornwall, Quebec, for the benefit of the Aluminium Company of America's power needs. Sifton opposed this hydroelectric development not only on the grounds that the power would mostly go to the United States, and the navigation of the river would be impeded, but also because the development would affect an area "internationally appreciated for its scenic beauty."[62] In this way, the commission represented national interests, but in line with the multipurpose ideas of conservation pushed by Pinchot and Roosevelt. (The project was later revived, but the dam was not built until the 1950s when the Saint Lawrence Seaway was being constructed.)

Canada did not act simply because Bryce told them to do so, nor because Pinchot in his whistle-stop visit had done so. The Canadians had their own interests in supporting international conservation. Laurier indicated privately that he wanted to see more forest reserves created in the Canadian West. As Edwards warned the prime minister, a United States that ran out of timber and coal would need Canadian alternatives, and the Canadian government did not want a repeat performance from profligate Americans.[63] Canadians also wanted to take advantage of American agricultural knowledge to develop the prairie states, which were later in being opened to settlement than the American ones and in competition with the latter for immigrants.[64]

Further complicating Canadian (and American) calculations were the vexed issues of the tariff and reciprocity.[65] From 1906 to 1909, the Republican Party had been, under Progressive Republican pressure against the dominance of large corporations, considering changes to tariff policy to create more competition and so lower the cost of living for American workers. But reduced tariffs threatened premature exhaustion of Canadian forests as lumber flowed south to replace depleted American equivalents. The negotiations could not be separated from conservation, since some Canadian provinces (and the new conservation commission) wished to protect supplies of those species useful for Canadian paper and pulp manufacturing. It therefore made sense, as Bryce pointed out, to deal with the issues of tariffs and conservation together. This was yet another reason for British and Canadian conservationists to support multilateral environmental diplomacy.[66] In the end, the Canadians rejected reciprocity in the 1911 election, displaying a surge of economic nationalism that matched American impulses to horde their own resources. Yet the climate for nation-centered conservation remained. Despite attacks upon it, the Canadian Conservation Commission continued to operate until the circumstances of World War I made the maximum use of raw materials more important than conservation, let alone preservation. In the end, the commission was abolished in 1921, not because it abandoned its prewar ideals, but because it "persisted in advocating conservation, an expression of early twentieth-century environmentalism, in a world which was no longer interested in this issue."[67] This was not much different from the eventual outcome in the United States, but in the meantime, Roosevelt had initiated a debate about conservation that ignited interest in many countries.

South of the US border, the response to Roosevelt revealed similar good intentions, though circumstances turned out to be radically different. Mexico wanted international action on conservation with the United States as a partner, but once again for its own reasons and with its own priorities. The North American meeting was its primary concern, not the proposed World Congress. Water issues lay at the heart of this choice. An agreement regarding the Colorado River was a pressing objective, where a faulty irrigation diversion in Mexican territory had in 1905 flooded the Salton depression to form the Salton Sea. Partly because the diversion had been undertaken by Americans to water the Imperial Valley, and with congressional action not forthcoming, Roosevelt determined that an international solution was necessary. In addition, an equitable agreement for use of the lower Rio Grande was still a Mexican objective, because prior appropriation of this essential asset upstream in Colorado and New Mexico had deprived Mexican farmers in the valley of adequate water supplies since 1889. Needing the cooperation of its powerful northern neighbor, Mexico welcomed the chance to join the North American forum because of "so many international rivers and mountains" shared between the two countries. Action "for the preservation of the waterways must be concurrent" across the continent, the Mexican president announced.[68]

Just as Pinchot practiced personal diplomacy to solicit Canadian support for global conservation, the chief forester bypassed the normal channels of the Department of State and sped in high drama to Mexico City to meet President Porfirio Díaz on 20 January 1909. Cooperation between the United States under Roosevelt and Mexico under Díaz was not unexpected. American corporations were already entrenching themselves in northern

Mexico, and the "Porfirian" regime sought US capital for economic development. Despite the Mexican government's reputation as authoritarian, many elite Americans, including Roosevelt's most influential officials, viewed Díaz as an effective modernizer and diplomatic ally. They hoped that economic progress would come in such a way that resources would be wisely used and available for American consumption on a more sustainable basis than had occurred in either country to that time. Contributing to a 1910 book by José G. Godoy, *Porfirio Díaz, President of Mexico*, Pinchot spoke of the Mexican glowingly: "He seems to me to have done more for his people than any other ruler now living has done."[69] Extravagant praise indeed this was, since the judgment notionally included Roosevelt. Secretary McGee of the Inland Waterways Commission, that other key conservationist in the administration, claimed that Díaz "admired" Roosevelt. He was "a sort of Mexican Roosevelt" who "shared the same . . . personal vigor."[70]

Cordially, Díaz offered his own support for conservation, and there is reason to take the Mexican leader's professions seriously. This was a mutual admiration society with humorous and even bumbling overtones, as the recorded interview with Díaz shows. Eyebrows were raised when Pinchot opined that Spanish was a beautiful language, but one that he proposed to master in just a few weeks. While Pinchot likened Díaz to TR, because both had "good intentions" for the development of their countries,[71] Díaz thought Roosevelt and he were "two strings tuned to the same key, that when one is struck, the other vibrates in harmony with it."[72] In discussion, Díaz expressed agreement with the NCC's concept of national vitality. The underlying or residual "vigor" of Mexico's original inhabitants needed a strong infusion of masculine forcefulness in the Roosevelt tradition, he told Pinchot: "While we must devote much attention to forests and waters and mines and the soils, we must not lose sight of the fact that the principal one of all the resources is man."[73] Díaz praised the Anglo-Saxon race but located the vigor that he coveted not in genetics but in the economic modernization of society. Technical education and industrial development revealed that "the degree of civilization of a race was measured by the per capita amount of power they employed," and on this score the United States had leadership, since "no other country on earth" could show such an example of the deployment of energy. Clearly, Díaz saw the United States as a model of technical, scientific, and economic development because of its plans for efficient use of resources. By following this model, Díaz hoped to "improve the Mexican race."[74]

Mexico was no more slavish in following the United States than Canada was, however. It had already been pursuing an agenda for conservation.[75]

Miguel Ángel de Quevedo, who was the most important Mexican delegate at the North American Conservation Conference, was an engineer known as *el apóstol del árbol* (the tree apostle). Until Pinchot traveled to Mexico City, he knew nothing of Quevedo's work for the restoration of the tree cover on the hills surrounding the capital. Working for the Mexican secretary of agriculture, Quevedo had looked to Europe rather than the United States for advice. Believing erroneously that Pinchot was interested only in lumber production, he traveled to France in 1907 because of the Mexican interest in wider issues of watershed protection and to Algeria to study the role of trees in stopping dune erosion. On the way he visited the forestry school at Nancy and those in other European cities and, in the following year, used his French contacts to establish the first Mexican equivalent.[76] Quevedo learned from the 1909 conference that Pinchot was not so narrow-minded as he had thought, but the Mexican delegation still believed that the American example was profitable to study only for northern Mexico, where the land approximated that of the US Southwest, and that many conservation problems could be solved only by attention to European precedents. That might have made a World Congress desirable, and Mexico did go along with the idea.[77] But Díaz fretted that a *World* Congress might undermine the possibility of focusing US attention on the borderland issues of irrigation and water supply. In Washington, the "Mexican delegates were especially emphatic" that a bilateral relationship continue, the *Washington Post* reported, and hoped that the proposed World Congress would not stop progress on the already-canvassed continental agenda.[78]

In any case, the Mexican contribution to Roosevelt's plans proved to be hypothetical because the revolution commencing in 1910 undermined Díaz's previous "good intentions,"[79] turning Mexicans to more pressing and immediate matters. Quevedo eventually had to leave the country in 1914 to escape assassination at the direction of Victoriano Huerta, the dictator who had seized power.[80] Under the Wilson administration, the United States did not recognize any negotiators from the Huerta regime, and the North American Conservation Conference agreement of 1909, where the nations vowed international cooperation on all resource issues, became a dead letter. Cross-border contacts had to be rekindled after World War I.[81]

Roosevelt had more sustained success elsewhere, especially within the British Empire, and not only in Canada. The president had shrewdly discerned the importance of conservation to Britain's far-flung possessions for the economic integration of that empire and for the separate interests of the self-governing dominions. For Britain itself, Roosevelt observed that, though they "might not have any large forests to conserve . . . the proceedings of the [World] Congress would bring out many valuable data and suggestions regarding the best means of planting and maintaining woodland, regarding modes of economizing supplies of coal and iron," and other similar matters.[82]

Internal British government advice persistently favored the idea of a congress. Bryce believed that a global stocktaking of resources might lead the United States to a more "rational" tariff arrangement in which Americans would "realize better than they now do the benefits of a freer interchange of all products, natural and manufactured, than existing Tariffs permit." The British had good reason to go along with the idea for the very reason that Roosevelt and Bryce set forth, namely, their dependence on the outposts of their empire in the strategic allocation of resources. As Bryce stated, "we should be directly interested at home, not to speak of our interest in conserving the natural resources of India, Australia, and the Crown Colonies."[83] Bryce was as keen on a World Congress as on the North American equivalent and saw the proposed meeting as a way of creating Anglo-American, as well as Canadian-American, rapprochement.

Though of course run by British officials, the government of India also wished to participate, and its attitudes reflected widespread concern over resource management among the colonial elite across the subcontinent. In semiofficial circles, no institution stood out more prominently than the *Times of India*, which was edited by Herbert Stanley Reed from 1907 to 1924. This was "one of the two most influential journals in India" during those years.[84] It channeled the opinions of many British on the subcontinent and some moderate Indian reformers. A remarkable piece appeared in that paper in early 1909. Titled "An Epic of Waste," it drew heavily upon the American jeremiad. The article parroted the *McClure's Magazine* article by Rudolf Cronau. But the *Times of India* took up this story with a telling twist. The British colonies had, at the start of "Caucasian settlement" in the 1600s, unparalleled natural wealth. In "a few moments in the career of humanity," however, Americans had stripped their land almost bare. The paper called the slaughter of the buffalo "a completed tragedy" but emphasized that much in the way of soil, water, and minerals as well as wildlife had gone without adding to the capital benefit of the future, and thus constituted "waste." The tragedy was most closely identified with "the disappearance of the forests," but this was no "self-contained evil." Deforestation had knock-on effects in torrential flooding that denuded soil and silted up navigable rivers. "Unscientific farming" robbed "the soil without repayment" and pushed the United States to the edge of dependency on others. It would soon cease "to be a self-feeding country," proclaimed the alarmed writer, who supported USGS critiques of the waste of coal and Andrew Carnegie's remarks on the squandering of the best iron-ore deposits. All this, the *Times of India* proclaimed, must "drive the United States to foreign sources for the elements of her organised industry," including "an impetus to the exploitation of Canada." Yet there was more. American profligacy would affect the entire world market for raw materials.[85]

Here, the *Times of India* drew without acknowledgment on Brooks Adams's geopolitics and offered a glimpse of the startling global implications if the United States hoovered up the world's resources. "The American people have written the history of other lands than their own in the epic of waste," the paper proclaimed. "[T]he next half-century must witness a tremendous inflow of commercial energy to all the countries that can still rank as undeveloped purveyors of raw material [*sic*]." Then came the intellectual coup de grace: "In this epoch no portion of the earth's surface may figure more prominent than the middle and northern latitudes of Asia, and the forces set in motion by economic necessity will produce transformations of a political and social nature with which the most skilful conjecture cannot assume to grapple."[86] This interpretation was straight out of Adams's manual for American geopolitical domination.

At the government level, India was a country primed for the conservation message of Theodore Roosevelt. Its bureaucracy had practically invented the science of tropical forestry, with the assistance of Pinchot's mentor, Dietrich Brandis. Indian Raj officials sought out American information on methods to implement agricultural reform, notably dryland farming and the use of fertilizers. Agricultural associations working with the assistance of the Indian Civil Service demonstrated an "immense amount of energy." Officials such as Bernard Coventry referred to American scientific agriculture as a model and advocated soil improvement.[87] All this was stimulated by the conservation debate in the United States and explains why the (British) government of India was keen to go to the proposed congress and pestered the British government at home on the issue.[88]

Equally apparent was the unexpected support for the idea in Australia. It was especially the Australian insistence on separate representation that caught Pinchot by surprise, though he readily acquiesced in the idea as a "natural" extension of his view on the need for all the world's resources to be considered.[89] The British dominions in the southwest Pacific had attitudes in line with the ex-president's, and the last year of Roosevelt's presidency and the first of his postpresidential life received strong coverage in the Australasian press. For Australia, the interests were in forests, irrigation, and mineral development. For New Zealand, the country's poor coal supplies made the energy issue of greater significance, and Roosevelt had spoken out on the waterpower problem. New Zealand was a strong candidate for hydroelectric experimentation.[90] As in so many other cases, Pinchot and Roosevelt's idea of efficient and statesman-like allocation of resources on a sustainable basis appealed to the white British dominions even as they picked and chose which aspects of Roosevelt's policies to emphasize in line with their own interests.

In all, nineteen countries quickly agreed to come to the proposed Hague meeting, and twenty-three had signed up by the end of 1909, leading State Department officials to conclude that positive action could be taken.[91] France indicated support through Roosevelt's friend Jean Jules Jusserand, the French ambassador to Washington, as did Germany, the latter already having a strong homegrown conservation movement.[92] Italy was also keen but wished to host the conference in Rome to coincide with the next meeting of the International Institute of Agriculture, in which organization many countries, including Canada and Australia, were already permanently represented. Canadian government delegates had been at the First Institute Assembly in 1908 and intended to draw this European support into the North American project. Sydney Fisher saw huge advantages from enlisting the many countries attending that meeting. The delegates at Rome "were just the stamp of men who would be interested in the movement for the conservation of natural resources, and would have at their command a vast fund of information which would be most useful," wrote the Canadian. In this and in many other ways, Roosevelt was able to tap preexisting bureaucratic and political networks abroad voicing conservationist concerns. These were concerns over which he had no control, and was not responsible for generating, but which resulted in support for the World Congress idea. The "cult of conservation" was "by no means wholly American." It was, as an astute contemporary observer put it, "altogether cosmopolitan."[93]

Acting at a most propitious time, Roosevelt had attempted to seize a global moment of enthusiasm for conservation, even as he only inadequately realized the extent of interest abroad. Ultimately, twenty-nine nations agreed to come. A remarkable opportunity for a pioneering conservation congress of global scope existed, but it became caught up in the shifting priorities of a new presidency. Taft stood in the way of American action not by opposing the proposal but by prevarication. When the first international conservation conference assembled in Europe four years later, the number of attending nations was less than the nineteen that initially responded to Roosevelt and Pinchot's call for the World Congress.[94] Taft's strict constructionism over the constitutional powers of the federal government on conservation and

11.2. Roosevelt strikes a vigorous pose and identifies his geopolitical orientation, 1903. LCPP.

his political timidity meant, at most, only tepid support for Roosevelt's and Pinchot's schemes. In March 1910 Taft had not abandoned the idea of the World Congress but told Ambassador Bryce that legislation for purely US conservation was a priority. It was clear that the new president did not want to have his hands tied by a Roosevelt-orchestrated international agreement.[95]

Bryce hurriedly jotted down Taft's thoughts as the portly president went on: "Must first get legisln. Thinks that safer than to rely on ex'tive power. Esp. as litig'n threatened." Taft even went so far as to insinuate to Bryce that special interests and monopolists themselves had effectively trapped the ex-president and Pinchot. That is, by seeking to withdraw resources from the marketplace to defer consumption to the future, the conservationist duo actually increased the value of those resources already in the market—those controlled by monopolists.[96]

No wonder, then, that The Hague initiative was eventually scuttled, yet the wait was agonizing. Not until January 1911 did the Department of State officially snuff out the idea, "owing to insufficient interest." Too few countries had agreed to attend, the line went. Those that had signed up, the State Department already claimed in a press release, were "not enthusiastic." None cared "enough to ask for further details." The diplomatic cables, not to mention the world's newspapers, show that this was a lie, while the Dutch hosts were clearly annoyed at the inconvenience to them of the procrastinating diplomacy since Taft's ascension to office.[97] Through Bryce, the British had repeatedly pressed Taft on the matter, and Australia, Canada, and India were among those that lobbied the Foreign Office for the right to attend. Bryce insisted to Taft that any "slackness or remissness on his part in regard to the International Congress would lay him open to severe strictures."[98] The subsequent history of the Progressive Republican insurgency against Taft would drive home the danger of abandoning the Roosevelt agenda.

Though the World Congress plan had been defeated, the bold bid inspired many abroad, including some who had not agreed to attend. Argentina was one such country. The ideas of Pinchot and Roosevelt were particularly appealing to liberal Argentinean leaders seeking to modernize their country economically and socially without transferring power to a radical democracy. The case was significant for showing the extent of international interest in the US model of conservation in Latin America. Conservation was useful in challenging the informal British imperial influence in those economies and as part of a growing interest in Pan-Americanism. The episode also further demonstrated the global appeal of utilitarian conservation alongside, not in opposition to, support for preservation of wild nature. In Argentina, American conservationists worked not as a simple outward projection of US concerns but in partnership with like-minded foreigners. Argentinean reformers solicited these American Progressives not only as technical experts but also as reformers interested in the creation of a racialized settler society—in line with Roosevelt's vision of how American empire should ideally operate. Latin Americans were not Anglo-Saxons, but both Progressive Americans and Argentinean reformers believed that they could participate in Anglo-style liberal economic development that would produce a new "race" of Pan-American settlers. The vision was never fully realized, of course, but it revealed much about conservation's global connections and consciousness in the wake of Roosevelt's presidency.[99]

The two decades before World War I marked a hopeful and yet rocky period of economic development for Argentina. Within the country's dominant Partido autonomista nacional, a split had emerged between conservatives and moderates. Working mainly under reformist president Roque Sáenz Peña, the Argentinean minister for public works (1908–13) and leading agricultural reformer Ezequiel Ramos Mexia was especially attracted to the US model of conservation.[100] Recognizing similarities with the western United States, which he believed had become prosperous through government policy on resource development, Ramos Mexia hired US geologists to survey the land for extension of the country's rail system deep into Patagonia. Though the railroads that dominated Argentina's economy were mostly British controlled, private, and export oriented by the 1890s, those now envisaged would be state owned. One of the leaders of the moderate, reformist segment of landowners who wished for greater integration in the global economy, Ramos Mexia proposed to use these improved networks to open up new land to agricultural settlement. This would provide a more democratic alternative, along US Homestead Act and National Reclamation Act lines, to existing large landed estates.[101] This thinking was congruent with Pinchot's. To this end Ramos Mexia authored the Irrigation Law of 1909 and tasked the American geologists with assessing the supplies of water and the engineering required.

Leading the American advisers was none other than Bailey Willis, who supplied the alarming pictures of environmental decay in China that Roosevelt used in his last annual message to Congress. Willis was a strong supporter of both national parks in the American West (he had helped plan Mount Rainier National Park) and the proclamation of southern national forests to protect both timber supplies and the watersheds of the Mississippi Valley.[102] Appointed "[w]ith the official approval of the U.S. Government" as commissioner of "Hydrologic Studies" in February 1911 on a two-year contract, he took with him several USGS officials and academics. Ramos Mexia's intention was to build a transcontinental railroad, starting from the Atlantic port of San Antonio, and the American party duly surveyed the northern Patagonia territory southwest of Buenos Aires for this plan.[103] As they went along, they broadened the scope of their work to include study of the "natural resources and possible industries that should eventually provide traffic for the transcontinental railroad."[104] The railroad would have its temporary terminus in present-day San Carlos de Bariloche. In the summer of 1913, the Americans undertook a reconnaissance of the land adjacent to the Rio Negros and the hill country of the Argentinean Andes and classified thirty-one thousand square kilometers of land for grazing, agriculture, or forestry, just as the United States had done under Pinchot's leadership for the American West.[105]

Willis's recommendations to the Argentinean government embodied wise-use conservation. He lectured his new employers from the standard Pinchot script. "Exploitation and conservation are by many considered to be contradictory terms, exploitation being taken to mean exhaustive utilization for immediate profit, and conservation representing the idea of preservation for future use." But this was not Willis's view. Conservation meant using "that which is ripe" and leaving the immature "in the case of all living things." Resources that did not "grow," like soils, must be conserved to prevent waste and promote "their highest utilization."[106] Citing the example of the US Reclamation Service and its relevance for arid areas, Willis argued that water was "the most important factor among the natural resources." Like Pinchot, Willis argued that "the people" were "anxious" to "retain control of the inexhaustible power which the falling streams can be made to yield" through hydroelectricity. There was also a need for "scientific forestry" controlled by "judicious, well-trained, far-sighted men, for the harvest of their planting is to be gathered by the next generation."[107]

Despite these clear commitments to Roosevelt and Pinchot's conservation platform, Willis did not distinguish between preservationist and utilitarian wings. In Patagonia he found the forests, lakes, and glaciers enchanting and enthusiastically supported Argentinean efforts to create a national park because "Nowhere have I seen Nature more majestic."[108] Explorer and scientist Francisco Pascasio Moreno had already donated the nucleus of the site in 1903. This, at Lago Nahuel Huapi in the foothills of the Andes, was to be developed "because of the value attaching to the scenic beauty and the environment of the lake and mountain region, as elements which will vitally affect the well-being and culture of the Argentine Nation."[109] Though Willis waxed lyrical on the area's scenic values and urged that "no exploitation, even of the water powers, which would injure that beauty or affect adversely the enjoyment of the National Park should ever be permitted,"[110] he sensed immediately a rare opportunity to plan an entire regional settlement on Progressive conservation principles. Yet the model was as much European

11.3. Bailey Willis (at left) with assistant Otto Lugenbuehl and the deputy director of the Comisión de estudios hidrológicos, Emilio Frey (Argentina), surveying Lake Nahuel Huapi, site of the proposed national park, c. 1911. Public domain.

as American. The area surveyed he compared in potential to Switzerland, which it resembled in aesthetics and "material future."[111] To this end, Willis completed a multiple-use plan for the industrial and agricultural development of the region, based on tourism, nature protection, cheap and renewable hydroelectric power for industry, and small-scale agricultural settlement nurtured by irrigation through dam construction and canalization.[112]

Willis and Ramos Mexia's shared vision was also a racial one, based on ideas of a white settler society model similar to that of the United States. Predicting that the area would become a "country of independent prosperous farmers,"[113] Willis recommended importing whites from central Europe—because Germans, Slovenes, and Swiss had already migrated to the area and seemed a promising "racial" group for his purposes. Willis advocated a campaign in Europe and North America to entice such hardy and resourceful people. Settlement would not rely on the indigenous people, who were in a hierarchical and marginalized relationship with the Argentinean elite of large landowners across the country. Indeed, Ramos Mexia's plans could easily be construed as a way for the nation to "consolidate its control" of the sparsely settled region after the violent removal of Indians from Patagonia under General Julio Roca, an event with some uncanny similarities to the North American experience. Willis was unconsciously complicit in this national consolidation of white settler society through liberal developmentalism. But neither Ramos Mexia nor Willis believed the area should be dominated by large-scale farming as practiced elsewhere in Argentina. Through Progressive conservation, Willis predicted that a "higher type of man" would evolve in this new environment with the right government assistance and would contribute to a "Pan-American" race shared between the United States and its Latin American neighbors.[114]

Willis's influence over Ramos Mexia was tied to the latter's career and, for this reason, was ultimately limited in impact. Together the two had planned a future for Argentina along the path of Progressive conservation.[115] However, in 1913 the national treasury's ability to borrow overseas deteriorated. During the Balkans crisis, procuring loans in Europe became more expensive. The existing railroad companies opposed Ramos Mexia's scheme bitterly, and private landowners in areas already settled resisted the possible diversion of state assistance away from the export-oriented economy based on staples such as beef and wheat.[116] This economic and political opposition took advantage of the financial crisis and Ramos Mexia was forced to resign in mid-1913. Willis's ambitious scheme for a combined industrialized and scenic development of the Lago Nahuel Huapi area could no longer be implemented. Now elected under a "universal" male suffrage (actually the extension by secret-ballot provisions of an earlier fraud-ridden law), the Argentinean government took a leftward turn in 1916, leaving moderates such as Ramos Mexia on the outside.[117] Willis returned to the United States, to a chair at Stanford University. Yet the episode showed that American conservation pioneered under Roosevelt and Pinchot was attractive to modernizers in that nation, as it also was in Brazil,[118] and in ways not dissimilar to the demands for conservation as "development" of resources across the Pacific in Australia and New Zealand. All was not in vain in Argentina, however. The plan for a national park at Lago Nahuel Huapi was eventually implemented in 1934, and the area became a focus for Argentinean tourism rather than industry.

Though initially missing from the respondents to Roosevelt's 1909 proposal while the battle between old-guard conservatives and more liberal reformers was still unresolved, by 1913 the now more progressive Argentina supported the idea of a World Congress on Conservation. When the Conférence internationale pour la protection de la nature was held that year at Berne, Switzerland, the Argentinean delegate sat alongside the US representative. But the conference owed nothing directly to American influence, or to Taft, who himself was an ex-president by that time. Woodrow Wilson's emissary was the minister to Switzerland, Pleasant Smith. Not known as a conservationist, Smith took no active part in the deliberations, which were orchestrated by Swiss zoologist Paul Sarasin.[119] Roosevelt and Pinchot had been scooped in their campaign for a global meeting, but it is unlikely they would have found the proceedings entirely satisfactory, since the conference focused on zoological and proto-ecological issues of species depletion rather than national resource allocation as a whole. Pinchot continued up to his death in 1946 to call for such a comprehensive meeting, but it was not to be held during his lifetime. Even the 1913 conference could achieve little because international circumstances quickly changed. The conference's "permanent" secretariat to advise governments on management and protection of endangered species was established but cut short because the outbreak of World War I shattered the apparent unity of the European representatives. Roosevelt's global conservation diplomacy had ended without tangible result, but the internal consequences for the United States were profound. To those we must now turn.

TWELVE

"A Senseless and Mischievous Fad?": From Alarm to Sobriety as a Nation Takes Stock

Amid the daily struggle of politics, the glaring headlines of the yellow press, and the muckraking journalism of *McClure's Magazine*, one newspaper that stood out in the time of Roosevelt for calm resolve was the *Christian Science Monitor*. Founder Mary Baker Eddy had herself felt the force of the popular penchant for sensationalism when she was attacked in Joseph Pulitzer's *New York World*. Eddy's paper followed through on the mission that she set. When it approached the topic of conservation, the paper avoided the alarmism of many competitors.[1] It did not claim to be anticonservation but, in 1910, presented a sober and conservative judgment of events since the 1908 Conference of Governors. Through two years of agitation the movement had grown at "full steam" to become "so strong that to oppose it would court public disfavor" in the East and Midwest. Yet in other sections of the country, there were respectable people who regarded conservation as "a senseless and mischievous fad." "Alarming assessments" of the diminution of resources had found their way into print and political favor and served the cause of people discontented with the rising cost of living. Conservation directed anger at "the trusts" that corralled resources and created "a fresh antagonism toward capital." Though the hyperbole seemed mischievous, and despite "unwarranted assertions and exaggerated claims of those overzealous in the cause," alarm was not without its virtues, the paper conceded. The nation had discovered its "reckless" path of "extravagance." It had come to realize the need to "make a new start along safer and saner lines." In future, conservation would "not be taken to mean stagnation" but would denote orderly development.[2]

All this was perceptive. Conservation had not just galvanized an intellectual elite but captured the attention of a broad swath of the American public as well. Alarm had had its moment, and its results. The *Monitor* described this post-Roosevelt trajectory, in which conserving nature became "entangled with issues of a partisan and factional nature." The quest for conservation, including its global goals, did not end when Roosevelt left office. Pinchot created the NCA as a voluntary association to continue the work started by the NCC, and he became deeply involved in the running of successive national conservation congresses from 1909 until 1916.[3] He also went to extraordinary lengths to ensure that Taft remained supportive, including the provision of speeches and data for the new president to deliver. Pinchot hoped that Taft would be TR's man.

Within months of the election, however, Taft was in rancorous dispute with Roosevelt's right-hand man of old. The issues were ostensibly internal to the United States, but they had international ramifications. Pinchot was still chief forester, and with Roosevelt had dreamed up the proposed World Congress in part as a weapon to extend Roosevelt's postpresidential influence. As we have seen, the conference was delayed and eventually canceled. With Roosevelt in Africa hunting, and then on a triumphal tour of Europe until June 1910, Pinchot was the eyes and ears of the ex-president in Washington. As he watched and listened, the matters agitating him concerned the "trusts" and the political agenda of the Taft cabinet.

In the swirl of political life, most important of all the sources of conflict was the Ballinger affair. Taft did not retain the pro-Roosevelt James Garfield as secretary of the interior and turned instead to Richard Ballinger. Secretary Ballinger was a Seattle mayor and lawyer who had served briefly as the head of the General Land Office. In Pinchot's view, the office was a den of corruption and inefficiency, and he distrusted its appointees. Believing that Roosevelt's administration had wrongly withheld large sections of the public lands from development, Ballinger reopened millions of acres for entry. These included rich coal deposits in Alaska subsequently obtained by the Morgan-Guggenheim trust.[4] Through a proxy, Louis R. Glavis, the chief of the Field Division of the Department of the Interior, Pinchot launched an attack on Ballinger in August 1909. Taft referred the matter to Attorney General George Wickersham, who cleared Ballinger of wrongdoing. But Pinchot and his allies refused to remain silent and, through the pen of Glavis, excoriated Ballinger once more in a *Collier's Magazine* article.[5] When Taft dismissed Glavis for insubordination, Pinchot wrote a letter of objection read into the proceedings of the Senate by Jonathan P. Dolliver on 6 January 1910. As a result, Taft immediately sacked Pinchot.[6] Though a special congressional committee investigated the whole affair, Ballinger was eventually cleared once more.[7] Not deterred, Pinchot took Taft's actions to be disloyal to Roosevelt's conservation legacy. Roosevelt was at first conciliatory, since he knew that Pinchot had a tendency to go off half-cocked. But the controversy ultimately produced a deep rift between Taft and the ex-president and led to plans to have Roosevelt run as a candidate against Taft's reelection—announced in February 1912. Roosevelt's Osawatomie, Kansas, speech on the New Nationalism on 31 August 1910, started this process. Therein, he made conservation one of the strongest pillars of a transparently political platform. As Taft stated, if Roosevelt suddenly died and an autopsy was conducted, "1912" would be found written on his brain.[8]

The Ballinger affair had internal conservationist origins. It was tied in some measure to Roosevelt's personal longing to set the boundaries of conservation policy by remote control while no longer president and, even more, to Pinchot's inclination to interpret the former president's wishes. But the controversy is similarly intelligible through Roosevelt and Pinchot's cherished assumption of a world struggling for control of the planet's natural wealth. The white heat of the Ballinger affair reflected the ever-present geopolitical thinking of the Roosevelt presidency. The affair was significant because Pinchot's allies claimed that the Alaska coalfields would, in effect, be the "last great national storehouse of natural resources" for the United States in its public lands. Thus stated Representative Edward H. Madison of Kansas in the Ballinger hearings. These never-to-be-duplicated Alaskan treasures rightfully belonged "to all the people of the United States" and should not be subject to "both fraud and monopoly."[9]

Because Pinchot charged that Ballinger was also improperly dispensing valuable waterpower sites on public lands, the Ballinger affair resonated with the question of public power supplies as well.[10] The growing interest in Europe and Canada as well as the United States in the potential of hydroelectricity lay behind this anxiety. Whether derived from fossil fuels or not, the Progressives saw energy as critical to the survival of modern society. As Progressive reformer Frederic C. Howe explained, "The problem of energy is the problem of civilization." Social, as much as economic, progress in the matter of democracy and class harmony depended on sustainable use. Progressives therefore emphasized the need to save waterpower from the private ownership that had already befallen coal.[11] For Pinchot and congressmen supporting him in the Ballinger affair, "great progress" had recently been made "in the use of electricity as power, and the means of its easy and cheap transmission" from the point of generation was "equally well known." Germany, Switzerland, and Italy had seen to that. Because of the "added value of water power," said Representative Madison in the Ballinger hearings, it was only a question of time before great demand from the western streams developed regarding this source. At its core, the Ballinger affair involved the question of whether public or private interests were to prevail across the West in the struggle for control of energy.[12] This affair and its implications increasingly preoccupied the Progressive conservation forces from 1910. In turn, the need to mobilize to defeat the apparently apostate Taft placed huge emphasis on galvanizing the national and local roots of Progressivism, at the expense of international contacts and cooperation.

Though Taft had taken a position against Roosevelt's brand of political intervention through enhanced executive authority, he was not truly anticonservationist. Progressive "insurgents" saw Taft's actions as "betrayal," yet the Taft administration was, historian James Penick concludes, "in so many ways an extension of the preceding regime." Even as "two competing views of the national interest" developed, "both . . . claimed to be progressive."[13] The true clash was over the use of executive commissions and rule by executive fiat. Was this a legitimate strategy in conservation policy, or should legislative control within specific acts of Congress prevail? Taft was, another authority avers, "not against conservation, [and] he merely favored a policy based on valid statutes."[14] This complex relationship between conservation and Progressivism cannot be denied. Pinchot sharpened and distorted the difference in the interests of his own partisan ambitions for his beloved chief, and for himself.[15] During his presidency, Taft declared several national monuments, supported the creation of Glacier National Park, withdrew public lands for oil conservation, and aided the passage of a number of important congressional conservation measures. One of these was the Weeks Act of 1911, which extended national forests to the eastern states, provided for interstate compacts for water and forest conservation to protect watersheds, and included measures against destructive forest fires. Taft's administration also completed the negotiations with Canada, Britain, Russia, and Japan, begun under Roosevelt in 1908, over limits to pelagic fur sealing.[16]

Nevertheless, Taft's speeches, as well as his executive style, revealed his reservations about Roosevelt's high-handed actions and the emotional temper of the conservation movement that Pinchot now effectively led. The new president's stance challenged the idea of a bureaucratic and centrally administrated state that Pinchot had come to embrace, as well as the all-encompassing nature of the conservation juggernaut.[17] In his speech to the Saint Paul convention of the National Conservation Congress in September 1910, Taft stated: "the time has come for a halt in general rhapsodies over conservation, making the word mean every known good in the world." Such appeals did not in Taft's view help with the "specific course" that legislators should take. Rather, practical remedies were needed. "Real conservation

12.1. Roosevelt handing his policies to Taft. *Puck* cartoon, 24 February 1909. LCPP.

involves wise, non-wasteful use in the present generation, with every possible means of preservation for succeeding generations." As Taft shrewdly observed, the problem was "how to save and how to utilize, how to conserve and still develop." Otherwise, any program would be counterproductive and "certain to arouse the greatest opposition to conservation as a cause."[18] Given that Pinchot and Roosevelt had both denied being against wise development, the adversarial political process artificially exaggerated the gap.[19]

The peculiarities of the American political system were important in bringing the conservationist crusade to a head in this way, yet equally important in its dissipation. Taft's response and his attempts to carve out a measure of support for conservation while liberating himself from the legacy of Roosevelt's increasingly poisonous relations with Congress were indicative of this tricky institutional context. Pre–World War I political circumstances shaped the American trajectory of environmental alarmism and conservation consciousness. The tussle between congressional and

Rooseveltian presidential power and, later, the rivalry with Taft encouraged and reflected TR's frenetic activity to secure his legacy in the latter days of his presidency and after. This personal context explains in part the climate of alarm.[20] This conjuncture raised both the fears and the hopes of many Progressives, giving the broader Progressive "movement," through conservation, greater definition, form, and impetus than it would otherwise have obtained. The same processes made longer-term coherence and continuity of national conservation initiatives virtually impossible.[21]

The idiosyncrasies of post-1909 Progressive politics also complicated the global perspective. Before 1910, Pinchot was vital in channeling industrial, scientific, and reform efforts toward international cooperation. After Roosevelt left office, Pinchot continued to push the international agenda, unilaterally sending to consulates around the world copies of the 1908 Conference of Governors' proceedings and conducting a sizable correspondence with foreign individuals and officials in favor of the World Congress idea. He set up a virtual foreign relations unit within his own Forest Service for this purpose.[22] After Pinchot's dismissal on 7 January 1910, that campaign necessarily lost its official imprimatur and bureaucratic wherewithal, and efforts had to turn inward to defend national Progressivism and, still later, to prepare for the 1912 election to defeat Taft with TR's third-party candidacy.[23] Pinchot adviser Harry Slattery told industrialist C. A. Grasselli in July 1912 that other NCA activities had to be stopped "in order to enable the officers of the Association to take up the more important work in Conservation immediately ahead."[24]

Organizationally, the position of the Progressives within some of the national associations previously sympathetic to conservation weakened, especially after Pinchot's dismissal from government. He now lacked the political power that Roosevelt's presidential backing had provided. The leaders of formerly aligned groups saw opportunities to abandon the drive to balance the interests of private enterprise against public control. Notably, the American Forestry Association[25] and the National Rivers and Harbors Convention fell into the hands of prodevelopment people at the expense of what Slattery and Pinchot regarded as Progressive conservation orthodoxy. Pinchot had been affiliated with the Rivers and Harbors Convention, but by 1915–16 both he and Slattery were convinced that it had become the mouthpiece of the "pork barrel river and harbor system" with "strong backing from the railroads and water power people."[26] In 1916 Pinchot lost control of the National Conservation Congress at the last of its meetings to be held.[27]

In addition to debilitating political splits, friction grew within the conservation movement between preservationists and the supporters of a more utilitarian approach. The proximate cause was Hetch Hetchy. The proposed dam in the Hetch Hetchy Valley in Yosemite National Park did not get its final go-ahead until 1913, well after Roosevelt's presidency, but when pressed to take sides in 1908, Roosevelt had supported the utilitarians.[28] The controversy had pitted concern over the physical integrity of national parks such as Yosemite against providing San Francisco access to publicly owned water and power.[29] The preservationist critique of the project, inspired by John Muir, *Century Magazine* editor Robert Underwood, and others, became another complication to add to the political controversy between Progressive Republicans favoring conservation and the regular Republicans whose conservation credentials Pinchot, Roosevelt, and Muir alike considered inadequate. Moreover, an articulate and increasingly well-organized national park lobby was taking shape and agitating for a park service to prevent future Hetch Hetchy land grabs. Its leaders also sought land from national forests for future park preservation, much to the irritation of Pinchot's entourage. Now Pinchot's efficiency-oriented conservation forces were fighting on two fronts.[30]

Beyond this festering two-front political conflict, the urgency of the conservation agenda also came into question through an apparent surge in resource availability. The prognosis for a number of key natural resources improved after 1910. This was especially the case for oil. USGS expert David Talbot Day had been found in error concerning the absolute depletion of petroleum. In 1912 the *Los Angeles Times* could gloat that "doubt has become conviction that the supply is abundant for years." Better still, "more exact and intelligent field investigations" that were "generally accepted" by the industry gave scope for optimism. Not until American entry to World War I in 1917 produced a large spike in oil usage by the military did the first true oil crisis occur. Subsequent concern over postwar industrial shortages in 1919–21 helped change attitudes toward the unlocking of corralled assets taken from the public lands. Thus, the war experience and its aftermath provided the wider political and economic context for the improper release of navy oil reserves for public entry in the Teapot Dome scandal of 1921–22, although, by that time, Americans had become used to periodic alarms over energy. To be sure, oil use soared in the twenties with the rise of mass automobile ownership, and demand intermittently exceeded supply. But further exploration and discoveries both in the United States and in Venezuela left the nation with a glut of the black gold and record low prices by the end of the decade. Oil crisis? What oil crisis? Or so it seemed. Alarm and excessive optimism had taken turns dominating public perception.[31]

In a similar fashion, coal shortages did not come to pass, except during the surging demand of World War I. Anthracite coal production reached an "absolute peak" in the early 1920s. Despite the generally booming economy, demand for coal declined during the remainder of the decade. Even bituminous coal, to which consumers were switching because of anthracite shortages, saw prices decline and production levels stagnate from 1923. Overall, coal production never again returned to such heights per capita, nor did it need to do so. This outcome reflected not only the switch to oil and gas but also other changes.[32]

Along with new discoveries of otherwise scarce commodities, technological innovations in the use of key resources also helped dissipate alarm. The Progressive Era fear of waste ironically ebbed because it encouraged more efficient production. Greater efficiency in the use of a wide range of minerals was achieved, in many cases for the first time in American history.[33] The mining industry's output benefited from the general alarm, thanks to work that began under the Roosevelt administration in 1904, when Joseph A. Holmes had become head of the USGS Technologic Branch. He was a close ally of W J McGee and the forestry chief. In fact, newspapers called him "Doctor Holmes, the Pinchot of the Mines."[34] A North Carolinian geologist and Democrat who nevertheless became loyal to Roosevelt and his conservation program, Holmes knew that coal, oil, and natural gas were all subject to huge waste. Recognizing the difference between renewable and nonrenewable resources, Holmes had argued in 1908 that "by far the most serious problem before the Nation" was "the approaching exhaustion of its mineral resources," for which there could be "no re-creation with the seasons and the exhaustion of which when once accomplished is a permanent exhaustion."[35]

Holmes's warning was never unqualified alarmism. Because of market forces, Holmes believed, private enterprise would remain central to improved efficiency. Scarcity itself would create the incentives for business to economize on energy use. But he also recognized market failure, outlining the seemingly endless race between supply and demand, where supply was often overtaken, and where innovation did not occur until stocks were almost exhausted. Here a role for government in limiting waste was vital. At the 1908 governors' conference he stressed in Rooseveltian fashion the need for a *national* policy to regulate the wise use of minerals, led by the federal government in collaboration with the states.[36]

12.2. Joseph Holmes, known as the "Pinchot of the Mines," was a fuel expert in the Technologic Branch, USGS, from 1904 to 1910 and chief of the Bureau of Mines from 1910 to 1915. Courtesy USGS.

At the USGS, Holmes had begun experiments to improve efficiencies in coal-consuming engines and furnaces during the Saint Louis World's Fair, where he observed technologically advanced machinery at the Belgian exhibit. He received permission to tour Europe twice to study European breakthroughs. He also agitated for an extension of the Technologic Branch's work and joined forces with the mining industry, links that he cultivated extensively. Under the Organic Act of 1910, he became chief of the Bureau of Mines, the new government section in the Department of the Interior, which included the old Technologic Branch.[37] Backed by McGee, Holmes engineered this appointment by personally lobbying congressmen.[38] Taft dithered over the appointment for several months, despite the North Carolinian's suitability, because he feared Holmes was a Pinchot ally. Skillfully, Holmes tied the legislation to broader Progressive interest in worker health and growing knowledge regarding the inferior working conditions of American miners vis-à-vis those of European countries.[39] Mine safety was therefore the law's priority, but the new bureau inherited the program for technological innovation in the use of minerals as well. The original act's scope was extended in 1913 by an amendment to provide explicitly for investigations of "mineral fuels and unfinished mineral products" in order to discover "their most efficient mining, preparation, treatment and use."[40] Subsequently, Holmes not only tested more efficient boilers but also supported a shift to electricity that improved the conversion rate of coal to usable energy.[41] Likewise, he continually encouraged the search for better methods for the recovery of coal and gas at the source of their extraction.[42] The Bureau of Mines also explored ways of heating federal buildings less wastefully to provide an example for the entire economy. In all these processes the federal government cooperated with businesses and state governments.

Ovens and furnaces gradually improved as a result of these innovations, though domestic sources of pollution in industrial cities such as Pittsburgh were harder to reduce. The drop in coal consumption in the 1920s was partly due to the implementation of such research on a broader, industrial scale.[43] As the economic historian Harold Williamson has noted, "Improvements in technology had made possible a reduction in the wastes involved in production, had increased the effective use of materials, and had made possible the exploitation of lower grades of resources or of substitutes."[44] Holmes was a pioneer in these efforts through federal government action, with the backing of Pinchot. In this way, alarmism was itself a stimulus to government and business-generated innovation that eased the threat of shortages.

This efficiency drive was not limited to the United States, as Holmes's trips to learn from Belgian and German scientists indicated. Roosevelt and Pinchot's calls for an end to waste were echoed elsewhere, promoting both cooperation within regional trading blocs and transnational scientific collaboration. The inception of a British Royal Commission on Natural Resources did for the British Empire what Roosevelt had proposed for the whole world. Action first came from discussions between the British dominions and the metropolitan government, leading to a resolution of the 1911 Imperial Conference in London. As with the enthusiasm for Roosevelt's Hague proposal, much of the impetus came from the white settler societies that were now fledgling nations. At the imperial conference, it was Canada's prime minister Laurier who pressed for the appointment of a Royal Commission, with support from Australia and South Africa. In all these places, Roosevelt's call for a World Congress on Conservation had raised hopes, and the dominion governments continued the agitation. Beginning in 1912, the commission visited South Africa, Canada, Australia, and New Zealand to assess the resource situation. The process dragged on because of the onset of World War I, and the final report, published in 1918, reflected the wartime, rather than the prewar, experience. The final recommendations called for "scientific development of the resources of the Empire" through coordination of research. As in the dominions' responses to the proposed conservation congress at The Hague, the emphasis was strongly utilitarian.[45] One spin-off was the creation in Australia of the Council for Scientific and Industrial Research in 1926. This has been called "a by-product of the Indian summer of neo-mercantilism and the idea that the British nations should combine economically to yield a high level of self-sufficiency."[46] The council was to sponsor scientific research for resource development, paralleling the Canadian Conservation Commission in this respect. In the 1920s this British initiative also bore fruit in the British Empire Forestry Conferences, the first being held in London in 1920.[47]

A second international legacy was the drive within the Western Hemisphere for further regional engagement on conservation. Here, Americans took the lead. At the 1910 National Conservation Congress in Saint Paul, Roosevelt's address focused on the importance of the North American Conservation Conference of 1909 and the necessity to build upon its example. He stated:

> It is clear that unless the governments of our southern neighbors take steps in the near future by wise legislation to control the development and use of their natural resources, they will probably fall into the hands of concessionaires

> and promoters, whose single purpose, without regard to the permanent welfare of the land in which they work, will be to make the most possible money in the shortest possible time. There will be shameful waste, destructive loss, and shortsighted disregard of the future, as we have learned by bitter experience here at home. Unless the governments of all the American Republics, including our own, enact in time such laws as will both protect their natural wealth and promote their legitimate and reasonable development, future generations will owe their misfortunes to us of today.[48]

Punctuated with rapturous applause as it was, the ex-president's speech made clear that international conservation campaigns were not to be discarded simply because Roosevelt had left office. They had become the centerpiece of all his future hopes for the nation and the world.

Under Pinchot's influence, the 1910 congress itself adopted an international conservation theme. The director of the Bureau of Latin American Republics, John Barrett, was made a vice-president of the meeting and gave a stirring speech on the need for cooperative development of Latin American resources: "I have the honor of representing here today some twenty nations as showing their interest in this great Conservation movement which is sweeping over the wide world." Barrett invoked the pioneering role of Roosevelt at the North American conference and believed the time was right for Latin America and the United States to act together on this issue. Because the countries to the south were "in the infancy of the development of their natural possibilities," they could benefit from the mistakes of the United States. "By taking wise steps," they could provide themselves with a lasting "natural endowment which this and other older lands have squandered."[49] The result would be permanent provision "for ourselves and for all men who are to come."[50] Of course, this cooperation would further open Latin America to US companies seeking not only export markets but also the abundant raw materials needed for American manufacturing and consumption. Yet to Barrett's mind this would be a process of trade, not military intervention or imperialist exploitation.[51]

With the coming of the world war in 1914, Pinchot abandoned any attempt at a truly global conservation meeting for the war's duration and, with allies in the Department of State and elsewhere, concentrated on Barrett's agenda of hemispheric action. This tactic was compatible with the Wilson administration's approach to maintaining neutrality toward the combatant states and with Barrett's long-standing efforts to develop American trade in Latin America, especially in view of the disruption to British trade dominance caused by the war itself. US government officials sympathetic

12.3. A jaunty Gifford Pinchot on Capitol Hill, c. 1913–14. LCPP.

to conservation pushed this topic at the Pan-American Scientific Congress, which met in Washington in the winter of 1915–16. The congress devoted an entire section and a separate published volume to "the conservation of natural resources, irrigation, agriculture and forestry."[52] Conservation thus became more closely aligned to US foreign policy and to the new "internationalism" in conservation policy that Roosevelt and Pinchot had pioneered.

The congress furthered Pinchot's agenda of utilitarian international conservation on a regional level. At the section meetings, many prominent US conservation officials, such as Henry Graves, Pinchot's replacement as chief of the Forest Service, spoke. Raphael Zon, the compiler of the sections on global forests for the *NCC Report*, discussed South American forest resources. A map of the "commercially valuable forest regions of South America" detailed forest boundaries as determined by W. H. Lamb of the US Forest Service.[53] Zon gave special attention to resource management through scientific surveys to inform policy development as Pinchot had urged, but the entire conference showed the conservation interest, with emphasis upon using education to promote wise use.[54]

The final report of the meeting was remarkable for its illustration of the Washington bureaucracy's loyalty toward global conservation. Retelling the story of the attempted World Congress, the report favorably portrayed the pioneering effort of Secretary of State Robert Bacon in aid of that proposal and showed how the US government decided, in the absence of a world meeting, to campaign for insertion of conservation into the Pan-American proceedings, making that meeting a site for a preliminary inventory of resources for the Americas. The Washington networks that Pinchot had cultivated worked hard for this result. Key figures within the American delegation included William Phillips, a Harvard-educated Department of State official under both Roosevelt and Wilson, who became chairman of the congress's executive committee; and James Brown Scott, secretary of the Carnegie Endowment for International Peace. Phillips had favored the World Congress idea and was "Mr Bacon's friend," while Scott edited *Robert Bacon: Life and Letters*[55] and served as the reporter-general for the Executive Committee.[56] Woodrow Wilson appointed Barrett secretary-general.[57]

Barrett's preference had always been for integration of Latin American and US markets in what, following John Hay, he called complementary trade, where the manufactured goods of the United States were exchanged for the raw materials of Latin America. This strategy favoring American capitalism's outreach for cheap resources paralleled the efforts of Britain and its dominions to coordinate their resource management within their own imperial sphere through the Royal Commission on Natural Resources, efforts that culminated in the Imperial Preference Scheme of 1932. By 1914–18, the Western world was already being dividing into a nascent set of trading blocs.[58] While Barrett indefatigably championed Latin American development on "peaceful" lines and discounted its turbulent revolutionary episodes as temporary aberrations in the expansion of capitalism in the region,[59] the speeches of the Latin Americans were subtly different. They emphasized that "America" meant all of the Americas. Bolivian delegate Ignacio Calderon effused, along lines also laid down by Cuban and Chilean members, that "[t]he Almighty has endowed this New World with lavish gifts of abundant resources, which we are prepared to let the rest of mankind share." To Calderon, this should not be a US but a Pan-American movement in which the continent as a whole would lead the "upward movement of the nations pushing forward in the path of justice and progress."[60]

Closely related to the Latin American orientation was the expanding network of tropical forestry research that built upon experience in the Philippines. From Bureau of Forestry alumni in the Philippines, a nucleus of tropical expertise emerged by the 1920s. Former bureau chief George Ahern worked in the Tropical Plant Research Foundation in Washington, DC (founded in 1924) on "the promotion of forestry in the Caribbean," while Harry N. Whitford returned to an academic position at Yale and "introduced the study of tropical forestry there."[61] Whitford undertook pioneering research on a tropical softwood, the Paraná pine (*Araucaria angustifolia*) of southern Brazil, said to compose "the most extensive coniferous forest in the Southern Hemisphere."[62] Tom Gill, an American influenced by the Philippine-based forester Hugh Curran, was to become prominent in this tropical work after 1926 and took part in the post–World War II efforts of the Food and Agricultural Organization of the United Nations (FAO) to help developing countries practice sustainable forestry. Much of Gill's experience was to be profoundly international rather than American centered: "[W]e are passing beyond the era when the influence of any one country dominates," Gill later reflected concerning forestry.[63]

Continued American input to the movement for international conservation also came through the work of David Lubin, a Polish-born merchant. As a commercial farmer in California in the 1880s, Lubin had helped found the California Fruit Growers' Union. Very interested in the Jewish diaspora, he then became the director of the International Society for the Colonization of Russian Jews in 1891, hoping to settle Jewish migrants on farms in California and Mexico. Turning to the improvement of the farmers' lot on a global stage, he campaigned tirelessly in Europe for what he called an international "chamber of agriculture." In 1906 the International Institute of Agriculture (IIA), the forerunner of the FAO, was established in Rome. Sponsored by the Italian government, the IIA sought to help the world's farmers with technical advice. Lubin had already moved to Europe, and because he had been so instrumental in persuading the Italians of the proposed IIA's value, Roosevelt made him the permanent US delegate from the organization's inception. Through Lubin's efforts, the IIA promoted rural credit schemes and cooperative farming techniques that echoed American contemporary interests.[64]

Roosevelt's Country Life initiatives bore fruit indirectly through the IIA. Horace Plunkett's work for the Commission on Country Life started the process, and Congress took up two tangible areas of activity: the one, practical education for farmers through land-grant college extension services; the other, rural credit schemes and cooperatives. These causes had gained impetus during the Taft years, stimulated not only by internal American political pressures but also circuitously by Lubin's international organizational work. Though educational extension was achieved in the Smith-Lever Act of 1914,[65] farm credit schemes and protection of rural cooperatives took longer. An English-language pamphlet published in 1911 by the IIA, *An Outline of the European Co-operative Credit Systems*, summarized farm credit measures in Europe and gained a "wide circulation in all English speaking countries, above all in the United States."[66] Stimulated by this literature, the Southern Commercial Congress (a meeting of prominent business and political leaders at Nashville in 1912) plumped for an agricultural credit system.[67] With all major candidates in the 1912 presidential election favoring rural credit reforms, the Southern Commercial Congress decided to send an "American Commission" of farmers and scientific advisers to Europe "to study on the spot the co-operative credit institutions."[68] Official federal government representatives were added to the traveling party under an act of Congress as a "United States Commission." Aware that the credit question was difficult to disentangle from other rural issues, the two technically separate commissions also surveyed the wider problems of agricultural cooperation.[69] The *Times* of London saw the initiative as an "awakening of a new American consciousness" in conservation and credited Theodore Roosevelt with its origins. Even as many aspects of his policies might "pass away," the *Times* believed, his stand on rural life and "the conservation of the natural resources of his country" seemed "likely to prove permanently productive."[70]

In May 1913 the commissioners began work in Rome, when the Fourth General Assembly of the IIA was in session. After visiting Italy, the Americans traipsed across Europe, finishing up in Ireland in August.[71] The United States Commission's three-volume report recommended agricultural banks to provide long-term credit, and short-term credit arrangements to meet "recurring needs" through rural credit associations run by the farmers themselves.[72] With support from the Wilson administration, Congress eventually passed the Federal Farm Loan Act (1916), which set up such a plan, but it was more successful in long-term credit provision through a "quasi-public system of regional mortgage banks" than in short-term credit arrangements.[73] Through the roundabout route of Lubin and the IIA, Plunkett's ideas had achieved a partial fulfillment, but the broader cooperative dream that Plunkett had championed presented a thornier prospect, since it clashed with the antimonopoly impulse in Progressive thought. The Clayton Antitrust Act of 1914 had contained a general exemption for farmers but did not address the specifics of what cooperatives were able to do. Only when farmers sought desperately to protect agriculture during the post–World War I glut of food on international markets did Congress explicitly exempt cooperatives from antitrust action, through the Capper-Volstead Act of 1922.[74] At last the Country Life vision achieved some practical success, but the dream of a holistic approach to the renaissance of farming was not fulfilled, and the demographic drain from the land continued.

On the national level, the Roosevelt legacy persisted in many such concrete achievements. Most obvious were the tangible residues in a greatly enhanced national forest system; national parks, monuments, and wildlife refuges that became foundational to American conservation policy and intrinsic to American culture; and the inception of a national irrigation system that was to reach large proportions in later decades. More important than a catalog of policies and activity, however, is the way Roosevelt used conservation to strengthen the American nation-state within the international state system. His effort was part of the larger process of Progressive Era state formation. Conservation was not the only way that the American state was built, but it was important in shaping its precise contours and emphasis on natural resources.[75] In recent historiography, including environmental history, there is a tendency to denigrate "the state" as an impersonal, bureaucratic, and oppressive entity. Critics have documented how colonized peoples, including indigenous Americans, were changed or oppressed by the growth of a big state, including its conservation aspects, whether through national parks, national forests, irrigated allotments on Indian land, or the environmental effects of dams.[76] These critics have an important point, though Roosevelt's paternalistic approach in the case of Indian irrigation policy was an attempt to avoid the market economy's effect in a more systematic expropriation of tribal land. In the case of Philippine forestry, congressional parsimony and the need to adapt to the realities of the Philippine social structure and economy heavily compromised the modernizing, state-centered drive of American colonialism.[77]

Critics have also linked this state development to empire building that required bureaucracies and taxes to support colonies, foreign bases, and powerful armed forces. In the wake of 1898, American anti-imperialists identified the intervention in the Spanish-American War and the taking of insular colonies as a move away from traditional republican values and toward the creation of such a state. Academic historians have generally accepted that Roosevelt's "large policy" and overseas "expansion" were at least functionally part of this process, while other, rasher commentators have excoriated Roosevelt on this score.[78] Rather than domestic imperatives determining empire's path, however, the new global engagement spurred the conservation reform that helped to shape the distinctive American state. Given the emergence of the United States as a major industrial power in a world of imperialist rivalries, the real issue was neither the development of stronger nation-state structures nor the existence of American empire as fact. It was what kind of state and what kind of empire the United States was to become. That turned on questions of intergenerational and international equity and the many contingencies of politics, diplomacy, and war after Roosevelt left office.

EPILOGUE

The Present, the Future, and the Power of Contingency in Human Life

In the middle of the Potomac River, west of the Lincoln Memorial in Washington, DC, stands Theodore Roosevelt Island. Though perhaps little known outside the nation's capital and certainly internationally, the island serves as a metaphor for Roosevelt's conservation, its residual effects, and its mixed legacy. Set in the midst of a vast urban and suburban expanse, it provides a tranquil escape from the bustle of nearby Georgetown's main street. Its tidewater swamps and forested areas provide a green refuge from the nearby city, and the island has its fair share of wildlife and flora. At its center stands an imposing statue of Roosevelt within a circular memorial, flanked by four large tablets inscribed with Roosevelt's views on boyhood, manhood, nation, and nature. Though the last three are directly relevant to this study, the legacy of Theodore Roosevelt's conservation policy is perhaps best reflected in "Nature." "There is delight in the hardy life of the open," we are informed. "There are no words that can tell the hidden spirit of the wilderness, that can reveal its mystery, its melancholy and its charm." No surprise for Roosevelt scholars here, but at the same time the inscription reads: "The nation behaves well if it treats the natural resources as assets which it must turn over to the next generation increased and not impaired in value." This assertion comes closest to TR's message for the future. The island itself provides a testimony too. It is a reclaimed and reforested space, not a "natural" one. Its resources are social, ecological, and cultural, not economic. At the same time the legacy is contingent and fragile. The island's tranquillity is broken by distant sirens and sounds of jet aircraft and military helicopters using the Potomac River flight path to avoid air traffic over built-up areas. It is no wilderness there, but nature is deeply impressed upon the mind. The central tension in Progressive Era conservation is contained within the inscriptions. Alongside the need for repair of nature for future

13.1. Roosevelt Monument, Theodore Roosevelt Island, Washington, DC. Author photograph.

generations, Roosevelt affirms that "[c]onservation means development as much as it does protection." But how were these two principles to be balanced or resolved?

For a contemporary observer of conservation's progress immediately before World War I, the Roosevelt legacy would have seemed finely poised between success and failure. "Wise" conservation had made great strides within academic and government circles, but the political conflicts that Pinchot and Roosevelt faced from 1910 to 1912 forced compromise and clarification that weakened the insistence upon intergenerational equity at the heart of the ex-president's plans. Under political pressure in 1910, Pinchot had assured readers of his *Fight for Conservation* that economic development would not stop under Progressive policies. "There may be just as much waste in neglecting the development and use of certain natural resources as there is in their destruction," he confessed.[1] Roosevelt, too, was forced to emphasize time and again that conservation would not halt present-day growth. At the National Conservation Congress of 1910 in Saint Paul, he agreed with Taft on this point: "one of the difficulties that we have to meet, in connection with the fight for Conservation, is that our aim is continually misrepresented—that the effect [*sic*] is constantly made to show that we are anxious to retard development." If there were "any doubt whether the conditions are liberal enough to the men who are to do the developing," Roosevelt told his audience, he would resolve "the doubt in favor of liberality to those men."[2]

Yet the details of any such compromise were difficult to specify. The conundrum worried the institutional economist and historian Lewis C. Gray, who devised the first attempt at discounting principles for the use of natural resources over time. A disciple of prominent political economist Richard Ely at the University of Wisconsin, Gray argued that the rate of interest at which capital could be invested provided the most convenient (albeit rough) rule of thumb for measuring the necessary discount for future generations. Yet Gray refrained from a purely economic and statistical calculation. To state the obvious, he had no econometric method with which to allocate the legitimate claims of the future. That would await the ruminations of a mathematical economist, Harold Hotelling, in the early 1930s.[3] Gray's view did not rest on economic formulae but on an argument that governments (led by far-sighted and knowledgeable people like Roosevelt and Pinchot) would intervene directly in the market to withdraw resources in the public interest using their own judgment.[4] Where necessary, public lands containing resources such as coal and oil ought to be withheld from the market by governments to raise the price of those resources and therefore encourage a more circumspect use. Thus would intergenerational equity be best preserved. Economists have barely noticed that Gray initiated within economics an ethical debate over the use of resources, but in doing so, he followed the public discourse of the Progressive conservationists, and his work is utterly inconceivable without them and the political debates over resource use that they generated.[5]

The contradiction in Progressive conservationists' stance between economic and ethical calculations did not come to the fore at the time and was subsequently resolved by abandoning the ethical side. By the 1920s, neoclassical economists of resource depletion came to rely purely upon the market. These later developments rested on technical economic expertise in a way that Progressive Era policies did not.[6] Then, theoretical discussion of private versus public interest in the regulation of natural resources had rested on Roosevelt's interventionist state. A public interest would measure, restrain, and allocate resources for development in an orderly fashion, based on a socially constructed idea of efficiency, not markets.[7]

Intellectually, the Progressives continued during World War I to advocate an ethical and political solution to the intergenerational problem that figured so prominently in the Roosevelt-Pinchot approach. In 1917 Ely and coauthors published *The Foundations of National Prosperity: Studies in the Conservation of Permanent National Resources*, in which Ely endorsed Gray's work as "[o]ne of the best articles on the economic theories involved in conservation."[8] He agreed with Gray that "the real heart of the conservation problem" lay in "the conflict between the present and future," noting that the primary problem of conservation "expressed in economic language" was "the determination of the proper rate of discount on the future" for natural resources.[9] Ely concurred that this problem could never be resolved through quantification. Instead, it depended on "questions of individual and social philosophy. What is the purpose of existence? The source and nature of ethical obligation? Our duties to posterity? All questions of the gravest import and beyond the range of economic science." Ely professed to use "the ethical notions and sentiments of normal men, or perhaps those somewhat above the average man," to "find guiding principles and helpful suggestions" in economic theory.[10] Exactly how economic theory helped remained unclear, however, since the book was taken up with normative and classificatory statements of the particular conditions under which resources could be cataloged and used prudently—whether renewable or not, whether in high demand or not, whether socially useful or not. This formulation, like that of Roosevelt and Pinchot, cast a high-minded elite at the center of moral judgments on the future of the United States and was at odds with the rough-and-tumble of a political system that refused to allow such an elite to decide resource allocation at the expense of the give-and-take of interest-group politics. An intellectual impasse over intergenerational equity persisted.

That said, the practical exigencies of war were already undermining the Progressive compromise. The first impact of the wartime environment was felt even before the United States entered the conflict in 1917. Labor shortages, increased factory output, and rising prices raised concerns over waste during 1915–16 and produced new opportunities to push the efficiency version of conservation. Preparedness rhetoric dominated at the 1916 National Conservation Congress in Washington.[11] That meeting saw the first day devoted to conservation, efficiency, and the war. Congress president E. Lee Worsham of Atlanta stated, "The situation which America faces calls for conservation, development and utilization of the country's resources. There must be no incompetency, no useless expenditure of either energy, time or substance." The congress also attempted to view conservation positively through a substitution of new sources of power to supply the need for nitrate-based munitions. Munitions (and fertilizer that could help feed the crops that would nourish the troops) could be manufactured from nitrogen in the atmosphere itself, but only with the aid of enormous amounts of additional energy at a time when coal resources were stretched thinly. Power companies and the federal government alike were increasingly attracted to utilizing hydroelectric power for these wartime purposes. Discussions at the 1916 meeting included "Water Power Resources and the Manufacture of Atmospheric Nitrogen." Undertaken by the government in 1918, the Muscle Shoals hydroelectric power station project took up the manufacturing challenge presented by such agitation.[12]

Pinchot attempted to use the threat of war in 1916 to advance Progressive policies, of which government ownership of power stations was one. He told the press that resource battles were not new in the United States, but the European conflict had "shown that the economic struggle" conducted by means of clashes over resource use was "the most important aspect of modern war." Pinchot remarked on the paradox that "while all the world is concentrating Government ownership of resources," American legislators were "about to dispose of all our remaining waterpowers [*sic*], oil, coal, gas, and similar resources with a develop-at-any-price principle."[13] The balance of interests was, however, changing. Oil was becoming more important. Pinchot aide Harry Slattery claimed that an "oil grab" was in progress: "Everyone agrees that the Standard Oil monopoly" was "the real cause" of shortages. Oil was now, "on account of the preparedness angle and the high price of gasoline," the "most important conservation question before the country." Americans were already consuming 1.75 million barrels a year to fuel ships, and war would add "two or three times that amount," even without further expansion of the US Navy. Britain already used six million barrels a year.[14]

The war itself advanced conservation through economies in consumption, with daylight saving time introduced and recycling of glass and metal encouraged.[15] The Conservation Division of the War Industries Board took the lead.[16] War also enabled Herbert Hoover's Food Administration to emphasize eating less of certain staples such as wheat so that grain could be preserved for the troops and the Allies. (Pinchot briefly worked in this administration, but he argued with Hoover over the power of the meat packers and left.)[17] Given the prewar anxieties about coal, it is not surprising that many attempts to redirect raw-material use came through the US Fuel Administration (established in 1917). Forced to rationalize coal supplies, this agency gave priority to freight transport and troop use over purely private consumption. Harry A. Garfield, a friend of Woodrow Wilson's and Fuel Administration head, issued an "idle Mondays" order in January 1918, closing nonessential industries for five consecutive days beginning

18 January and every Monday thereafter to 25 March.[18] After lobbying from manufacturers, however, the Senate requested a postponement, and the order was suspended.[19] The measures followed the Rooseveltian idea of state direction and national coordination of conservation measures but, as in so many cases, faced resistance from consumers, industry lobby groups, and political factions.

While the war encouraged restraint in nonessential consumption of food, coal, and other items—themes that accorded with ideas of intergenerational equity—short-term thinking took precedence in the wartime mobilization to maximize production. As a consequence of these pressures, the war exposed the contradictions in Progressive approaches, which had aimed to use efficiency to reduce the strain on resources. More efficiently produced goods would lower costs and hence encourage greater demand in response to price falls. In economics this is called the "Jevons effect" or "paradox," after the English economist of the coal industry who first theorized the position.[20] Such a response was likely in peacetime, but the very different and abnormal war conditions, by emphasizing maximum production, glaringly exposed the contradictions of efficiency in resource use *versus* conservation. This made measures to rationalize labor processes through scientific management and transport more immediately relevant than longer-term initiatives such as energy efficiency. The trend was clear in the US Fuel Administration, which necessarily aimed at using every ton of coal possible.[21]

During the war, Pinchot's constant harping on the need to conserve resources came under attack for the same reason. Clarence Barron, president of Dow Jones and publisher of the *Wall Street Journal*, criticized him for putting out "propaganda" that "directly" decreased production when it was "so much needed in the war time." Not that Pinchot personally flinched. He replied that "particularly in view of what has happened in Europe in the last three years . . . the one great lesson taught by the European War is this: national efficiency in war depends directly on the wise common control of the natural resources."[22] By the end of the war, Pinchot felt politically marginalized in the debates over waterpower and mineral rights on public land, even as the wise-use idea had prevailed across so many fronts. By this time, conservationists were thoroughly alienated from the Wilson administration and chafed at Secretary of the Interior Franklin Lane's "siding . . . with special business interests."[23] The 1920s saw continued erosion of the power of those wishing to reserve natural resources in public control for future use. Soon-to-be president Warren G. Harding redefined conservation in 1920 as "putting our resources conserved to *immediate* use, whether by private or public capital." This might well have served as a gloss for the alienation of the US Navy's Teapot Dome oil reserves leased illegally to private interests in 1922. Harding's lax attitude toward conservation already placed him on the slippery slope to that scandal.[24]

By the mid-1920s, conservation had also been weakened strategically by its wartime equation with the saving of grain through prohibition of alcohol under the Lever Act of 1917. Grain for Europe and the troops meant none for liquor. Prohibition thereby identified conservation with a program of moral restraint. This approach was consistent with the interest in "national vitality" during the Roosevelt presidency. Recall that the eugenicist-inclined Irving Fisher of Yale was a strong advocate of Prohibition as an instrument of such "vitality" in his section of the 1909 *NCC Report*.[25] Religious fundamentalists could rejoice in that version of conservation and, through the Anti-saloon League and Woman's Christian Temperance Union, many did. Political figures who stood for old-stock Protestant and rural values, such as Secretary of State Bryan, had become "conservationists" in a rhetorical maneuver that stretched the scope of the conservation crusade almost to breaking point.[26]

Alcohol served as a convenient target through which to discharge anxieties over material consumption and to channel the moral cause that Pinchot and Roosevelt had emphasized long before. Though the ex-president was no Prohibitionist, Pinchot found the cold-water crusade attractive to broaden his political base in Pennsylvania when he turned his ambitions to state politics,[27] a smart move in the short term. The evangelical influence on conservationist thought was already apparent at the grassroots level. But this identification was not purely reflective of cultural anxieties or political calculation. The intellectual justifications for considering waste a moral matter were deeply embedded in Progressive economic thinking. Waste had assumed the garb of a moral, as much as an efficiency, problem. In the Progressive Era, the social roots of moral restraint were not far from the surface of intellectual life and permeated the higher levels of economic thought. Thomas N. Carver, the David A. Wells Professor of Political Economy at Harvard University and president of the American Economic Association for 1916, wrote in *The Foundations of National Prosperity*: "Vice is, after all, nothing in the world except waste of human energy. Nothing is vice except that which wastes or dissipates human energy, and everything is vice which does waste or dissipate human energy. This brings vice definitely within the scope of the economist's study." Not surprisingly in view of wartime agitation, Carver found "probably no vice" more "reprehensible from a rational point of view than drinking." Ely endorsed this position, noting that Carver preached a "stern gospel" but a fair one, since "measures that give temporary relief and lead in the end to softness, flabbiness and degeneration deserve only condemnation."[28]

This ascetic alignment with the cold-water crusade was understandable in view of wartime calls for a moral campaign against German autocracy, but it was disadvantageous in the longer term. Conservation became identified with wartime moral sacrifice, which ill-suited both the mood of disillusionment with war after 1919 and the intensified pattern of economic consumption that American capitalism pushed in the 1920s through mass advertising. If Prohibition was a conservation measure, its cultural politics came under sustained attack from 1922, as Prohibition itself did. As in the case of drinking behavior, broader social change quickly made the Carver-Ely position on waste seem highly ideological, unrealistic, and old-fashioned.[29]

That is not to say that conservation died in the 1920s, any more than Progressivism did.[30] Hoover became secretary of commerce and in many ways followed in the tradition of Progressive conservation. He pushed further the rationalizing and business-oriented aspects of Progressive Era efficiency—in arguments for management of fisheries and other wildlife, as well as water and other inanimate resources. In 1924 he "only half-jokingly" explained in an interview that some waterfalls were in the wrong place, and that they might well appeal to the public with only half the amount of water they had: "We could save water and we could also have waterfalls in better locations if we handled the subject . . . with the aid of human intelligence added to the resources of the nation." The great engineer surmised: "the sheet of water used to produce the scenic effect could be much thinner." In the end, "scenically as well as industrially we can be better off through the civilizing of our rivers."[31] This was the spirit of the late W J McGee reincarnated.

But if there was continuity in—indeed magnification of—the wise-use philosophy, the international and intergenerational critiques were marginalized in the evolving discourse over waste. Instead, in moves that consolidated wartime trends, conservation became identified with rationalizing production and, especially, with labor productivity. *The Report of the Committee on Elimination of Waste in Industry* (1921) was written by fifteen engineers appointed by Hoover in his position as president of the Federated Association of Engineering Societies before he joined Harding's administration. Studying six areas of industry, the commission did not inquire into industrial resource depletion. The viewpoint of engineers was geared toward eradicating idle hours, intentionally restricted production by managers, and losses due to ill health, industrial accidents, and the like, all measured against commonsense benchmarks established in industries.[32] The approach was a version of Frederick Winslow Taylor's scientific management, which had come increasingly into vogue by 1920.[33] Domestically, this type of conservation centered on minimizing waste in the production process. Thereby the price of products could be lowered, the consumer economy extended, and the standard of living raised. This had been a concern of Roosevelt's and a reason for seeking efficiency in coal production and distribution, but it did not dominate Roosevelt's thinking as it did Hoover's.

The goal of equity between generations would, to be sure, resurface later in US history and draw upon the Roosevelt-Pinchot legacy. In the 1920s Lewis Gray was a Bureau of Agricultural Economics official, and within the USDA he advocated the application of Progressive conservation that regulated access to the public lands. This work provided the connecting link to many New Deal agricultural reforms.[34] Journalist Stuart Chase drew upon Gray and the idea of prudent and minimal use of nonrenewable resources by the present generation in his Depression era book, *Rich Land, Poor Land* (1936), an important if historiographically neglected statement of New Deal conservation's logic.[35] But Gray himself lamented the absence of national planning on Progressive principles in the 1920s, and Chase recognized the change from pre–World War I in the fragmented and defensive posture of conservation. He announced that the conservation movement of the 1920s was "still a living force, but somehow strangely shrunken from the great days when Roosevelt was its King Arthur and Pinchot its Launcelot." It lacked leadership and cohesion. Despite Theodore Roosevelt's efforts, the "embattled front of sturdy individualism" had "not been broken."[36]

Neither was international equity an issue in the 1920s, with the United States beginning a shift from surplus producer of raw materials to net importer. In contrast to the prewar initiatives for combined internal reform and cooperative international action, the development of foreign markets and resources in the interest of American corporations took on new importance, especially in Latin America.[37] Internationally, "conservation" now included intensifying the search for secure foreign supplies of raw materials as a possible alternative to internal conservation. Though the United States manifestly produced an oil surplus and exported this commodity, it also imported substantial amounts of different types of oil. Intercorporate competition, repeated scares of shortage, and specialist oil demands spurred further exploitation of Venezuelan oil fields in the 1920s.[38] Experimentation with rubber plantations in the Philippines, Liberia, and Brazil (needed for the booming automobile market) also proceeded apace. The Republican administrations of the 1920s reversed expectations of a move toward

Philippine independence and renewed the call for intensified exploitation of tropical forest resources there. The former Philippine Bureau of Forestry official Harry Whitford played a part in developing rubber production abroad. Meanwhile, coffee and tropical fruit imports rose, and the transformation of Central America into banana republics continued.[39] On the minerals front, the domestic balance sheet was still largely positive, but the gap was narrowing in some cases. The nation began to import slightly more iron ore than it exported, and the prewar surplus of domestic copper production declined due to greatly increased internal demand, with lessening US exports a result. US copper-mining corporations expanded key operations in Peru (1915) and Chile (1923) to augment domestic supplies and those from Mexico. Only in lumber production, however, was there "a classic case of increasing scarcity" on the horizon, and not until the late 1940s did resource shortage concerns for a whole range of strategic raw materials become apparent.[40]

There was another path, not followed. At the end of the war Pinchot had revived the idea of a world conservation congress; this was backed by Wilson's adviser on natural resources, the geologist Charles K. Leith, who had written chapters for *The Foundations of National Prosperity*. Leith accompanied Wilson to Versailles as part of The Inquiry, Wilson's brain trust on the conditions of peace. His views resembled those of Pinchot and the English economist John Maynard Keynes: that the postwar settlement should ideally prevent resource conflicts from contributing to future wars and economic upheaval. Advocating "an internationalist solution for the global minerals problem," Leith discovered that the president's plate was very full with great-power political squabbles. Wilson "exhibited no interest in substantive economic or natural resource issues."[41] When the United States rejected the League of Nations, this rebuff of Leith was merely reinforced. Subsequent international action within the Republican administrations of the 1920s was narrowly construed. It centered on trade access to foreign raw materials, prompted by concern over the proliferation of tariff restrictions and consolidation of trading blocs in world commerce. Absence from the League of Nations did not help but was not the decisive influence. Within the league, Europeans initiated a new conservationist agenda concerning, among other things, pollution on the high seas, even as the international results were, in the end, "minimal." Yet no US president from 1919 to 1933 showed the slightest interest in holding—or participating in—a world conservation congress.[42]

All this was a different world from the one that TR inhabited. It was a world in which colonial nationalism was ever more assertive, especially within the British Empire, and in which other empires had either crumbled or were being challenged. US formal imperialism was increasingly marginalized in American thinking, even as economic investment and financial intervention in the affairs of Latin America grew. Colonialism was far from over, but Woodrow Wilson's liberal internationalism and his principle of free commercial exchanges between nations were potentially more compatible with these transitions than Roosevelt's brand of great-power interventionism. Yet it was a world whose implications on the level of resource development Roosevelt had glimpsed as he watched the American frontier close, the European powers expand into the "uncivilized" world through colonial grabs, and fossil fuel consumption spike as US industrialization proceeded. He had seen the United States quickly turn from an economy founded on extensive settlement of land to one based on the intensive transformation of raw materials.[43] Roosevelt had not entered the presidency with a clear view on this transformation and its implications for national and international conservation, but he belatedly developed one through experience. His views went far beyond national parks and wildlife appreciation; they encompassed nascent ideas of environmental sustainability in a set of interlocking policies for national control over forests, rivers, soils, fuel, and water. Understood as habitability, the notion of a sustainable society remained a goal in the process of formulation, but Secretary of State Bacon's 19 February 1909 call for "repairing the injuries" to natural environments contained within it the fundamental perception of the globe as a closed energy system. It pronounced the recycling of energy as Roosevelt's hope for the future.[44] His presidency ended with many schemes still incompletely articulated, or thwarted, but the ambitiousness of—and complex motives behind—these attempts to come to grips with fundamental problems of unsustainable resource use have not been fully appreciated. The audacity of his plans as they took shape in the last few years of his tenure only makes full sense in the context of a struggle over what kind of empire the United States should become: one that provided international leadership on resource conservation or one that sought perpetual and global extension of its resource-based abundance.

Roosevelt's presidency witnessed a rare opportunity in American history to bring the threads of sustainability and nature conservation together on a national level and to begin these conversations on an international level too. Americans did not repudiate his legacy entirely, or suddenly. It remained a benchmark for the future. To the extent that they did take a different path, they did so not for any one reason, or in a conscious process of decision making, but through the incremental forces and events analyzed in the last chapter and this. Changing calculations of actual resource shortage, new technologies, the contingencies of war, and the transformation of US society to a mass consumer economy all mattered, but so too did political partisanship, institutional constraints, and personal rivalry figure in the outcome. As Rhys Isaac has written, "History is the most particularizing of the social sciences; it must stand tall to remind the others of the power of contingency in human life. . . . This is not just a . . . scholars' debate; it is an affirmation of the possibilities of changing the disposition of things. The future is always being made by the present generation. . . . [But] however we read the signs of doom and gloom, we cannot predict the future. The shape of the world to come remains to be made by human action in circumstances that can never be foretold."[45] That applied to Roosevelt as it applies to us.

NOTES

NOTES TO CHAPTER ONE

1. Rudolf Cronau, "A Continent Despoiled," *McClure's Magazine* 32 (April 1909): 639–64; Rudolf Cronau, *Our Wasteful Nation: The Story of American Prodigality and the Abuse of Our National Resources* (New York: Michael Kennerley, 1908), 11 (quote). Cronau had first visited the United States in 1881, and he served the *Cologne Gazette* from 1893 to 1899. On Cronau, see H. Glenn Penny, *Kindred by Choice: Germans and American Indians since 1800* (Chapel Hill: University of North Carolina Press, 2013), chap. 4.
2. Emerson Hough, "The Slaughter of the Trees," *Everybody's Magazine* 18 (May 1908): 579–92, quoted in Cronau, *Our Wasteful Nation*, 130; Frank Luther Mott, *A History of American Magazines*, vol. 5 (Cambridge, MA: Harvard University Press, 1968), 82. For the long history of the jeremiad "simultaneously lamenting a declension and celebrating a national dream," see Sacvan Bercovitch, *The American Jeremiad* (Madison: University of Wisconsin Press, 1978), esp. 180.
3. *New York Times*, 13 February 1909, BR85.
4. *New York Times*, 20 December 1908, SM11.
5. "A Nation's Prodigal Waste," quoted in *American Conservation* 15 (February 1909): 106.
6. "Among the New Books," *Chicago Daily Tribune*, 18 March 1909, 9. See also *New York Sun*, 2 January 1909, 7.
7. Mary Huston Gregory, *Checking the Waste: A Study in Conservation* (Indianapolis: Bobbs-Merrill, 1911); Cronau, "Continent Despoiled"; Charles R. Van Hise, *The Conservation of Natural Resources in the United States* (New York: Macmillan, 1910).
8. See, e.g., "Report of National Conservation Commission," in *Chautauquan* 55 (June 1909): 53–104; "Statements by Men Prominent in National Affairs upon the Subject of Conservation," ibid., 115–20; Clara E. Fanning, comp., *Selected Articles on the Conservation of Natural Resources*, Debaters Handbook Series (Minneapolis: H. W. Wilson, 1913); Conservation, Reports [Daughters of the American Revolution], National Conservation Committee, fl. (folder) Conservation General PS, Gifford Pinchot Papers, LC; Harry Slattery, "From Roosevelt to Roosevelt, 1900–1946, Part 1," 25, fl. Autobiography, box 28, Harry A. Slattery Papers, David M. Rubenstein Rare Book and Manuscript Library, Duke University, Durham, NC.

9. "Mexican Timber Lands," *Wall Street Journal*, 1 October 1907, 8; ibid., 4 November 1907, 7; Ian Tyrrell, *True Gardens of the Gods: Californian-Australian Environmental Reform, 1860–1930* (Berkeley: University of California Press, 1999), 80; Samuel P. Hays, *Conservation and the Gospel of Efficiency: The Progressive Conservation Movement, 1890–1920* (Cambridge, MA: Harvard University Press, 1959), 127–32, 140; Slattery, "From Roosevelt to Roosevelt." Beginning as early as 1900, the jeremiads over waste preceded the concerted campaign of the Roosevelt administration from 1907 to 1909 and came from professional groups and the wider public as well as from the Washington bureaucracy. See, e.g., "The Coming Exhaustion of Nature's Resources," *Engineering News* 45 (31 January 1901): 81–82.
10. On the role of hunting in Progressive Era conservation, see Thomas R. Dunlap, "Sport Hunting and Conservation, 1880–1920," *Environmental Review* 12 (Spring 1988): 51–60; John F. Reiger, *American Sportsmen and the Origins of Conservation*, 3rd ed. (Corvallis: Oregon State University Press, 2001), chap. 7.
11. This description was used by Matt Donnelly in his *Theodore Roosevelt: Larger than Life* (North Haven, CT: Shoestring Press, 2003), 1.
12. James Garfield diary, 4 January 1909, box 8, James R. Garfield Papers, LC; Theodore Roosevelt to Garfield, 23 May 1908 and 24 February 1909, box 119, ibid. ("intimacy"); Reiger, *American Sportsmen*, 33–34; Kathleen Dalton, "Theodore Roosevelt, Knickerbocker Aristocrat," *New York History* 67 (January 1986): 39–65 at 40, 50 (quotes).
13. *Report of the National Conservation Commission, February, 1909: Special Message from the President of the United States, Transmitting a Report of the National Conservation Commission, with Accompanying Papers . . .* , 3 vols. (1909; repr., New York: Arno Press, 1972), hereafter *NCC Report*; Gifford Pinchot to Dr. J. T. Rothrock, 5 May 1909, fl. May 1909, box 489, Pinchot Papers.
14. Daniel T. Rodgers, "In Search of Progressivism," *Reviews in American History* 10 (December 1982): 113–32. As Rodgers explains, adherents of Progressivism used "three languages of discontent," which can be put in three "rough but serviceable" categories: focusing on antimonopolism, "social bonds and the social nature of human beings," and social efficiency (ibid., 123). For a more recent interpretation, see Michael McGerr, *A Fierce Discontent: The Rise and Fall of the Progressive Movement in America, 1870–1920* (New York: Oxford University Press, 2005).
15. *New York Times*, 13 February 1909, BR85; George L. Knapp, "The Other Side of Conservation," *North American Review* 191 (April 1910): 465–81.
16. For the range of arguments, see Fanning, *Selected Articles*, esp. Leslie M. Scott, "Why East and West Differ on the Conservation Problem," *Independent* 68 (31 March 1910): 697–99, in ibid., 131–35; "Are We Conservation-Crazy?," *Literary Digest* 41 (26 November 1910): 967–68, in ibid., 135–37; "Living on Our Capital," *Wall Street Journal*, 24 September 1907, 1; "The Exhaustion of Oil," *Los Angeles Times*, 7 August 1909, II4 ("comforts").
17. Hays, *Conservation and the Gospel of Efficiency*, 256–60; *Proceedings of a Conference of Governors in the White House, Washington, D.C., May 13–15, 1908* (Washington, DC: Government Printing Office, 1908), 162–63; Edwin L. Norris quoted in "Are We Conservation-Crazy?," in Fanning, *Selected Articles*, 136; Gregory Randall Graves, "Anti-conservation and Federal Forestry in the Progressive Era" (PhD diss., University of California, Santa Barbara, 1987), esp. 171–72; Knapp, "The Other Side of Conservation"; Richard White, *"It's Your Misfortune and None of My Own": A History of the American West* (Norman: University of Oklahoma Press, 1994), 407–9. George

Leonard Knapp was born in Dover, Minnesota, on 6 April 1872 and died in Lafayette, Indiana, on [8?] May 1950. He was a writer, at one time for the *Rocky Mountain News* and at another for a railroad newspaper, *Labor*. *Niagara Falls Gazette*, 9 May 1950, 9.

18. Governor Frank Gooding of Idaho, a sheep rancher and conservative Republican, stated in 1908: "that which lies wholly within the borders of one State, can, in my judgment, with the passage of proper laws by Congress, be best done by the People of that State. We need the strong arm of the general Government in the initiation of this great work, but, if it is to be successful, the State must be made an interested party in the administration and development of its own resources." *Conference of Governors*, 169.
19. Knapp, "The Other Side of Conservation," 467.
20. Ibid., 465.
21. Henry Carey, *The Slave Trade, Domestic and Foreign* (Philadelphia: L. Johnson, 1853), 44.
22. *New York Times*, 13 February 1909, BR85.
23. "America's Profligacy with Her Heritage," *New York Times*, 20 December 1908, SM11 (quote); *New York Times*, 13 February 1909, BR85.
24. "Living on Our Capital."
25. For the role of technology in American culture, see, e.g., Alan Trachtenberg, *Brooklyn Bridge* (1965; Chicago: University of Chicago Press, 1979), chap. 1; John F. Kasson, *Civilizing the Machine: Technology and Republican Values in America, 1776–1900* (New York: Penguin, 1977).
26. Char Miller, *Gifford Pinchot and the Making of Modern Environmentalism* (Washington, DC: Shearwater Press, 2001), 153–54.
27. George Grinnell, "The Adirondacks," *Forest and Stream* 21 (17 January 1884): 489; Reiger, *American Sportsmen*, 85.
28. Richard Grove, *Green Imperialism: Colonial Expansion, Tropical Island Edens and the Origins of Environmentalism, 1600–1860* (Cambridge: Cambridge University Press, 1995); Clarence Glacken, *Traces on the Rhodian Shore: Nature and Culture in Western Thought from Ancient Times to the End of the Eighteenth Century* (Berkeley: University of California Press, 1976). For the nineteenth-century British Empire, see James Beattie, *Empire and Environmental Anxiety: Health, Science, Art and Conservation in South Asia and Australasia, 1800–1920* (Basingstoke, Eng.: Palgrave Macmillan, 2011).
29. George Perkins Marsh, *Man and Nature; or, Physical Geography as Modified by Man* (New York: Charles Scribner, 1864); Ian Tyrrell, "Acclimatisation and Environmental Renovation: Australian Perspectives on George Perkins Marsh," *Environment and History* 10 (May 2004): 153–66; Donald Pisani, "Forests and Conservation, 1865–1890," *Journal of American History* 72 (September 1985): 340–59; Richard T. Ely, *Introduction to Political Economy*, rev. ed. (1889; New York: Eaton and Mains, 1901), 81–83 (quote at 82).
30. E.g., Nathaniel Southgate Shaler, ed., *The United States of America*, vol. 2 (New York: D. Appleton, 1894), 454, quoted in Harold F. Williamson, "Prophecies of Scarcity or Exhaustion of Natural Resources in the United States," *American Economic Review* 35 (May 1945): 97–109 at 99–100; Nathaniel S. Shaler, *Man and the Earth* (New York: Fox, Duffield, 1905), 128 (quote); Frank A. Fetter, *Source Book in Economics Selected and Edited for the Use of College Classes* (New York: Century, 1912), 91–101; D. N. Livingstone, "Nature and Man in America: Nathaniel Southgate Shaler and the Conservation of Natural Resources," *Transactions of the Institute of British Geographers*, n.s., 5, no. 3 (1980): 369–82.

31. Harold K. Steen, *The U.S. Forest Service: A History* (Seattle: University of Washington Press, 1977), 26–27, 34–37; Organic Act of 1897, 30 Stat. 34–36; codified U.S.C. vol. 16, sec. 551, an amendment to the Sundry Civil Appropriations Act; Tyrrell, *True Gardens*, 51.
32. Pisani, "Forests and Conservation, 1865–1890"; Sidney Fine, *Laissez-Faire and the General Welfare State: A Study of Conflict in American Thought, 1865–1901* (Ann Arbor: University of Michigan Press, 1956), 365; *Report of the Committee Appointed by the National Academy of Sciences upon the Inauguration of a Forest Policy for the Forested Lands of the United States: To the Secretary of the Interior, May 1, 1897* (Washington, DC: Government Printing Office, 1897), 8.
33. Ely, *Introduction to Political Economy*, 83 (quote). Moreover, the 1890s critique concerned the need for Americans to catch up with European forestry, not alarm over a global crisis. *Report of the Committee Appointed by the National Academy of Sciences*; Gifford Pinchot, "Government Forestry Abroad," *Publications of the American Economic Association* 6 (May 1891): 7–54.
34. Oscar Kraines, "The President versus Congress: The Keep Commission, 1905–1909; First Comprehensive Presidential Inquiry into Administration," *Political Research Quarterly* 23 (March 1970): 5–54 at 4, 33 (quote); Harold T. Pinkett, "The Keep Commission, 1905–1909: A Rooseveltian Effort for Administrative Reform," *Journal of American History* 52 (September 1965): 297–312 at 298; Bruce J. Schulman, "Governing Nature, Nurturing Government: Resource Management and the Development of the American State, 1900–1912," *Journal of Policy History* 17, no. 4 (2005): 375–403.
35. Douglas Brinkley, *The Wilderness Warrior: Theodore Roosevelt and the Crusade for America* (New York: HarperCollins, 2009).
36. Henry Nash Smith, *Virgin Land: The American West as Symbol and Myth* (Cambridge, MA: Harvard University Press, 1950).
37. Roosevelt sent Arthur C. Veatch to Australia to study this matter. Arthur C. Veatch, *Mining Laws of Australia and New Zealand* (Washington, DC: Government Printing Office, 1911), 11–22; John D. Leshy, *The Mining Law: A Study in Perpetual Motion* (Washington, DC: Resources for the Future, 1987), 243; Theodore Roosevelt, "Seventh Annual Message," 3 December 1907, in Gerhard Peters and John T. Woolley, American Presidency Project, http://www.presidency.ucsb.edu/ws/?pid=29548 (accessed 28 June 2010). The "filed intention to take up a homestead claim" was known as an "original entry." E. Louise Peffer, *The Closing of the Public Domain: Disposal and Reservation Policies, 1900–50* (Stanford, CA: Stanford University Press, 1951), 13. "Public" entry was the process whereby the public lands were transferred to private ownership through the General Land Office.
38. Peffer, *Closing of the Public Domain*, 147–48; Coal Lands Act, chap. 270, 35 Stat. 844 (3 March 1909); Hays, *Conservation and the Gospel of Efficiency*, 68–69; Debra Donahue, *The Western Range Revisited: Removing Livestock from Public Lands to Conserve Native Biodiversity* (Norman: University of Oklahoma Press, 1999), 22; Graves, "Anticonservation," 152; Paul Gates, *History of Public Land Law Development* (Washington, DC: Government Printing Office, 1968), 724–26.
39. Kathleen Dalton, *Theodore Roosevelt: A Strenuous Life* (New York: Knopf, 2002), 242–46; David M. Wrobel, *The End of American Exceptionalism: Frontier Anxiety from the Old West to the New Deal* (Lawrence: University Press of Kansas, 1993).
40. Brinkley, *Wilderness Warrior*; Paul R. Cutright, *Theodore Roosevelt: The Making of a Conservationist* (Urbana: University of Illinois Press, 1985).

41. Theodore Roosevelt to William T. Hornaday, 29 December 1908, series 2, reel 353, Theodore Roosevelt Papers, LC (quotes); "Saving of America," *Washington Post*, 25 February 1909, 1; Theodore Roosevelt to William T. Hornaday, 1 February 1909, Miscellaneous Correspondence, 1883–1970, MS Am 1454, Theodore Roosevelt Collection, Houghton Library, Harvard University, Cambridge, MA, http://pds.lib.harvard.edu/pds/view/22396592?n=41.
42. See, e.g., "A Man of Action: Mr. Roosevelt's Record," *Times of India*, 26 March 1909, 7 (quote); "Man of Action," *Star* (Christchurch, N.Z.), 3 May 1909, 2; *West Gippsland Gazette* (Warragul, Vic.), 11 May 1909, 5.
43. Roosevelt to Hornaday, 29 December 1908, series 2, reel 353, Roosevelt Papers.
44. G. R. Searle, *The Quest for National Efficiency: A Study in British Politics and Political Thought, 1899–1914* (Oxford: Blackwell, 1971), 64–65, 66; R. A. Lowe, "Eugenicists, Doctors and the Quest for National Efficiency: An Educational Crusade, 1900–1939," *History of Education* 8 (December 1979): 293–306; Anna Davin, "Imperialism and Motherhood," *History Workshop Journal* 5 (Spring 1978): 9–65; Linda Simpson, "Imperialism, National Efficiency and Education, 1900–1905," *Journal of Education* 16 (January 1984): 28–36.
45. Michel Girard, "Conservation and the Gospel of Efficiency: Un modèle de gestion de l'environnement venu d'Europe," *Histoire sociale/Social History* 33, no. 45 (1990): 63–79.
46. Daniel T. Rodgers, *Atlantic Crossings: Social Politics in a Progressive Age* (Cambridge, MA: Belknap Press of Harvard University Press, 1998). But see Girard, "Conservation and the Gospel of Efficiency."
47. Kurkpatrick Dorsey, *The Dawn of Conservation Diplomacy* (Seattle: University of Washington Press, 1998). Calls for the study of a wider environmental diplomacy have come from Mark Lytle, "Research Note: An Environmental Approach to American Diplomatic History," *Diplomatic History* 20 (Spring 1996): 279–300; and Thomas Robertson, "'This Is the American Earth': American Empire, the Cold War, and American Environmentalism," *Diplomatic History* 32 (September 2008): 561–84. These calls concentrate on recent decades in American history.
48. For the most authoritative account of this changing historiography, see Paul A. Kramer, "Power and Connection: Imperial Histories of the United States in the World," *American Historical Review* 116 (December 2011): 1348–91.
49. Walter LaFeber, *The New Empire: An Interpretation of American Expansion* (Ithaca, NY: Cornell University Press, 1963), and the leader of the so-called Williams school of US foreign relations history, William Appleman Williams, have stressed "expansion" loosely defined and equated with "empire." They have also focused on the drive for overseas markets. William Appleman Williams, "The Frontier Thesis and American Foreign Policy," *Pacific Historical Review* 24 (November 1955): 379–95 at 381, 382. See also William Appleman Williams, "Brooks Adams and American Expansionism," *New England Quarterly* 25 (June 1952): 225–28. For critiques of overemphasis on trade expansion, see John A. Thompson, "William Appleman Williams and the 'American Empire,'" *Journal of American Studies* 7 (April 1973): 91–104, esp. 99, 101; William H. Becker, "American Manufacturers and Foreign Markets, 1870–1900: Business Historians and the 'New Economic Determinists,'" *Business History Review* 47 (Winter 1973): 466–81.
50. "Foreign Trade in Forest Products," *Forestry and Irrigation* 9 (September 1903): 461; "The Four Islands," *Irrigation Age* 15 (December 1899): 87–88. For the broader demand for tropical products, see Richard P. Tucker, *Insatiable Appetite: The United States*

and the Ecological Degradation of the Tropical World (Berkeley: University of California Press, 2000).

51. *Greater America: Address by Hon. David J. Hill, Ll. D., Assistant Secretary of State, Delivered at the Annual Banquet of the Rochester Chamber of Commerce, December 8, 1898* (Washington, DC: Judd and Detweiler, 1898); George Marvin, "The Greater America," *World's Work* 28 (May 1914): 22–30; George Clarke, *Greater America: Heroes, Battles, Camps,* Dewey *Islands, Cuba, Porto Rico* (Boston: Little, Brown, 1899); H. Addington Bruce, *The Romance of American Expansion Illustrated* (New York: Moffat, Yard, 1909); *Greater America: The Latest Acquired Insular Possessions* (Boston: Perry Mason, 1909); "A Plea for Greater America," *New York Times,* 9 December 1898, 2 (quote); Archibald R. Colquhoun, *Greater America* (New York: Harper and Brothers, 1904). For the impact, see, e.g., Alfred W. McCoy and Francisco A. Scarano, eds., *Colonial Crucible: Empire in the Making of the Modern American State* (Madison: University of Wisconsin Press, 2009).
52. Fritz Fischer, *Griff nach der Weltmacht: Die Kriegszielpolitik des kaiserlichen Deutschland* (1961), translated as *Germany's Aims in the First World War,* introduction by Hajo Holborn and James Joll (New York: W. W. Norton, 1967). For German colonial conservation, see, e.g., Thaddeus Sunseri, "Reinterpreting a Colonial Rebellion: Forestry and Social Control in German East Africa, 1874–1915," *Environmental History* 8 (July 2003): 430–51.
53. Even the British at times dabbled in discriminatory trade practices. Marion Johnson, "Cotton Imperialism in West Africa," *African Affairs* 73 (April 1974): 178–87.
54. Thompson, "William Appleman Williams and the 'American Empire,'" 99, 101; Becker, "American Manufacturers and Foreign Markets"; Marc-William Palen, "The Imperialism of Economic Nationalism, 1890–1913," *Diplomatic History,* 2014 (Advance Access published 7 February 2014, doi:10.1093/dh/dht135).
55. Early in the Taft administration, this distinction would become clear in forest policy. The United States sought to shape the Payne-Aldrich tariff of 1909 to protect American lumber interests where forest supplies were still adequate in the United States and amenable to sustained-yield conservation, but favored free access to *Canadian* lumber where US supplies were exhausted, as in wood pulp supplies of Canadian spruce. Gifford Pinchot was implicated in this compromise. Gifford Pinchot to Sereno E. Payne (chair, Ways and Means Committee, US Congress), 10 March 1909, fl. Pa, box 474, Pinchot Papers; "Forester Pinchot's Views on the Lumber Tariff as Proposed in the Payne Bill," *Ohio Architect and Builder* 13 (April 1909): 33–35 at 33. On conservation of forests and the tariff issue, see also H. V. Nelles, *The Politics of Development: Forests, Mines and Hydro-electric Power in Ontario, 1849–1941,* rev. ed. (1974; Montreal: McGill-Queen's University Press, 2005); Bernhard Fernow, "The Outlook of the Timber Supply in the United States," pt. 2, *Forestry and Irrigation* 9 (May 1903): 226–27; Roscoe R. Hess, "The Paper Industry in Its Relation to Conservation and the Tariff," *Quarterly Journal of Economics* 25 (August 1911): 650–81; "President Roosevelt and Mr. Bryce on the Conservation of America," *Outlook* 91 (9 January 1909): 50.
56. However, Russia had a roughly comparable land empire. Jane Burbank and Frederick Cooper, *Empires in World History: Power and the Politics of Difference* (Princeton, NJ: Princeton University Press, 2010), chap. 9.
57. Michael Adas, "Improving on the Civilizing Mission? Assumptions of United States Exceptionalism in the Colonization of the Philippines," in Lloyd Gardner and Marilyn B. Young, eds., *The New American Empire: A 21st Century Teach-in on U.S. Foreign Policy* (New York: New Press, 2005), 153–81; Glenn May, *Social Engineering in*

the Philippines: The Aims, Execution, and Impact of American Colonial Policy, 1900–1913 (Westport, CT: Greenwood, 1980), 179–81.

58. Daiva K. Stasiulis and Nira Yuval-Davis, eds., *Unsettling Settler Societies: Articulations of Gender, Race, Ethnicity and Class* (London: Sage, 1995); James Belich, *Replenishing the Earth: The Settler Revolution and the Rise of the Anglo World* (Oxford: Oxford University Press, 2009). There is a huge literature developing on settler societies, mostly on the colonial exclusion of indigenous people. On the theory of settler colonialism, see Lorenzo Veracini, *Settler Colonialism: A Theoretical Overview* (Basingstoke, Eng.: Palgrave Macmillan, 2010). On settler states that consolidate settler colonial rule in racially based nation-states, see Frederick E. Hoxie, "Retrieving the Red Continent: Settler Colonialism and the History of American Indians in the U.S.," *Ethnic and Racial Studies* 31 (September 2008): 1153–67. Since Roosevelt was of Dutch American descent, he regarded himself as "Teutonic" rather than narrowly "Anglo-Saxon" in descent, but this does not negate his affiliation with what was, in the late nineteenth century, described as the Anglo-Saxon "race" in the British Empire. Theodore Roosevelt, *African Game Trails: An Account of the African Wanderings of an American Hunter-Naturalist* (London: John Murray, 1910), 37; Thomas G. Dyer, *Theodore Roosevelt and the Idea of Race* (Baton Rouge: Louisiana State University Press, 1980), 2–3. For more on Roosevelt's conceptions of race, see chapters 9 and 10, below.
59. Kramer, "Power and Connection," 1361. This settler formulation was similar to and probably influenced by John R. Seeley's vision of Great Britain's white colonies. See Amanda Behm, "Empire Divided: Seeley's *Expansion of England* in Context," *Storia della storiografia*, no. 61 (2012): 59–74; Teodoro Tagliaferri, "Legitimizing Imperial Authority: Greater Britain and India in the Historical Vision of John R. Seeley," *Storia della storiografia*, no. 61 (2012): 75–91.
60. E.g., Cutright, *Theodore Roosevelt*.
61. For the (huge) historiography on Roosevelt and his presidency, see, e.g., the three-volume survey by Edmund Morris, esp. *Theodore Rex* (New York: Random House, 2001); Dalton, *Theodore Roosevelt*; John Morton Blum, *The Republican Roosevelt* (Cambridge, MA: Harvard University Press, 1954); Serge Ricard, ed., *A Companion to Theodore Roosevelt* (Malden, MA: Wiley-Blackwell, 2012). Roosevelt's concept of relations with the wider world is surveyed through a European lens in Hans Krabbendam and John M. Thompson, eds., *America's Transatlantic Turn: Theodore Roosevelt and the "Discovery" of Europe* (New York: Palgrave Macmillan, 2012). Though many works have addressed the themes of Roosevelt's world politics, none captures Roosevelt's geopolitical context better than Howard K. Beale, *Theodore Roosevelt and the Rise of America to World Power* (Baltimore, MD: Johns Hopkins University Press, 1956).
62. Arthur F. Beringause, *Brooks Adams: A Biography* (New York: Alfred A. Knopf, 1955), 235; Brooks Adams, *America's Economic Supremacy* (New York: Macmillan, 1900); Brooks Adams, *The New Empire* (New York: Macmillan, 1902).
63. Mark Elvin, *The Retreat of the Elephants: An Environmental History of China* (New Haven, CT: Yale University Press, 2004); Theodore Roosevelt, "Eighth Annual Message," 8 December 1908, in Peters and Woolley, American Presidency Project, http://www.presidency.ucsb.edu/ws/?pid=29549 (accessed 11 December 2010); Bailey Willis, *Friendly China* (Stanford, CA: Stanford University Press, 1949), 162–63.
64. Among standard general works, Blum, *The Republican Roosevelt*, 111–12, best recognizes this point.
65. Cf. Dalton, *Theodore Roosevelt*, 244.

NOTES TO CHAPTER TWO

1. Stated the editor of *Conservation*, Frank G. Heaton: "The cult of conservation is by no means wholly American; it is altogether cosmopolitan." "Conservation, a World Movement," *Conservation* 14 (September 1908): 496–98 at 498. For the theory of transnational history used here, see Ian Tyrrell, "In the Shadow of the Nation? Space and Time in the Practice and Problems of U.S. Transnational History," in Udo Hebel, ed., *Transnational American Studies*, American Studies Monograph Series, 222 (Heidelberg: Winter, 2012), 75–96.
2. Theodore Roosevelt, *History as Literature* (New York: Charles Scribner's Sons, 1913), 100–104; "Mr. Roosevelt's Lecture," *New York Times*, 13 May 1910.
3. "The International Navigation Congress," *Engineering News*, 31 August 1911, 272.
4. *Proceedings of a Conference of Governors in the White House, Washington, D.C., May 13–15, 1908* (Washington, DC: Government Printing Office, 1908), 56.
5. E.g., see Henry J. van der Windt, "Biologists Bridging Science and the Conservation Movement: The Rise of Nature Conservation and Nature Management in the Netherlands, 1850–1950," *Environment and History* 18 (May 2012): 209–26.
6. "International Ornithological Congress [London]," *Forest and Stream* 64 (1 April 1905): 253; "The Protection of Birds," *Times* (London), 22 May 1900, 3; "The Protection of American Game," *Scientific American* 83 (11 August 1900): 82 (first quote); Sherman Strong Hayden, *The International Protection of Wild Life* (New York: Columbia University Press, 1942), 38; *Troisième congrès ornithologique international, Paris, 26–30 juin 1900: . . . IIIe* Congrès ornithologique international, *Compte rendu des séances* (Paris: Masson, 1901), 34 ("protection des oiseaux"), 49 ("Acclimatation"), 35–41, 85–86, 344–45.
7. See, e.g., William Dutcher to Countess Stephanie von Wedel, 8 December 1902; Dutcher to William G. Scott, City Treasurer's Office, Winnipeg, 19 June 1902, box A-10, National Audubon Society Records, NYPL.
8. James Wilson, "Importation of Birds and Animals," *Forest and Stream* 55 (21 July 1900): 48; "The Lacey Game Bill," ibid. 51 (24 December 1898): 509.
9. Hayden, *International Protection of Wild Life*, 36–37; Bernhard Gissibl, "German Colonialism and the Beginnings of International Wildlife Preservation in Africa," *GHI Bulletin Supplement* 3 (2006): 132; John MacKenzie, *The Empire of Nature: Hunting, Conservation and British Imperialism* (Manchester: Manchester University Press, 1989), 81, 202, 321; David K. Prendergast and William M. Adams, "Colonial Wildlife Conservation and the Origins of the Society for the Preservation of the Wild Fauna of the Empire (1903–1914)," *Oryx* 37, no. 2 (2003): 251–60.
10. *Journal of the Society for the Preservation of the Wild Fauna of the Empire* 3 (1907): 6; ibid. 2 (1905): 7; William T. Hornaday, "Discovery of a Big-Game Paradise," ibid. 4 (1908): 56–60; Theodore Roosevelt, "Extract from Message from the Hon. Theodore Roosevelt, President of the United States," ibid. 4 (1908): 8 (quote); Edward North Buxton to Theodore Roosevelt, 19 December 1902, and Roosevelt to Buxton, 8 December 1902, Theodore Roosevelt Papers, LC, Theodore Roosevelt Digital Library, Dickinson State University, http://www.theodorerooseveltcenter.org/Research/Digital-Library/Record.aspx?libID=0183699.
11. Thomas R. Dunlap, "Sport Hunting and Conservation, 1880–1920," *Environmental Review* 12 (Spring 1988): 51–60; cf. John F. Reiger, *American Sportsmen and the Origins of Conservation*, 3rd ed. (Corvallis: Oregon State University Press, 2001), chap. 7.

12. Hayden, *International Protection of Wild Life*, 38, 44. However, crocodile eggs were protected, and female and young elephants, rather than adult males, were given protection to preserve game stock.
13. MacKenzie, *Empire of Nature*; Karl Jacoby, *Crimes against Nature: Poachers, Squatters, Thieves, and the Hidden History of American Conservation* (Berkeley: University of California Press, 2001); Louis Warren, *The Hunter's Game: Poachers and Conservationists in Twentieth-Century America* (New Haven, CT: Yale University Press, 1997); Mark David Spence, *Dispossessing the Wilderness: Indian Removal and the Making of the National Parks* (New York: Oxford University Press, 1999).
14. Bernhard Gissibl, Sabine Höhler, and Patrick Kupper, eds., *Civilizing Nature: National Parks in Global Historical Perspective* (New York: Berghahn Books, 2012); Gissibl, "German Colonialism and the Beginnings of International Wildlife Preservation in Africa," 132; Paul Jepson and Robert J. Whittaker, "Histories of Protected Areas: Internationalisation of Conservationist Values and Their Adoption in the Netherlands Indies (Indonesia)," *Environment and History* 8 (2002): 129–72 at 135–38; Tom Mels, *Wild Landscapes: The Cultural Nature of Swedish National Parks* (Lund: Lund University Press, 1999), 81; E. J. Hart, *J. B. Harkin: Father of Canada's National Parks* (Calgary: University of Alberta Press, 2009).
15. George Kunz, "Foreign Regulations for the Conservation of Scenic and Historic Places and Objects," in *Sixteenth Annual Report of the American Scenic and Historic Preservation Society* (New York: J. B. Lyon, 1911), 555, hereafter ASHPS, *Sixteenth Annual Report, 1911* (all ASHPS reports are uniformly given this shortened citation except for 1895–1900, where the first reference identifies the specific title used at that time, followed by the same shortened citation); ASHPS, *Nineteenth Annual Report, 1914* (New York: J. B. Lyon, 1914), 316.
16. Caroline Ford, "Nature, Culture and Conservation in France," *Past and Present*, no. 183 (May 2004): 173–98.
17. Hugo Conwentz, "On National and International Protection of Nature," *Journal of Ecology* 2 (June 1914): 109–24 (quote at 122). See also Hugo Conwentz, *The Care of Natural Monuments with Special Reference to Great Britain and Germany* (Cambridge: Cambridge University Press, 1909), 174, 178.
18. Minutes of Trustees' Meeting, 25 May 1908, 108–12, box 10, ASHPS Records, NYPL; Ian Tyrrell, "America's National Parks: The Transnational Creation of National Space in the Progressive Era," *Journal of American Studies* 46 (January 2012): 1–25, 45–49.
19. "Ruins of the Southwest Protected," *Forest and Stream* 67 (8 September 1906): 373 (quote); Charles B. Hosmer Jr., *Presence of the Past: The History of the Preservation Movement in the United States before Williamsburg* (New York: G. P. Putnam's Sons, 1965); James M. Lindgren, *Preserving the Old Dominion: Historic Preservation and Virginia Traditionalism* (Charlottesville: University of Virginia Press, 1993), 116; James M. Lindgren, *Preserving Historic New England: Preservation, Progressivism, and the Remaking of Memory* (New York: Oxford University Press, 1995).
20. Hal Rothman, *Preserving Different Pasts: The American National Monuments*, chap. 3, http://www.cr.nps.gov/history/online_books/rothman/chap3.htm; "Ruins of the Southwest Protected," 373; L. H. Pammel, "Major John F. Lacey and the Conservation of Our Natural Resources," in L. H. Pammel, comp., *Major John F. Lacey: Memorial Volume* (Cedar Rapids: Iowa Park and Forestry Association, 1915), 36; John F. Lacey, "The Pajarito: An Outing with the Archaeologists," in *Lacey: Memorial*, 211–12; John F. Lacey, "Cliff Dwellers' National Park," in *Lacey: Memorial*, 222.

21. The Antiquities Act was to be important in enabling Roosevelt to bypass Congress on key nature conservation issues. See below, chapter 8.
22. Alexander Hume Ford, "Engineering Opportunities in the Russian Empire," *Engineering Magazine* 21 (April 1901): 29–42; Thomas Curtis Clarke, "European and American Bridge-Building Practices," ibid., 43–58; A. G. Charleton, "Gold Mining and Milling in Western Australia" (on "water problems and their solution"), ibid., 89–104; M. I. Pupin, "Transatlantic Communication by Means of the Telephone," ibid., 105–14. See also "The Development of the Submarine Cable," *Engineering News* 41 (6 April 1899): 220–21. On engineers' transnational activity, see Carroll Purcell, "Herbert Hoover and the Transnational Lives of Engineers," in Marilyn Lake, Penny Russell, and Desley Deacon, eds., *Transnational Lives: Biographies of Global Modernity, 1700–Present* (Basingstoke, Eng.: Palgrave Macmillan, 2010), 109–20; Bruce Sinclair, "The Power of Ceremony: Creating an International Engineering Community," *History of Technology* 21 (1999): 203–11 at 205 ("technical accomplishment"); Jessica Teisch, "'Home Is Not So Very Far Away': Californian Engineers in South Africa, 1868–1915," *Australian Economic History Review* 45 (July 2005): 139–60.
23. US wages had been low in the late nineteenth century due to bounteous unskilled migration, but there are indications that they were rising in the first decade of the twentieth century, narrowly overtaking those in Britain in 1906 for low-paid workers and by a much greater margin for highly paid, skilled workers. Peter Shergold, "Reefs of Roast Beef: The American Worker's Standard of Living in Comparative Perspective," in Dirk Hoerder, ed., *American Labor and Immigration History, 1877–1920s: Recent European Research* (Urbana: University of Illinois Press, 1983), 97; M. L. Holman, "The Conservation Idea as Applied to the American Society of Mechanical Engineers," *Journal of the American Society of Mechanical Engineers* 31 (January 1909): 40 (quote). Scientific management was not a factor in the calculations of Roosevelt or Pinchot. That system came into vogue in the decade after the Eastern Rate Case of 1910–11, which concerned rail freight. Samuel Haber, *Efficiency and Uplift: Scientific Management and the Progressive Era, 1890–1920* (Chicago: University of Chicago Press, 1964), 51–66.
24. Purcell, "Herbert Hoover"; Jessica Teisch, *Engineering Nature: Water, Development, and the Global Spread of American Environmental Expertise* (Chapel Hill: University of North Carolina Press, 2011), 84–86 (on Mead), 183–84 (on Hoover); John Hays Hammond, *The Autobiography of John Hays Hammond*, 2 vols. in one (1935; repr., New York: Arno Press, 1974); Elbridge Baldwin (associate editor of the *Outlook*) to Theodore Roosevelt, 4 June 1908, series 1, reel 83, Roosevelt Papers; "Political Notes: Unique," *Time*, 10 May 1926; *New York Times*, 18 November 1911, 1; Ian Tyrrell, *True Gardens of the Gods: Californian-Australian Environmental Reform, 1860–1930* (Berkeley: University of California Press, 1999), chap. 8.
25. "China's Coal Supply," *Wall Street Journal*, 13 May 1908, 2; Bailey Willis, "Mineral Resources of China, Part I," *Economic Geology* 3 (January–February 1908): 1–36; Eliot Blackwelder, "Bailey Willis: 1857–1949; A Biographical Memoir," *Biographical Memoirs* (National Academy of Sciences) 35 (1961): 333–50; "Deforestation in China" [1905], fl. 2, box 3, Bailey W. Willis Papers, Huntington Library, San Marino, CA.
26. Bailey Willis to Raphael Zon, 30 January 1907, and Zon to Willis, 26 January 1907, fl. 28, box 20, Willis Papers; *Northern Patagonia: Character and Resources*, text and maps by the Comisión de estudios hidrológicos, Bailey Willis, director, 1911–14 (New York: Scribner Press, 1914), vi; Bailey Willis, *Friendly China: Two Thousand Miles*

Afoot among the Chinese (Stanford, CA: Stanford University Press, [1949]); "Some Conditions in China," *Outlook* 87 (30 November 1907): 737–41.

27. Raphael Pumpelly et al., *Explorations in Turkestan: With an Account of the Basin of Eastern Persia and Sistan* (Washington, DC: Carnegie Institution, 1905); Ellsworth Huntington, *The Pulse of Asia: A Journey in Central Asia . . .* (Boston: Houghton, Mifflin, 1907); Ellsworth Huntington, "Climatic Change and Agricultural Exhaustion as Elements in the Fall of Rome," *Quarterly Journal of Economics* 31 (February 1917): 173–208; "Proposed Work of Carnegie Institution," *New York Times*, 30 December 1903, 6.
28. *Engineering News* 59 (21 May 1908): 569–70 at 569.
29. "The Coming Exhaustion of Nature's Resources," *Engineering News* 45 (31 January 1901): 82 (first quote), 81; Haber, *Efficiency and Uplift*, 14 ("larger").
30. Joseph Holmes to (Mrs.) Jane Sprunt Holmes, 30 August 1905, fl. 12, 1904–7, box 1, Joseph A. Holmes Papers, SHC; Holmes to Holmes, 4 March 1906, ibid.
31. "News and Notes," *Forestry and Irrigation* 8 (November 1902): 440–44; Biographical Information, Thomas H. Means Papers, Water Resource University of California, Bancroft Library, http://www.oac.cdlib.org/findaid/ark:/13030/tf5489n85p/admin/#bioghist-1.8.4 (quote); Charles C. Colley, "The Desert Shall Blossom: North African Influence on the American Southwest," *Western Historical Quarterly* 14 (July 1983): 277–90; Charles C. Colley, *The Century of Robert H. Forbes: The Career of a Pioneer Agriculturist, Agronomist, Environmentalist, Conservationist and Water Specialist in Arizona and Abroad* (Tucson: Arizona Historical Society, 1977), 10–20.
32. Pinchot's official title from 1905 was chief of the Forest Service and before that chief of the Division of Forestry (1898) and then chief of the Bureau of Forestry from 1901. I use "chief forester" to avoid clumsiness.
33. Char Miller, *Gifford Pinchot and the Making of Modern Environmentalism* (Washington, DC: Shearwater Press, 2001), 88–90; Gregory A. Barton, *Empire Forestry and the Origins of Environmentalism* (New York: Cambridge University Press, 2002), esp. 138–40, 143.
34. Brian Balogh, "Scientific Forestry and the Roots of the Modern American State," *Environmental History* 7 (April 2002): 199 (quote); Miller, *Gifford Pinchot*.
35. Gifford Pinchot to Theodore Roosevelt, 22 September 1905, fl. Theodore Roosevelt, box 98, Gifford Pinchot Papers, LC (quote); Charles H. Sherrill to Gifford Pinchot, 20 October 1905, and Pinchot to Sherrill, 23 October 1905, fl. "sc-sh," box 98, Pinchot Papers.
36. Wilcomb E. Washburn, *The Cosmos Club of Washington: A Centennial History, 1878–1978* (Washington, DC: Cosmos Club, 1978); Wallace Stegner, *Beyond the Hundredth Meridian: John Wesley Powell and the Second Opening of the West* (Boston: Houghton Mifflin, 1954), 242; Donald Worster, *A River Running West: The Life of John Wesley Powell* (New York: Oxford University Press, 2001), 437–40; diary entry, 20 January 1909, vol. 26, 1909, box 8, James R. Garfield Papers, LC ("jolly meeting"); Joseph Holmes to Jane Sprunt Holmes, 5 March 1906, 1 April 1906 ("splendid dinner"), fl. 12, J. A. Holmes to wife [*sic*], 1904–7, box 1, Holmes Papers; Polly Welts Kaufman, *National Parks and the Woman's Voice: A History* (Albuquerque: University of New Mexico Press, 2006), 31, 246n14.
37. Bruce J. Schulman, "Governing Nature, Nurturing Government: Resource Management and the Development of the American State, 1900–1912," *Journal of Policy History* 17, no. 4 (2005): 375–403 at 375.

38. For different European cases, all of which show interaction between amateurs and experts, though with varying trajectories, see Raf De Bont and Rajesh Heynickx, "Landscapes of Nostalgia: Life Scientists and Literary Intellectuals Protecting Belgium's 'Wilderness,' 1900–1940," *Environment and History* 18 (May 2012): 237–60 at 260; van der Windt, "Biologists Bridging Science."
39. Richard Hofstadter, *The Age of Reform: From Bryan to F.D.R.* (New York: Alfred A. Knopf, 1955), 186.
40. Rudolf Cronau, "A Continent Despoiled," *McClure's Magazine* 32 (April 1909): 639–64. On *McClure's*, see Peter Lyon, *Success Story: The Life and Times of S. S. McClure* (New York: Scribner, 1963), 159. In 1898, its circulation passed four hundred thousand. See also George Juergens, *News from the White House: The Presidential-Press Relationship in the Progressive Era* (Chicago: University of Chicago Press, 1981).
41. Theodore Roosevelt, "The Man with the Muck Rake," 15 April 1906, http://www.theodore-roosevelt.com/trspeeches.html (accessed 20 August 2011).
42. Edward Scripps to Gifford Pinchot, 27 September 1910, in "A New Platform" and "My Visit at San Clemente" (quote); "Disquisition by Scripps, E. W., The Growth of National Wealth, May 20, 1908" (discussing "laying up wealth for future generations"); Scripps to John F. Forward, 23 November 1908; "Disquisition by Scripps, E. W., Mr. Pinchot's Visit; A Scheme of Foresting and a Means of Financing It, June 16, 1908"; all in Ohio Digital Libraries Collection, http://dmc.ohiolink.edu/cgi/i/image/image-idx?page=index;c=scripps (accessed 23 July 2012).
43. "Edward W. Scripps," Ohio History Central, 1 July 2005, http://www.ohiohistorycentral.org/entry.php?rec=334 (quote) (accessed 23 July 2012); Vance H. Trimble, *The Astonishing Mr. Scripps: The Turbulent Life of America's Penny Press Lord* (Ames: Iowa State University Press, 1992), 315–16.
44. See, e.g., "Statements by Men Prominent in National Affairs upon the Subject of Conservation," *Chautauquan* 55 (June 1909): 115–21; Andrew Rieser, *The Chautauqua Moment: Protestants, Progressives, and the Culture of Modern Liberalism* (New York: Columbia University Press, 2003), 133 (Roosevelt quote); French Strother, "The Great American Forum: Chautauqua and the Chautauquas in Summer and the Lyceum in Winter," *World's Work* 24 (September 1912): 551–64.
45. W. E. B. Du Bois, *Autobiography* (New York: International Publishers, 1968), 249.
46. Lyman Abbott, *America in the Making* (New Haven, CT: Yale University Press, 1911), 110–14.
47. Obituary for Howland, *New York Times*, 28 February 1917, 10.
48. New York State Assembly, no. 35, *Thirty-First Annual Report of the Commissioners of the State Reservation at Niagara from October 1, 1913, to September 30, 1914* (Albany, NY: J. B. Lyon, 1915), 10, 12, 14, 15; Hamilton Holt, "William Bailey Howland," 12 March 1917, 433–34, and J. Horace McFarland to *Independent*, 2 March 1917, fl. Howland Memorial, box 4, American Civic Association General Correspondence, Horace McFarland Papers, PSA.
49. "The Marvel of Color-Photography," *World's Work* 18 (May 1909): 11536.
50. "The Calling of an International Conservation Congress," *World's Work* 17 (April 1909): 11416.
51. "The Problems of Country Life," *World's Work* 17 (February 1909): 11195–96; Frank N. Doubleday, "The Autobiography of 'Country Life in America,'" *Country Life in America*, Anniversary Issue, 21 (15 April 1912): 21–22; "Beautiful America: Conducted by J. Horace McFarland, President of the American League for Civic Improvement," *Ladies' Home Journal* 29 (March 1904): 15.

52. Edmund Morris, *Theodore Rex* (New York: Random House, 2001), 44–45, 123–24; Kathleen Dalton, *Theodore Roosevelt: A Strenuous Life* (New York: Knopf, 2002), 212; John M. Thompson, "Theodore Roosevelt and the Press," in Serge Ricard, ed., *A Companion to Theodore Roosevelt* (Madden, MA: Wiley-Blackwell, 2011), 223.
53. Dalton, *Theodore Roosevelt*, 211–13; George Kibbe Turner, quoted in Stephen Ponder, "News Management in the Progressive Era, 1898–1909: Gifford Pinchot, Theodore Roosevelt and the Conservation Crusade" (PhD diss., University of Washington, 1985), 4; Stephen Ponder, "Gifford Pinchot, Press Agent for Forestry," *Journal of Forest History* 31 (January 1987): 26–35.
54. George L. Knapp, "The Other Side of Conservation," *North American Review* 191 (April 1910): 465–481 at 465.
55. *Theodore Roosevelt: An Autobiography*, new ed. (1913; New York: Charles Scribner's Sons, 1922), 401 (quote); Miller, *Gifford Pinchot*, 305; Ponder, "News Management," 4–5.
56. *Forestry Pamphlets, History*, vol. 1, *U.S.F.S. Field Programs, 1904–1909* (n.p., n.d.), gives detailed personnel and designations. See, e.g., "Field Programme for April, 1906," 1; Harold K. Steen, *The U.S. Forest Service: A History* (Seattle: University of Washington Press, 1977), 83–86.
57. See, e.g., J. Pease Norton to Overton Price, 15 October 1908, file H–M, box 4, entry 24, Records Relating to the National Conservation Commission, 1908–9 Correspondence and Misc. A–Z, RG 95, NARA.
58. William Beinart and Peter Coates, *Environment and History* (London: Routledge, 1995), 45.
59. James P. Hull, "Fernow, Bernhard Eduard," in *Dictionary of Canadian Biography Online*, http://www.biographi.ca/009004-119.01-e.php?&id_nbr=8131&terms=death (quote) (accessed 9 December 2012).
60. Brett M. Bennett, "A Network Approach to the Origins of Forestry Education in India, 1855–1885," in Brett M. Bennett and Joseph Hodge, eds., *Science and Empire: Knowledge and Networks of Science across the British Empire, 1800–1970* (Basingstoke, Eng.: Palgrave Macmillan, 2011), 68–88 at 82.
61. Barton, *Empire Forestry*, esp. 138–40, 143. See also Miller, *Gifford Pinchot*, 81; Indra Munshi Saldanha, "Colonialism and Professionalism: A German Forester in India," *Economic and Political Weekly* 31 (25 May 1996): 1265–1273 at 1272; "Trip to the Philippines," typescript, dated 6 December 1941, insert 2a ("I had learned more than a little about forestry in Burma and in British India from Dr. Brandis and his successors. . . . For that reason I was better prepared than it seemed"), Book File, Breaking New Ground, Philippine Islands, box 1010, Pinchot Papers. See also Gifford Pinchot, "Government Forestry Abroad," *Publications of the American Economic Association* 6 (May 1891): 7–54, esp. 45.
62. For the Vienna congress of 1890, see *Internationaler land- und forstwirtschaftlicher Congress, Wien, 1890: Bericht* (Vienna: Verlag der k.k. Landwirthschaft-Gesellschaft, 1890).
63. See, e.g., Paul F. Kerr, "Memorial of George Frederick Kunz," *American Mineralogist* 18 (March 1933): 91–94 at 93. There is a large literature on international exhibitions. See Emily S. Rosenberg, "Transnational Currents in a Shrinking Word," in Emily S. Rosenberg, ed., *A World Connecting, 1870–1945* (Cambridge, MA: Belknap Press of Harvard University Press, 2012), 887–99; Robert W. Rydell, *All the World's a Fair: Visions of Empire at American International Expositions, 1876–1916* (Chicago: University of Chicago Press, 1987); Robert W. Rydell, John E. Findling, and Kimberly

D. Pelle, *Fair America: World's Fairs in the United States* (Washington, DC: Smithsonian Institution Press, 2000).

64. "For an International Congress," *Forester* 5 (December 1899): 288 (quote). The American Association for the Advancement of Science and the National Geographic Society also approved the plan. For German colonial forests in Africa, see Thaddeus Sunseri, "Reinterpreting a Colonial Rebellion: Forestry and Social Control in German East Africa, 1874–1915," *Environmental History* 8 (July 2003): 430–51.
65. E. Louise Peffer, *The Closing of the Public Domain: Disposal and Reservation Policies, 1900–50* (Stanford, CA: Stanford University Press, 1951), chap. 1.
66. W. Schlich, "The Outlook of the World's Timber Supply," *Transactions of the Royal Scottish Arboricultural Society* 16, pt. 3 (1901): 355–83 at 355.
67. See *Congrès international de sylviculture, Paris, 1900; Tenu à Paris du 4 août, 7 juin 1900 sous la présidence de M. Daubrée . . . compte rendu détaillé* (Exposition universelle internationale de 1900) (Paris: Imprimerie nationale, 1900), 30–45, 70–71; Jean-Yves Puyo, "Sur le mythe colonial de l'inépuisabilité des ressources forestières (Afrique occidentale française / Afrique équatoriale française, 1900–1940)," *Cahiers de géographie du Québec* 45, no. 126 (2001): 479–96 ("si les intervenants jugent exagérés les propos d'Alphonse Mélard, on n'écarte pas la menace d'une pénurie de bois 'à un moment donné' ").
68. For an acceptance of Mélard's warnings, see J. S. Gamble, "The International Congress of Sylviculture," *Transactions of the Royal Scottish Arboricultural Society* 16, pt. 2 (1900): 262–73, esp. 262–64. Schlich ("Outlook of the World's Timber Supply") drew on Mélard for the non–British Empire sources: "This pamphlet is drawn up on lines similar to those of my lecture given at the Imperial Institute in 1897, but Monsieur Mélard brings in a number of non-European countries with which I had not dealt, as not directly affecting the British Empire" (356). See also "Insufficiency of the World's Timber Supply," *Transactions of the Royal Scottish Arboricultural Society* 16, pt. 3 (1900): 384–87 (translation of A. Mélard, "L'insuffisance de la production des bois d'oeuvre dans le Monde," *Indian Forester*, September 1901); Val Shaw, "The Rape of the Redwood," *Overland Monthly and Out West Magazine* 39 (March 1902): 12.
69. *NCC Report*, 2:327, 333, 339, 340, 369. However, Raphael Zon did not necessarily accept all of the Frenchman's judgments (ibid., 339). See also Max Endres, *Handbuch der Forstpolitik mit besonderer Berücksichtigung der Gesetzgebung und Statistik* (Berlin: Julius Springer, 1905).
70. News Items, *Forester* 5 (March 1899): 50; "To Lumber in the Philippines," ibid. (June 1899), 148; "Preservation of Philippine Forests," *Washington Times*, reprinted in *Forester* 5 (August 1899): 190.

NOTES TO CHAPTER THREE

1. The impact of ideas is difficult to measure because of the problem of reception. For a summary, see John Conner, *Studying Media: Problems of Theory and Method* (Edinburgh: Edinburgh University Press, 1998), 21, 109. A sensible approach mediating between the power of ideas and political practice is John L. Campbell, "Institutional Analysis and the Role of Ideas in Political Economy," *Theory and Society* 27 (June 1996): 377–409 at 398, 400 (quote).
2. See *American Monthly Review of Reviews* 18–21 (1898–1900); e.g., Sylvester Baxter, "Java as an Example: How the Dutch Manage Tropical Islands," ibid. 19 (February 1899): 179–84; Samuel W. Belford, "Material Problems in the Philippine Islands," ibid. 19 (April 1899): 464–57; John S. Barrett, "The Value of the Philippines,"

ibid. 20 (August 1899): 205, regarding the "vast quantities" of "primeval forest" and "the undoubtedly extensive mineral wealth in the islands." See also fl. America in the Philippines and Far East, box 107, John Barrett Papers, LC; and chapter 4, below.

3. Benjamin Kidd, "The United States and the Control of the Tropics," *Atlantic Monthly* 82 (December 1898): 721–26; W. Alleyne Ireland, "European Experience with Tropical Colonies," ibid., 729–35; Carl Evans Boyd, "Our Government of Newly Acquired Territory," ibid., 735–41.
4. Alleyne Ireland, *Tropical Colonization: An Introduction to the Study of the Subject* (London: Macmillan, 1899), v; Alleyne Ireland, *The Far Eastern Tropics: Studies in the Administration of Tropical Dependencies: Hong Kong, British North Borneo, Sarawak, Burma, the Federated Malay States, the Straits Settlements, French Indo-China, Java, the Philippine Islands, . . .* (Boston: Houghton, Mifflin, 1905); Alleyne Ireland, *The Province of Burma: A Report Prepared on Behalf of the University of Chicago, . . .* (Boston: Houghton, Mifflin, 1907).
5. Benjamin Kidd, *The Control of the Tropics* (New York: Macmillan, 1898), 51.
6. D. P. Crook, *Benjamin Kidd: Portrait of a Social Darwinist* (Cambridge: Cambridge University Press, 1984), 125.
7. *Washington Times*, 20 February 1909, 2 (quote). Frank Vrooman appears to have virtually plagiarized the *Washington Times* on this point in his lecture at Oxford University delivered on 8 March 1909, seventeen days after the *Washington Times* piece. Frank Buffington Vrooman, *Theodore Roosevelt, Dynamic Geographer . . .* (London: Henry Frowde, Oxford University Press, 1909), 32.
8. Crook, *Benjamin Kidd*, 74–75. Roosevelt favored a Lamarckian form of evolutionary theory in which acquired characteristics, stimulated by environment, could be inherited. See Thomas Dyer, *Theodore Roosevelt and the Idea of Race* (Baton Rouge: Louisiana State University Press, 1980), 44–45; and below, chapter 9.
9. Kidd, *Control of the Tropics*, 46.
10. Michael Adas, *Dominance by Design: Technological Imperatives and America's Civilizing Mission* (Cambridge, MA: Harvard University Press, 2006), 153–54.
11. Bruce Stronach, *Beyond the Rising Sun: Nationalism in Contemporary Japan* (Westport, CT: Greenwood Press, 1995), 7.
12. William Elliot Griffis, "America in the Far East: II. The Anglo-Saxons in the Tropics," *Outlook* 60 (10 December 1898): 902–5 at 905.
13. William Elliot Griffis, *America in the East: A Glance at Our History, Prospects, Problems, and Duties in the Pacific Ocean*, 2nd ed. (1899; New York: A. S. Barnes, 1900), 51.
14. Ibid., 23.
15. Ibid.
16. Adas, *Dominance by Design*, 153–54.
17. Griffis, "America in the Far East," 905.
18. Griffis, *America in the East*, 208.
19. Ibid., 64 ("semi-slumber"), 213; William Martin, *A Cycle of Cathay or China, South and North with Personal Reminiscences . . .* , 3rd ed. (New York: Fleming H. Revell, 1900), 49 (final quote).
20. Griffis, "America in the Far East," 905.
21. Ibid., 902.
22. Ibid., 903 (quote), 904.
23. *American Monthly Review of Reviews* 18 (July 1898): 225 (quote); "Dr. Griffis, Friend of Japan, Dies," *New York Times*, 6 February 1928, 3; *Washington Post*, 8 September 1908, 3; "National Institute of Arts and Letters," *New York Times*, 24 January 1909, SM1.

24. Adas, *Dominance by Design*, 153–54. The William Elliot Griffis Papers lack evidence of direct contact or reciprocal influence. See Griffis Papers, in Japan through Western Eyes: Manuscript Records of Traders, Travellers, Missionaries and Diplomats, 1853–1941 (microfilm ed.), "Roosevelt and Japan," newspaper clipping, 1924, item 11, reel 73, box 50, pt. 5; ibid., fl. 12, reel 66, box 83, pt. 4; Theodore Roosevelt, *Japan's Part* (New York: Japan Society, 1918).
25. John Morton Blum, *The Republican Roosevelt*, 2nd ed. (Cambridge, MA: Harvard University Press, 1977), 33, 34 (quote).
26. Howard K. Beale, *Theodore Roosevelt and the Rise of America to World Power* (Baltimore, MD: Johns Hopkins University Press, 1956), 256–57.
27. Brooks Adams to Theodore Roosevelt, 17 July 1903; Roosevelt to Brooks Adams, 18 July 1903, Theodore Roosevelt Collection, MS Am 1971 (2), Houghton Library, Harvard University, in Theodore Roosevelt Digital Library, Dickinson State University, http://www.theodorerooseveltcenter.org/Research/Digital-Library/Record.aspx?libID=041333.
28. Worthington C. Ford, "Brooks Adams," in American Council of Learned Societies, *Dictionary of American Biography Base Set* (New York: Scribner, 1928–36), reproduced by History Resource Center, Farmington Hills, MI, http://www.faculty.fairfield.edu/faculty/hodgson/Courses/progress/BrooksAdamsBio.html (accessed 12 July 2012).
29. Brooks Adams, *The New Empire* (New York: Macmillan, 1902), vi; Brooks Adams, *The Law of Civilization and Decay: An Essay on History*, 2nd ed. (New York: Macmillan, 1896).
30. Quoted in Lawrence Janofsky, "The Brothers Adams," *Kenyon Review* 18 (Spring 1956): 303–9 at 307; David H. Burton, "Theodore Roosevelt and His English Correspondents: A Special Relationship of Friends," *Transactions of the American Philosophical Society*, n.s., 63, no. 2 (1973): 1–70 at 30.
31. Ford, "Adams" (quote); Timothy Donovan, *Henry Adams and Brooks Adams: The Education of Two American Historians* (Norman: University of Oklahoma Press, 1961), 151.
32. T. J. Jackson Lears, *No Place of Grace: Antimodernism and the Transformation of American Culture, 1880–1920* (Chicago: University of Chicago Press, 1981), 134.
33. Walter LaFeber, *The New Empire: An Interpretation of American Expansion* (Ithaca, NY: Cornell University Press, 1963), 85.
34. Beale, *Theodore Roosevelt and the Rise of America to World Power*, 257–59.
35. "Problems of Expansion," *Outlook* 67 (26 January 1901): 227–30 at 227. "[M]ost satisfactory" to Brooks was the review in the *Outlook*, "with which magazine, significantly," Roosevelt "had close relations." Arthur F. Beringause, *Brooks Adams: A Biography* (New York: Alfred A. Knopf, 1955), 240, 241. See also James T. Shotwell, *Political Science Quarterly* 18 (December 1903): 688–93.
36. "New Industrial Revolution," *Chicago Daily Tribune*, 5 February 1901, 6. See also *Independent* 53 (7 March 1901): 566.
37. Brooks Adams, "The Spanish War and the Equilibrium of the World," *Forum* 25 (August 1898): 641–51, reprinted in Brooks Adams, *America's Economic Supremacy* (New York: Macmillan, 1900), 23 (quote). Also reprinted in *America's Economic Supremacy* were Adams's "The New Struggle for Life among Nations," *McClure's Magazine* 12 (April 1899): 558–63; and "Russia's Interest in China," *Atlantic Monthly* 86 (September 1900): 309–17.
38. Adams, *America's Economic Supremacy*, 26.
39. Ibid., 47.

40. Charles Hirschfeld, "Brooks Adams and American Nationalism," *American Historical Review* 69 (January 1964): 371–92 at 387.
41. Beringause, *Brooks Adams*, 233.
42. Brooks Adams, "The Heritage of Henry Adams," in Henry Adams, *The Education of Democratic Dogma* (New York: Macmillan, 1919), 78.
43. Adams, *New Empire*, 211.
44. Beringause, *Brooks Adams*, 238; Beale, *Theodore Roosevelt and the Rise of America to World Power*, 256–58; Burton, "Theodore Roosevelt," 30.
45. Cf. LaFeber, *New Empire*; William Appleman Williams, "The Frontier Thesis and American Foreign Policy," *Pacific Historical Review* 24 (November 1955): 379–95 at 381, 382; William Appleman Williams, "Brooks Adams and American Expansionism," *New England Quarterly* 25 (June 1952): 225–28.
46. Bailey W. Willis, "Water Circulation and Its Control," *NCC Report*, 2:688.
47. Chase S. Osborn, *The Iron Hunter*, introduction by Robert M. Warner (Detroit, MI: Wayne State University Press, 2002), 10.
48. *Proceedings of a Conference of Governors in the White House, Washington, D.C., May 13–15, 1908* (Washington, DC: Government Printing Office, 1908), 367.
49. Ibid., 338.
50. Ibid., 26.
51. Ross Evans Paulson, *Radicalism and Reform: The Vrooman Family and American Social Thought, 1837–1937* (Lexington: University of Kentucky Press, 1964), 40–41.
52. Ibid., 100.
53. "Quit Preaching for Klondike," *New York Times*, 4 September 1897.
54. Paulson, *Radicalism and Reform*, 192.
55. Ibid., 195.
56. Frank B. Vrooman, "Uncle Sam's Romance with Science and the Soil, Part II: The Stream," *Arena* 35 (January 1906): 44 (quote); W. H. Dean, "Two Remarkable Brothers," *American Magazine* 80 (November 1915): 28–29.
57. Paulson, *Radicalism and Reform*, 192.
58. Vrooman, *Dynamic Geographer*, 6.
59. Ibid., 9.
60. *Daily Mail* (London), reprinted in "Man of Action," *Star* (Christchurch, N.Z.), 3 May 1909, 2.
61. Vrooman, *Dynamic Geographer*, 49.
62. Ibid., 20.
63. Frank B. Vrooman, "The Imperial Idea: From the Point of View of Vancouver," *Nineteenth Century and After* 73 (March 1913): 501–10; Frank B. Vrooman, "Business and Patriotism," *British Columbia Magazine* 8 (December 1912): 887–88; Frank B. Vrooman, "British Columbia and the Panama Canal," *British Columbia Magazine* 8 (August 1912): 631–38; Frank Vrooman to Gifford Pinchot, 13 September 1910, fl. P-O, box 470, Gifford Pinchot Papers, LC.
64. "A Man of Action: Mr. Roosevelt's Record," *Times of India*, 26 March 1909, 7 (quote); "Man of Action," *Star*, 2; *West Gippsland Gazette* (Warragul, Vic.), 11 May 1909, 5.

NOTES TO CHAPTER FOUR

1. *New York World*, quoted in "The Destruction of Cervera's Fleet: The Advance on Santiago," *Outlook* 59 (9 July 1898): 602.
2. "An Attack All Along the Line," *Century Illustrated Magazine* 56 (August 1898): 633 (quote); "The Danger to Forest Reserves," *Forest and Stream* 51 (21 May 1898): 1,

endorsed Pinchot as "by far" the best-equipped appointee; "Model Forest Culture," ibid. (31 December 1898): 1 ("so well-trained").

3. "Probable Revolt in Congress," *Chicago Daily Tribune*, 12 December 1898, 1.
4. "Professor Pinchot Coming," *San Francisco Chronicle*, 2 August 1899, 11; "Forests and Water: Their Relations Discussed by Experts," *Los Angeles Times*, 20 July 1899, 4.
5. "Washington Letter," *Journal of the American Geographical Society of New York* 30, no. 1 (1898): 255.
6. *Chicago Daily Tribune*, 12 December 1898, 1.
7. "Washington Letter," 255.
8. "Professor Pinchot Coming," 11; "Forests and Water," 4; "How to Save the Forests," *San Francisco Chronicle*, 5 August 1899, 5.
9. "Washington Letter," 255.
10. "The Far Philippines," *Los Angeles Times*, 11 June 1898, 10. See also "Gold in Plenty There," *Washington Post*, 4 June 1898, 7; W. Hazeltine Mayo, "What Shall Be Done about the Philippines?," *North American Review* 167 (October 1898): 385–92.
11. "The Philippines: An Unknown Empire," *Gunton's Magazine* 14 (June 1898): 384.
12. "Destruction of Cervera's Fleet," 602.
13. Howard L. Hyland, "History of U.S. Plant Introduction," *Environmental Review* 2, no. 4 (1977): 26–33 at 28.
14. W. H. Hodge and C. O. Erlanson, "Federal Plant Introduction: A Review," *Economic Botany* 10 (October–December 1956): 299–334 at 299.
15. Philip J. Pauly, *Fruits and Plants: The Horticultural Transformation of America* (Cambridge, MA: Harvard University Press, 2007), chap. 5.
16. Michael A. Osborne, *Nature, the Exotic, and the Science of French Colonialism* (Bloomington: Indiana University Press, 1994); Ian Tyrrell, *True Gardens of the Gods: Californian-Australian Environmental Reform, 1860–1930* (Berkeley: University of California Press, 1999); Hodge and Erlanson, "Federal Plant Introduction," 299; Hyland, "History of U.S. Plant Introduction," 27–28.
17. Hodge and Erlanson, "Federal Plant Introduction," 302.
18. Ibid., 306; David Fairchild, "Our Plant Immigrants," *National Geographic* 17 (April 1904): 179–201 at 182.
19. Hodge and Erlanson, "Federal Plant Introduction," 318 (quote); Hyland, "History of U.S. Plant Introduction," 30; Raymond G. McGuire, Raymond J. Schnell, and Walter Gould, "A Century of Research with USDA in Miami," *Proceedings of the Florida State Horticultural Society* (1999), http://www.ars-grin.gov/mia/Pages/Chapman/1999.htm (accessed 19 June 2012). The extensive nature of the original work is revealed in David Fairchild's manuscript "Southern Trip January to April 1917 Including Account of Effects of Freeze February 3, 1917," http://ufdc.ufl.edu/AA00003176/00001 (accessed 19 June 2012). See also David Fairchild, *The World Was My Garden: Travels of a Plant Explorer* (New York: C. Scribner's Sons, 1938), 115.
20. Hodge and Erlanson, "Federal Plant Introduction," 324 (quotes).
21. William Elliot Griffis, "America in the Far East," *Outlook* 60 (10 December 1898): 904.
22. Alphonse de Candolle, "The Origin of Cultivated Plants," in *NCC Report*, 3:212–26.
23. Fairchild, "Our Plant Immigrants," 180.
24. Kristin L. Hoganson, *Consumers' Imperium: The Global Production of American Domesticity, 1865–1920* (Chapel Hill: University of North Carolina Press, 2007), chap. 3.

25. John Stevenson, "Plants, Problems, and Personalities: The Genesis of the Bureau of Plant Industry," *Agricultural History* 28 (October 1954): 155–62 at 160; Randall E. Stross, *The Stubborn Earth: American Agriculturalists on Chinese Soil, 1898–1937* (Berkeley: University of California Press, 1987), 21. See also David Fairchild, "Reminiscences of Early Plant Introduction Work in South Florida," *Proceedings of the Florida State Horticultural Society* 51 (1938): 11–33; Marjory Stoneman Douglas, *Adventures in a Green World: The Story of David Fairchild and Barbour Lathrop* (Coconut Grove, FL: Field Research Projects, 1973); Fairchild, *The World Was My Garden*, chap. 7; "Barbour Lathrop," *New York Times*, 18 May 1927, 25; "Barbour Lathrop, Capitalist, Leaves $1,750,000 Estate," *Chicago Daily Tribune*, 9 October 1927, 4.
26. Also a member of the Board of Trustees of the National Geographic Society of America, Fairchild called American farmers "the best in the world." He was well aware of the global market connections of interest to them. They were not "peasants who blindly follow in the footsteps of their forebears" but kept in touch with plant industries "all over the world" through the daily press. Fairchild, "Our Plant Immigrants," 180.
27. John Eliot, *The Green Roosevelt: Theodore Roosevelt in Appreciation of Wilderness* (Amherst, NY: Cambria Press, 2010), 3.
28. Fairchild, "Our Plant Immigrants," 180, 201.
29. Ibid., 201.
30. Fairchild, *The World Was My Garden*, 115.
31. Fairchild, "Our Plant Immigrants," 187 (quote); Guy N. Collins, *The Mango in Porto Rico*, Bureau of Plant Industry Bulletin 28 (Washington, DC: Government Printing Office, 1903), 26, 29, 31–32; Seaman A. Knapp, *Recent Foreign Explorations, as Bearing on the Agricultural Development of the Southern States*, Bureau of Plant Industry Bulletin 35 (Washington, DC: Government Printing Office, 1903), 40–44, for the Philippines; E. N. Reasoner, "Tropical Fruits, Other Than Pineapples," *Proceedings of the Florida State Horticultural Society* 15 (1902): 62–65; Fairchild, *The World Was My Garden*, 223–25.
32. Fairchild, "Our Plant Immigrants," 198.
33. Knapp, *Recent Foreign Explorations*, 40.
34. Elmer D. Merrill, *Report on Investigations Made in Java in the Year 1902*, Department of the Interior, Forestry Bulletin 1 (Manila: Bureau of Public Printing, 1903), 8–9.
35. For the inception of those problems, see Warren Dean, *Brazil and the Struggle for Rubber: A Study in Environmental History* (New York: Cambridge University Press, 1987), 5 and chaps. 3–4.
36. Fairchild, "Our Plant Immigrants," 198 (quotes); Getachew Metaferia, *Ethiopia and the United States: History, Diplomacy, and Analysis* (New York: Algora, 2009), 15–18.
37. David G. Fairchild, *Persian Gulf Dates and Their Introduction into America*, Bureau of Plant Industry Bulletin 53 (Washington, DC: Government Printing Office, 1903), 23, cited in Charles C. Colley, "The Desert Shall Blossom: North African Influence on the American Southwest," *Western Historical Quarterly* 14 (July 1983): 277–90 at 286.
38. Colley, "Desert Shall Blossom," 288.
39. Gifford Pinchot, foreword to George Ahern, *Deforested America* (Washington, DC: n.p., 1928), 5. The official extent of the Philippine forests was 40 million acres, or 16.1 million hectares. Bankoff gives a much higher figure of 55 million acres (20.7 million ha.) for 1903, taking special note of the Mindanao forests. Greg Bankoff, "One Island Too Many: Reappraising the Extent of Deforestation in the

Philippines prior to 1946," *Journal of Historical Geography* 33, no. 2 (2007): 314–34 at 324. Raphael Zon, "Foreign Sources of Timber Supply," *NCC Report*, 2:366, gave 49 million as his appraisal.

40. "The Philippines: An Unknown Empire," *Gunton's Magazine* 14 (June 1898): 384. Editor George Gunton, a "staunch Republican," had been an adviser to Governor Theodore Roosevelt. Jack Blicksilver, "George Gunton: Pioneer Spokesman for a Labor–Big Business Entente," *Business History Review* 31 (Spring 1957): 1–22 at 6.
41. Pinchot, foreword to Ahern, *Deforested America*, 5; Michael Williams, *Deforesting the Earth: From Prehistory to Global Crisis* (Chicago: University of Chicago Press, 2003), 387–88; Bankoff, "One Island Too Many," 324; *NCC Report*, 2:366.
42. Greg Bankoff, "Gifford Pinchot and the Birth of Tropical Forestry in the Philippines," in Alfred W. McCoy and Francisco A. Scarano, eds., *Colonial Crucible: Empire in the Making of the Modern American State* (Madison: University of Wisconsin Press, 2009), 487.
43. Ibid. (quote); Gifford Pinchot to William Howard Taft, 20 July 1903, file 1991/72, box 240, entry 5, RG 350, NARA. The Philippine Commission was an all-appointed body with a majority of American members.
44. For a US Forest Service appraisal, see Theodore S. Woolsey Jr., "Management and Natural Reproduction of Chir Pine near Dehra Dun," *Forestry and Irrigation* 12 (April 1906): 183–89.
45. Gifford Pinchot to W. Cameron Forbes, 15 October 1910, fl. 1910 Oct., box 467, Gifford Pinchot Papers, LC (Ahern "has long had a habit of keeping me more or less posted on the situation of the Philippine Bureau of Forestry"); Lawrence Rakestraw, "George Patrick Ahern and the Philippine Bureau of Forestry, 1900–1914," *Pacific Northwest Quarterly* 58 (July 1967): 142–150 at 144.
46. George Ahern to Dean Worcester, 6 May 1901, file 1991/29, box 240, entry 5, RG 350, NARA.
47. Gifford Pinchot, "The Forester and the Lumberman," *Forestry and Irrigation* 9 (April 1903): 176–78 at 178.
48. Roy Crandall, "The Riches of the Philippine Forests," *World's Work* 16 (May 1908): 10228.
49. Eber Smith, "A Valuable Philippine Cabinet Wood," *Forestry and Irrigation* 10 (August 1904): 381–82 (Narra); Barbara Goldoftas, *The Green Tiger: The Costs of Ecological Decline in the Philippines* (New York: Oxford University Press, 2006), 51–52; Raphael Zon, "South American Timber Resources and Their Relation to the World's Timber Supply," *Geographical Review* 2 (October 1916): 256–66 at 257.
50. Pinchot, "The Forester and the Lumberman," 178.
51. *Imperata cylindrica* is a tufted, tall, perennial grass with hard, creeping rhizomes.
52. In the Philippines, the word is today written as *kaingin*. In India, the practice is called *kumri*. William Beinart and Lotte Hughes, *Environment and Empire* (Oxford: Oxford University Press, 2007), 119.
53. Joseph D. Cornell, "Slash and Burn," http://www.eoearth.org/article/Slash_and_burn (accessed 30 June 2010); Michael R. Dove, "Smallholder Rubber and Swidden Agriculture in Borneo: A Sustainable Adaptation to the Ecology and Economy of the Tropical Forest," *Economic Botany* 47, no. 2 (1993): 136–47; Beinart and Hughes, *Environment and Empire*, 127–28; David Henley, "Swidden Farming as an Agent of Environmental Change: Ecological Myth and Historical Reality in Indonesia," *Environment and History* 17 (November 2011): 525–54; Greg Bankoff, "'Deep Forestry':

Shapers of the Philippine Forests," *Environmental History* 18 (July 2013): 523–56 at 531–32.

54. "Report of the Secretary of the Interior, 1910," in *Annual Report of the Philippine Commission to the Secretary of War, 1910 (in One Part)* (Washington, DC: Government Printing Office, 1911), 105 (quote); William M. Maule, "Parang and Cogonales in the Philippines," *Forestry and Irrigation* 12 (July 1906): 313–17.
55. Barrington Moore, "Forest Problems in the Philippines" [pt. 1], *American Forestry* 16 (February 1910): 75–81 at 80.
56. Goldoftas, *Green Tiger*, 51–52; M. Patricia Marchak, *Logging the Globe* (Montreal: McGill-Queen's University Press, 1995), 184–85; Barrington Moore, "Forest Problems in the Philippines" [pt. 2], *American Forestry* 16 (March 1910): 149–53 at 151 (quote).
57. Moore, "Forest Problems in the Philippines" [pt. 1], 78–80.
58. Ibid., 80; Marchak, *Logging the Globe*, 184–85. The areas of cogon grass in turn encouraged locust plagues. *Report of the Philippine Commission to the Secretary of War, July 1, 1913 to June 30, 1914* (Washington, DC: Government Printing Office, 1915), 114; Antonio Racelis, "Forestry Education in the Philippines," *Journal of Forestry* 31 (April 1933): 455–61 at 455.
59. Ahern, however, blamed parsimonious US support of Philippine forestry. George Ahern to Gifford Pinchot, n.d., fl. 1910, October (2), box 467, Pinchot Papers.
60. Department of the Bureau of Forestry, *The Forest Manual; The Forest Act (No. 1148); Extracts from Other Laws of the Philippine Commission Relating to the Forest Service and the Forest Regulations Prepared in Accordance with the Provisions of the Forest Act* (Manila: Bureau of Public Printing, 1904), 15, 45. On private land where title could be proven, a forestry official or a municipal officer had to give permission for use, after instructing *caingin* cultivators on measures to prevent the spread of fire (ibid., 46). For five years under the revised rules, residents could take without license and free of charge "forest products, earth and stone, for personal use" except for class 1 timber trees. Timber for sale or export continued to be taxed at the normal government rate, but the regulation did away with the red tape previously needed "to obtain permission to cut a few cubic feet of wood for personal use." W. J. Hutchinson, "Reorganization of the Philippine Bureau of Forestry," *Forestry and Irrigation* 12 (February 1906): 89 (quotes). See also C. A. Schenck, *Forest Policy*, 2nd ed., rev. and enl. (Darmstadt: C. F. Winter, 1911), 112; H. N. Whitford, "The Vegetation of the Lamao Forest Reserve," *Philippine Journal of Science* 1, no. 4 (1906): 373–431. The attempt to regulate *caingins* rather than to ban them followed British Indian practice. Maule, "Parang and Cogonales," 315; Moore, "Forest Problems in the Philippines" [pt. 2], 149. The "homesteads" were being implemented by 1914. *Report of the Philippine Commission to the Secretary of War, July 1, 1913 to June 30, 1914*, 115, 125.
61. "Report of the Secretary of the Interior, 1910," 105.
62. *Report of the Philippine Commission to the Secretary of War, July 1, 1913 to June 30, 1914*, 115.
63. The situation is summarized in Ahern to Pinchot, n.d., fl. 1910, October (2), box 467, Pinchot Papers; "Report of the Secretary of the Interior, 1910," 105 (quotes). The American case was comparable to that of the Germans in East Africa. They, too, tried to introduce commercial crops and sought to prevent indigenous incursions on forested areas. But the fear of rebellions in 1905–6 caused them to ease the forest-clearing restrictions for the peasantry. Thaddeus Sunseri, "Reinterpreting a Colonial

Rebellion: Forestry and Social Control in German East Africa, 1874–1915," *Environmental History* 8 (July 2003): 430–51.

64. George Ahern to G. B. Sudsworth, 29 November 1911, file 1191/128, box 242, entry 5, RG 350, NARA; George Ahern to F. E. Olmsted, 14 April 1910, file 1191/100, ibid.; Donald M. Matthews, "Lumbering in the Philippines," with Acting Director of Forestry W. F. Sherfesee to Bernhard Fernow, 26 August 1911, file 1191/119, ibid.; H. N. Whitford, "Studies in the Vegetation of the Philippines. I. The Composition and Volume of the Dipterocarp Forests of the Philippines," *Philippine Journal of Science* 4 (December 1909): 699–725; Richard Tucker, *Insatiable Appetite: The United States and the Ecological Degradation of the Tropical World* (Berkeley: University of California Press, 2000), 380–82.
65. George Ahern to Dean Worcester, 6 May 1901, file 1991/29, box 250, entry 5, RG 350, NARA. Cf. Michael Adas, *Dominance by Design: Technological Imperatives and America's Civilizing Mission* (Cambridge, MA: Belknap Press of Harvard University Press, 2006); Tucker, *Insatiable Appetite*, 363–64; Bankoff, " 'Deep Forestry.' "
66. Rakestraw, "Ahern," 149.
67. W. Forsythe Sherfesee, "Organization and Activities of the Chinese Forest Service" [15 July 1916], fl. Forests, Foreign, Miscellany, box 563, Pinchot Papers. For Brazil, see Roy Nash, *The Conquest of Brazil* (New York: Harcourt, Brace, 1926), 66–68. Roy Nash, a Yale graduate in forestry (1909), began his tropical experience in the Philippines and later headed the Los Baños Forestry School. Warren Dean, *With Broadax and Firebrand: The Destruction of the Brazilian Atlantic Forest* (Berkeley: University of California Press, 1995), 285; "A Summary of the Career of Tom Gill, International Forester: An Interview Conducted by Amelia R. Fry" (Berkeley: Regional Oral History Office, 1969), 7, 11–16, 25–26. Hugh Curran, another forestry official in the Philippines, worked on Venezuelan forestry as "the grand old man of tropical forestry" ("Summary of the Career of Tom Gill," 26). See also "Personalities," *Forestry Quarterly* 13 (September 1915): 434; W. F. Sherfesee, "A Forest School in the Philippines," *American Forestry* 17 (September 1911): 517–21; Racelis, "Forestry Education in the Philippines," 455–61; Rakestraw, "Ahern," 148–49; Tom Gill, "America and World Forestry," in Henry Clepper and Arthur B. Meyer, eds., *American Forestry: Six Decades of Growth* (Washington, DC: Society of American Foresters, 1960), 283–87. On Argentina, see *Northern Patagonia: Character and Resources*, text and maps by the Comisión de estudios hidrológicos, Bailey Willis, director, 1911–14 (New York: Scribner Press, 1914), 365 (hereafter Willis, *Northern Patagonia*); Warren Dean, "Forest Conservation in Southeastern Brazil, 1900 to 1955," *Environmental Review* 9 (Spring 1985): 54–69; Stuart McCook, " 'The World Was My Garden': Tropical Botany and Cosmopolitanism in American Science, 1898–1935," in McCoy and Scarano, eds., *Colonial Crucible*, 499–507; Whitford, "Studies in the Vegetation of the Philippines," 722.
68. Wilhelm Klemme, "The Philippine Forestry Service," *Forestry and Irrigation* 10 (April 1904): 158–64 at 160.
69. Zon, "Foreign Sources," 367 (quote), 352; Bankoff, "One Island Too Many"; Luis J. Borja, "The Philippine Lumber Industry," *Economic Geography* 5 (April 1929): 194–202.
70. John Gifford, "A Porto Rico Forest Reserve," *Forestry and Irrigation* 11 (January 1905): 38–40 at 39, 40. Gifford typically favored mixed, multipurpose use, as did other Progressives. The "best returns from the sale of timber and other forest products, consistent with the maximum protection of the watersheds," should be the goal, but

the reserve "should also be made accessible to the public for its scenic attractions" (ibid., 40, 38).

71. The insular government had already extended the size of the reserve in 1903. Douglas Brinkley, *The Wilderness Warrior: Theodore Roosevelt and the Crusade for America* (New York: HarperCollins, 2009), 444–46. Gifford did not acknowledge indebtedness to the Spanish government's reserve in his official report, *The Luquillo Forest Reserve in Porto Rico*, Bureau of Forestry Bulletin 54 (Washington, DC: Government Printing Office, 1905), 33 (quotes).
72. Gifford, *Luquillo Forest Reserve*, 32–33; "News and Notes: Working for Porto Rico's Forests," *Forestry and Irrigation* 11 (February 1905): 50; Brinkley, *Wilderness Warrior*, 447; William D. Durland, "Forest Regeneration in Porto Rico," *Economic Geography* 5 (October 1929): 369–81; Ariel E. Lugo, "Preservation of Primary Forests in the Luquillo Mountains, Puerto Rico," *Conservation Biology* 8 (December 1994): 1122–31.
73. "Forestry in Our Island Territories," *Forestry and Irrigation* 9 (December 1903): 574 (quote). In April 1898, the departing chief forester, Bernhard Fernow, had been sent to Hawaii to make "preliminary explorations" regarding forest laws in the prospective acquisition. "Science Notes," *Scientific American* 127 (23 April 1898): 263.
74. William L. Hall, "Hawaiian Forests: A Description of the Island Forests Based on Recent Observations," *Forestry and Irrigation* 9 (December 1903): 582–88 at 586 (quote); W. R. Castle (a Honolulu lawyer and financier), "Forest Conditions of the Hawaiian Islands," *Forestry and Irrigation* 8 (January 1902): 37–39 at 37. See also W. L. Hall, *The Forests of the Hawaiian Islands*, Bureau of Forestry Bulletin 48 (Washington, DC: Government Printing Office, 1904), 1–29.
75. Hall, "Hawaiian Forests," 586 (quote), 588; Thomas R. Cox, "The Birth of Hawaiian Forestry: The Web of Influences," *Pacific Historical Review* 61 (May 1992): 169–92.
76. Jean-Yves Puyo, "Sur le mythe colonial de l'inépuisabilité des ressources forestières (Afrique occidentale française / Afrique équatoriale française, 1900–1940)," *Cahiers de géographie du Québec* 45, no. 126 (2001): 479–96; Val Shaw, "The Rape of the Redwood," *Overland Monthly and Out West Magazine* 39 (March 1902): 738–42 at 741; Zon, "Foreign Sources," 327, 333; Max Endres, *Handbuch der Forstpolitik mit besonderer Berücksichtigung der Gesetzgebung* . . . (Berlin: Julius Springer, 1905).
77. Gifford, "A Porto Rico Forest Reserve," 39.
78. Zon, "South American Timber Resources," 257.
79. Willis, *Northern Patagonia*, 9–11.
80. John Gifford, "The Planting of Exotic Trees in Southern Florida," *Forestry and Irrigation* 8 (March 1902): 116–21 at 116 (first quote); John Gifford, "Southern Florida," ibid. 10 (September 1904): 413.
81. John Gifford, "The Everglades of Florida and the Landes [*sic*] of France," *Conservation* 15 (August 1909): 453–62 at 462; Brinkley, *Wilderness Warrior*, 734–35; J. E. Dovell, "Thomas Elmer Will, Twentieth Century Pioneer," *Tequesta* 8 (1948): 22–23; Pauly, *Fruits and Plants*, 212–13.
82. Gifford, "Planting of Exotic Trees," 118.
83. Brinkley, *Wilderness Warrior*, 736.
84. *Report of the Committee Appointed by the National Academy of Sciences upon the Inauguration of a Forest Policy for the Forested Lands of the United States; To the Secretary of the Interior, May 1, 1897* (Washington, DC: Government Printing Office, 1897), 9–16; Gifford Pinchot, "Government Forestry Abroad," *Publications of the American Economic Association* 6 (May 1891): 54; Harold K. Steen, *The U.S. Forest Service: A History* (Seattle: University of Washington Press, 1977), 31–32.

85. Brinkley, *Wilderness Warrior*, 422; Char Miller, *Gifford Pinchot and the Making of Modern Environmentalism* (Washington, DC: Shearwater Press, 2001), 159; "New Regulations for Forest Reserves," *Outlook* 81 (16 September 1905): 103; *San Francisco Call*, 17 February 1905, 4.
86. Steen, *U.S. Forest Service*, 99–100.
87. Pinchot to Roosevelt, 22 November 1902, in Gifford Pinchot, *Breaking New Ground* (New York: Harcourt, Brace, 1947), 228.
88. Gifford Pinchot, "The Foundations of Prosperity," *North American Review* 188 (November 1908): 740–752 at 740.
89. "American Forestry Association, Twenty-Second Annual Meeting Was Held at Washington, D.C., December 9, 1903, Minutes of the Meeting," *Forestry and Irrigation* 10 (January 1904): 13; Alfred Gaskell, "Forestry at the World's Fair," ibid. 10 (September 1904): 400.
90. John F. Lacey, "Address before the Forestry Congress, Washington, DC, January 2–6, 1905," in L. H. Pammel, comp., *Major John F. Lacey: Memorial Volume* (Cedar Rapids: Iowa Park and Forestry Association, 1915), 95.
91. See, e.g., Francis Newlands to Gifford Pinchot, 31 July 1907, fl. 105, and Newlands to Pinchot, 23 November 1907, fl. 112, box 11, Francis Griffith Newlands Papers, Sterling Memorial Library, Yale University; Brinkley, *Wilderness Warrior*, 806.
92. Samuel P. Hays, *Conservation and the Gospel of Efficiency: The Progressive Conservation Movement, 1890–1920* (Cambridge, MA: Harvard University Press, 1959), 39–40; *1905 "Use Book": U.S. Department of Agriculture, Forest Service, Gifford Pinchot, Forester; The Use of the National Forest Reserves, Regulations and Instructions, Issued by the Secretary of Agriculture, to Take Effect July 1, 1905* (Washington, DC: Government Printing Office, 1905), 30–31.
93. Actually, 194 million acres. Brian Balogh, "Scientific Forestry and the Roots of the Modern American State," *Environmental History* 7 (April 2002): 199.
94. Department of the Bureau of Forestry, *The Forest Manual; The Forest Act (No. 1148); Report of the Chief of the Bureau of Forestry, Philippine Islands for the Period from September 1, 1903, to August 31, 1904* (Manila: Bureau of Public Printing, 1904), 7.
95. *1905 "Use Book,"* 8.
96. Greg Bankoff, "Breaking New Ground? Gifford Pinchot and the Birth of 'Empire Forestry' in the Philippines, 1900–1905," *Environment and History* 15, no. 3 (2009): 369–93 at 383.
97. William Howard Taft to Elihu Root, 4 September 1903, file 1991/72, box 240, entry 5, RG 350, NARA.
98. Rakestraw, "Ahern," 149–50; Pinchot, foreword to Ahern, *Deforested America*, 6.
99. Bankoff, "Breaking New Ground?," 369–93; Greg Bankoff, "First Impressions: Diarists, Scientists, Imperialists and the Management of the Environment in the American Pacific, 1899–1902," *Journal of Pacific History* 44, no. 3 (2009): 261–80.
100. Taking the area to be from 43 to 194 million acres. See http://www.theodore roosevelt.org/life/conNatlForests.htm (accessed 12 June 2012).
101. Brinkley, *Wilderness Warrior*, 676–78; Steen, *U.S. Forest Service*, 100; Gary Hines, *Midnight Forests: A Story of Gifford Pinchot and Our National Forests* (Honesdale, PA: Boyds Mills Press, 2005). For these proclamations, see no. 707 and following items at http://www.theodore-roosevelt.com/trproclamations.html.
102. Gifford Pinchot to William Howard Taft, 18 September 1908, fl. 1908, Ta, box 117, Pinchot Papers.

103. "Sustained annual, periodic, or intermittent workings are those under which the amount of wood cut is so regulated that the productive capacity of the forest does not decrease, but produces a sustained yield." *Terms Used in Forestry and Logging, Prepared in Cooperation with the Society of American Foresters*, Bureau of Forestry Bulletin 61 (Washington, DC: Government Printing Office, 1906), 27; *Forestry Pamphlets, History*, vol. 1, *U.S.F.S. Field Programs, 1904–1909* (n.p., n.d.), e.g., 111 and throughout, for ubiquitous references to "foreign language" publications; Arthur F. McEvoy, *The Fisherman's Problem: Ecology and Law in the California Fisheries* (Cambridge: Cambridge University Press, 1986), 158; C. G. J. Petersen, "What Is Over-fishing?," *Journal of the Marine Biological Association* 6 (December 1903): 587–95; "Colorado School of Forestry," *Conservation* 14 (December 1908): 676 ("sustained productiveness"); Woolsey, "Management and Natural Reproduction of Chir Pine," 182; *Proceedings of a Conference of Governors in the White House, Washington, D.C., May 13–15, 1908* (Washington, DC: Government Printing Office, 1908), 321 ("protection," "avarice"); cf. the fully worked out statement in W. W. Ashe, "The Tree That Does Not Yield a Profit," *Scientific Monthly* 31 (October 1930): 319–27.
104. Garrett Hardin, "The Tragedy of the Commons," *Science* 162 (13 December 1968): 1243–48.

NOTES TO CHAPTER FIVE

1. Brooks Adams, *The New Empire* (New York: Macmillan, 1902), xi.
2. *Theodore Roosevelt: An Autobiography*, new ed. (1913; New York: Charles Scribner's Sons, 1922), 407.
3. J. A. Holmes, "Production and Waste of Mineral Resources and Their Bearing on Conservation," *Annals of the American Academy of Political and Social Science* 33 (May 1909): 202–14 at 204 (first two quotes); memorandum accompanying amendment to provide appropriation for investigations in behalf of the mining industry [18 March 1907], fl. 21, Letters, Memoranda, etc. by J. A. Holmes, 1905–10, box 2, Joseph A. Holmes Papers, SHC; "A Barbarous Waste of Coal," *Boston Globe*, 9 October 1910, fl. 1, Biography: Joseph Austin Holmes, box 1, Holmes Papers (final quote).
4. Edward Atkinson, "Coal Is King," *Century Illustrated Magazine* 56 (April 1898): 828–30.
5. *Message regarding Conservation of Mineral Fuels in Public Lands, February 1907*, 59th Cong., 2nd Sess., S. Doc. No. 310, at 4 (1907).
6. *Proceedings of a Conference of Governors in the White House, Washington, D.C., May 13–15, 1908* (Washington, DC: Government Printing Office, 1908), 7.
7. Ibid., 60.
8. F. E. Saward, "The World's Need of Coal and the United States' Supply," *Engineering Magazine* 20 (October 1900): 1.
9. *Conference of Governors*, 12.
10. Charles R. Van Hise, "The Future of Man in America," *World's Work* 15 (June 1909): 117–18.
11. *Conference of Governors*, 47.
12. J. A. Holmes, in ibid., 440.
13. Thomas C. Chamberlain, "Soil Wastage," in ibid., 76.
14. Van Hise, "Future of Man in America."
15. Israel White, "The Waste of Our Fuel Resources," in *Conference of Governors*, 26, 27; Israel Charles White, *Memoir of Israel C. White*, Herman L. Fairchild, Fredrick H.

Armstrong Collection, w.wvculture.org/HiStory/businessandindustry/whiteic03 .html (accessed 17 January 2012).

16. See fl. Speech Material, Oil, PS, box 929, Gifford Pinchot Papers, LC; Israel White, "Waste of Our Fuel Resources," 35; *Preliminary Report on the Operations of the Coal-Testing Plant of the United States Geological Survey at the Louisiana Purchase Exposition, St. Louis, Mo., 1904* (Washington, DC: Government Printing Office, 1905), 17; Peter A. Shulman, "'Science Can Never Demobilize': The United States Navy and Petroleum Geology, 1898–1924," *History and Technology: An International Journal* 19, no. 4 (2003): 365–95 at 369–71; *Investigation of the Department of the Interior and of the Bureau of Forestry*, vol. 1, *Report of the Committee*, 61st Cong., 3rd Sess., S. Doc. No. 719, at 60 (1910–11).
17. *Conference of Governors*, 47.
18. Ibid., 53.
19. Ibid., 47 (quote); Waldemar Lindgren, "Resources of the United States in Gold, Silver, Copper, Lead, and Zinc," *NCC Report*, 3:521–57 at 525.
20. US Department of the Interior, US Geological Survey, Daniel E. Sullivan, John L. Sznopek, and Lorie A. Wagner, "20th Century U.S. Mineral Prices Decline in Constant Dollars, Open File Report 00–389," pubs.usgs.gov/of/2000/of00–389/ (accessed 22 November 2012). However, there were short-term fluctuations. Copper and aluminum prices spiked around 1905–9, then declined before rising during World War I (ibid., 5, 7).
21. *Conference of Governors*, 19.
22. White, "Waste of Our Fuel Resources," 26–27.
23. "Coal Specialist Resigns," *Wall Street Journal*, 11 May 1915, 8.
24. Edward W. Parker, "The Supply of Anthracite Coal in Pennsylvania," *Century Illustrated Magazine* 55 (April 1898): 830; "Oil and Fuel," *Current Literature* 31 (December 1901): 736.
25. "Increase in Coal Prices; Government Report Shows That Anthracite Costs Over 39 Per Cent. More Than Five Years Ago," *New York Times*, 1 February 1904, 1, regarding a "steady and considerable rise in the price of both anthracite and bituminous coal over the last five years."
26. Edward Wheeler Parker, "How Long Will Our Coal Supplies Meet the Increasing Demands of Commerce?," typescript [1906?], fl. 28, box 1, Bailey W. Willis Papers, Huntington Library. Though the document appears to have been composed in 1906, it was delivered as a speech and published in 1907.
27. Ibid. (quotes); *Report of Proceedings of the American Mining Congress, Tenth Annual Session, Joplin, Mo., November 11–16, 1907* (Denver, CO: American Mining Congress, 1908), 239–46.
28. Bailey Willis, "Coal Resources of the United States," fl. 27, box 2 [1906?], Willis Papers (quotes). On the international discourse over fuel shortage, see Nuno Luis Madureira, "The Anxiety of Abundance: William Stanley Jevons and Coal Scarcity in the Nineteenth Century," *Environment and History* 18 (August 2012): 395–420.
29. Marius Campbell and Edward W. Parker, "The Coal Fields of the United States," *NCC Report*, 3:249; Henry Gannett, "Estimates of Future Coal Production," in ibid., 443–46.
30. Julie A. Tuason, "The Ideology of Empire in *National Geographic Magazine*'s Coverage of the Philippines, 1898–1908," *Geographical Review* 89 (January 1999): 34–53 at 43.
31. Henry Gannett, *The Building of a Nation: The Growth, Present Condition and Resources of the United States, with a Forecast of the Future* (New York: H. T. Thomas, 1895), 238–39.

32. White, "Waste of Our Fuel Resources," 26.
33. David T. Day, "The Petroleum Resources of the United States," *NCC Report*, 3:446–64; David T. Day, "The Petroleum Resources of the United States," *American Review of Reviews* 39 (January 1909): 49–56; Diana Davids Olien and Roger M. Olien, "Running Out of Oil: Discourse and Public Policy, 1909–1929," *Business and Economic History* 22 (Winter 1999): 36–66 at 42 (quotes).
34. "Texas Oil Prices," *Wall Street Journal*, 14 September 1903, 7.
35. Day, "Petroleum Resources," *NCC Report*, 3:463.
36. The paper poured scorn on Day's prediction of exhaustion in twenty-seven years. "Oil Production, Uses, Waste," *Los Angeles Times*, 14 February 1909, II4 (quotes); "The Exhaustion of Oil," *Los Angeles Times*, 7 August 1909, II4.
37. Day, "Petroleum Resources," *American Review of Reviews* 39 (January 1909).
38. Ralph Arnold and V. R. Garfias, *Geology and Technology of the California Oil Fields; Reprinted from Bulletin No. 87, March, 1914, American Institute of Mining Engineers (New York Meeting, February, 1914)* (New York: n.p., 1914), 393–94, 423; Dan T. Boyd, "Oklahoma Oil: Past, Present, and Future," *Oklahoma Geology Notes* 62 (Fall 2002): 98.
39. John Mason Hart, *Empire and Revolution: The Americans in Mexico since the Civil War* (Berkeley: University of California Press, 2002), 155–56 (quote at 155).
40. Day, "Petroleum Resources," *NCC Report*, 3:462; David Day to Gifford Pinchot, 10 September 1910, fl. Conservation 1910-Oct., box 467, Pinchot Papers.
41. Tom McCarthy, "The Coming Wonder? Foresight and Early Concerns about the Automobile," *Environmental History* 6 (January 2001): 46–74 at 49; Olien and Olien, "Running Out of Oil"; Jonathan C. Brown, "Why Foreign Oil Companies Shifted Their Production from Mexico to Venezuela during the 1920s," *American Historical Review* 90 (April 1985): 362–85.
42. Day to Pinchot, 10 September 1910, box 467, Pinchot Papers.
43. Day, "Petroleum Resources," *NCC Report*, 3:461–62; Shulman, "'Science Can Never Demobilize,'" 371.
44. Day, "Petroleum Resources," *NCC Report*, 3:464.
45. E. D. Stodder to Gifford Pinchot, 1 November 1908, and encl., fl. 1908 St-Sy, box 117, Pinchot Papers; "Will Generate Electricity by Ocean Waves," *Los Angeles Herald*, 17 March 1907, 6; M. O. Leighton, "Undeveloped Water Powers," *NCC Report*, 3:159–77; Frederic C. Howe, "The White Coal of Switzerland," *Outlook* 94 (22 January 1910): 151–58; Robert W. Righter, *The Battle over Hetch Hetchy: America's Most Controversial Dam and the Birth of Modern Environmentalism* (New York: Oxford University Press, 2006), 57–61, 166; W. C. Shaffer to Francis Newlands, 17 February 1908, fl. 123, 1908 Feb. 15–19, box 13, Francis Griffith Newlands Papers, Sterling Memorial Library, Yale University ("ethanol").
46. "Texas Oil Prices," 7.
47. White, *Memoir of Israel C. White*; White, "Waste of Our Fuel Resources," 26, 27 (quote).
48. *NCC Report*, 1:142.
49. Ibid., 183; Mary Huston Gregory, *Checking the Waste: A Study in Conservation* (Indianapolis: Bobbs-Merrill, 1911), 266; Cyril K. Moresi, "Conservation of Louisiana's Mineral Resources," *Louisiana Conservation Review* 3 (October 1933): 21.
50. *Conference of Governors*, 26; David T. Day, "Natural-Gas Resources of the United States," *NCC Report*, 3:465–75 at 474; White, *Memoir of Israel C. White*.

51. "Increase in Coal Prices," 1; Parker, "How Long Will Our Coal Supplies Meet the Increasing Demands of Commerce?"
52. For summaries and charts, see William J. Baumol and Alan S. Blinder, *Economics: Principles and Policy* (Mason, OH: South-Western/CENGAGE Learning, 2009), 370–72; Sullivan, Sznopek, and Wagner, "20th Century U.S. Mineral Prices Decline in Constant Dollars"; Price Fishback, *Soft Coal, Hard Choices: The Economic Welfare of Bituminous Miners* (New York: Oxford University Press, 1992), table 31, 20–22.
53. For the high cost of household use, see Scott Nearing, *Anthracite: An Instance of Natural Resource Monopoly* (Philadelphia: John C. Winston, 1915), 87–90. According to the Douglas Index, from 1897 to 1913 the cost of living rose 37 percent, and the price of coal rose from $3.74 to $5.31, or 42 percent. See *Historical Statistics of the United States* (1976 ed.), 208, 212, 165.
54. Aaron Sakolski, "Are We Approaching the End of Our Economic Resources?," *Independent* 62 (28 March 1907): 726–30; David Stradling, *Smokestacks and Progressives: Environmentalists, Engineers and Air Quality in America, 1881–1951* (Baltimore, MD: Johns Hopkins University Press, 1999), 12–13; Fishback, *Soft Coal, Hard Choices*, 20–22.
55. "Increase in Coal Prices," 1.
56. *Conference of Governors*, 65.
57. R. Moses, "Ascending Coal Prices," *New York Times*, 22 April 1907, 8 (quote); "Harriman Roads Indicted," *Washington Post*, 21 November 1907, 4; "Fear Advance in Coal Price," *Chicago Daily Tribune*, 18 April 1907, 2; "Keeping Up Pig Iron Prices," *New York Times*, 29 January 1900.
58. Frank Julian Warne, "A Ton of Anthracite," *Outlook* 82 (21 April 1906): 889.
59. "No World-Wide Advances Found in Cost of Living," *New York Times*, 22 August 1912 (quotes); "Exorbitant Coal Prices," *Outlook* 73 (3 January 1903): 5; "The Duty on Coal," *Outlook* 73 (3 January 1903): 5.
60. "Strike Jumps Coal Prices," *Chicago Daily Tribune*, 3 November 1904, 3; Ray Stannard Baker, "Capital and Labor Hunt Together," *McClure's Magazine* 21 (September 1903): 2–15.
61. "Relieving the Northwest," *New York Times*, 9 January 1907, 1; "Americans in Control of Canadian Mines," *San Francisco Chronicle*, 4 March 1901, 3.
62. David C. Burley, "James Jerome Hill," *Canadian Dictionary of Biography*, http://www.biographi.ca/009004-119.01-e.php?&id_nbr=7445 (accessed 18 September 2012). The Northern Securities case left Hill bewildered: "It really seems hard when we know that we have led all Western companies in opening the country and carrying at the lowest rates, that we should be compelled to fight for our lives against the political adventurers who have never done anything but pose and draw a salary."
63. "Duty on Coal," 5.
64. Theodore Roosevelt to Winthrop Murray Crane (governor of Massachusetts), 22 October 1902, in Elting E. Morison, ed., *The Letters of Theodore Roosevelt*, vol. 3 (Cambridge, MA: Harvard University Press, 1951), 360. For the British parallel, see W. Stanley Jevons, *The Coal Question*, 2nd ed. (London: Macmillan, 1866), vi.
65. "Coal Miners' Weak Cause," *Los Angeles Times*, 15 April 1906, VI14; *New York Times*, 11 April 1906, 5.
66. *Washington Post*, 15 January 1907, 2; *NCC Report*, 1:111; "Minority Report," in *Investigation of the Department of the Interior*, vol. 1, *Report of the Committee*, 151 (quote).
67. White, "Waste of Our Fuel Resources," 27.

68. "The Institute of Social Service," *Outlook* 72 (25 October 1902): 472 (quotes).
69. "Carroll Wright Dies in Worcester," *New York Times*, 21 February 1909.
70. William H. Tolman, *Industrial Betterment* (New York: Social Science Press, 1900); William H. Tolman, *Social Engineering: A Record of Things Done by American Industrialists Employing Upwards of One and One-Half Million of People* (New York: McGraw, 1909), iii ("mutuality"), 2, 3–5.
71. Andrew Carnegie, "Introductory," in Tolman, *Social Engineering*, v.
72. *Conference of Governors*, 22
73. Stradling, *Smokestacks and Progressives*, 54–55, 76–77; E. D. Simon and Marion Fitzgerald, *The Smokeless City* (London: Longmans, 1922), 49.
74. Ernest Morrison, *Horace McFarland: A Thorn for Beauty* (Harrisburg: Pennsylvania Historical and Museum Commission, 1995); J. Horace McFarland, "Shall We Have Ugly Conservation?," *Outlook* 91 (13 March 1909): 594–99.
75. On Adirondack preservation, see Philip G. Terrie, *Forever Wild: A Cultural History of Wilderness in the Adirondacks* (Syracuse, NY: Syracuse University Press, 1994); Frank Graham Jr., *The Adirondack Park: A Political History* (Syracuse, NY: Syracuse University Press, 1984).
76. Theodore Roosevelt, "First Annual Message," 3 December 1901, http://millercenter.org/scripps/archive/speeches/detail/3773 (accessed 11 December 2010).
77. ASHPS, [Fourth] *Annual Report of the Society for the Preservation of Scenic and Historic Places and Objects, . . . 1899* (Albany, NY: State Printers, 1899), 11, 12 (quote); "News of the Railroads," *New York Times*, 9 July 1898.
78. Stradling, *Smokestacks and Progressives*, 17, 32.
79. Charles R. Van Hise, *The Conservation of Natural Resources in the United States* (New York: Macmillan, 1910), 35 (first quote); *Conference of Governors*, 433 (second quote).
80. *Conference of Governors*, 35, 84, 179, 199–200, 217, 229, 230.
81. Ibid., 12 (quote), 84, 179, 199–200, 217, 229, 230; Joshua David Hawley, *Theodore Roosevelt, Preacher of Righteousness* (New Haven, CT: Yale University Press, 2008), 72–73.
82. *Conference of Governors*, 8 (quote); Lewis Cecil Gray, "The Economic Possibilities of Conservation," *Quarterly Journal of Economics* 26 (May 1913): 497–519.
83. *Conference of Governors*, 64.
84. Gifford Pinchot, cited in "Conservation of Resources," *Times of India*, 21 January 1909, 10.
85. E.g., Atkinson, "Coal Is King," 830.
86. George Otis Smith, *Our Mineral Reserves: How to Make America Industrially Independent*, Department of the Interior, US Geological Survey, Bulletin 599 (Washington, DC: Government Printing Office, 1914).
87. Theodore Roosevelt, reported in James Bryce to Sir Edward Grey, 8 January 1909, no. 53, FO 414/214, *Part VI, Further Correspondence Respecting Proceedings for the Settlement of Questions between the United States and Canada, January to June 1909*, 34, National Archives, UK (digitized correspondence). For India, see "An Epic of Waste," *Times of India*, 19 April 1909, 6.
88. Svante Arrhenius, *Worlds in the Making: The Evolution of the Universe* (New York: Harper and Row, 1908). Arrhenius's theories were (mostly later) challenged by those who emphasized water vapor, not carbon dioxide. F. B. Mudge, "The Development of the 'Greenhouse' Theory of Global Climate Change from Victorian Times," *Weather* 52 (January 1997): 13–17 at 15–16.

89. *Washington Post*, 23 April 1899, 19; *Nebraska State Journal*, 15 June 1898, 4; "A Startling Scientific Prediction," *Wellington Evening Post* (N.Z.), 30 July 1898, 4.
90. Thomas C. Chamberlin, "A Group of Hypotheses Bearing on Climatic Changes," *Journal of Geology* 5 (September–October 1897): 653–83; Bailey Willis, "Climate and Carbonic Acid," *Popular Science Monthly* 59 (July 1901): 254.
91. Parker, "How Long Will Our Coal Supplies Meet the Increasing Demands of Commerce?"
92. Van Hise, *Conservation of Natural Resources*, 33.
93. Ibid., 373.
94. Ibid., 374.
95. Willis, "Climate and Carbonic Acid," 254, foreshadowed this shift.
96. Parker, "How Long Will Our Coal Supplies Meet the Increasing Demands of Commerce?"
97. Van Hise, *Conservation of Natural Resources*, 10.

NOTES TO CHAPTER SIX

1. An Act Appropriating the Receipts from the Sale and Disposal of Public Lands in Certain States and Territories to the Construction of Irrigation Works for the Reclamation of Arid Lands, chap. 1093, 32 Stat. 388 (17 June 1902) (Newlands Act); Francis Newlands, "Irrigation in the West," *New York Times*, 2 March 1902 (quote); William D. Rowley, *Reclaiming the Arid West: The Career of Francis G. Newlands* (Bloomington: Indiana University Press, 1996).
2. Donald J. Pisani, *From the Family Farm to Agribusiness: The Irrigation Crusade in California, 1850–1931* (Berkeley: University of California Press, 1984), 189, 358; Donald J. Pisani, *Water and American Government: The Reclamation Bureau, National Water Policy, and the West, 1902–1935* (Berkeley: University of California Press, 2002); Samuel P. Hays, *Conservation and the Gospel of Efficiency: The Progressive Conservation Movement, 1890–1920* (Cambridge, MA: Harvard University Press, 1959), 26.
3. Henry Nash Smith, *Virgin Land: The American West as Symbol and Myth* (Cambridge, MA: Harvard University Press, 1950); Donald Worster, *Rivers of Empire: Water, Aridity, and the Growth of the American West* (New York: Pantheon Books, 1985), 150; Ian Tyrrell, *True Gardens of the Gods: Californian-Australian Environmental Reform, 1860–1930* (Berkeley: University of California Press, 1999), chap. 7; Lawrence B. Lee, "William E. Smythe and San Diego, 1901–1908," *Journal of San Diego History* 14 (Winter 1973): 10–24.
4. Laura L. Lovett, *Conceiving the Future: Pronatalism, Reproduction, and the Family in the United States, 1890–1938* (Chapel Hill: University of North Carolina Press, 2007), 113; "Concerning Empire," *Century Illustrated Magazine* 56 (August 1898): 633.
5. G. D. Rice, "Irrigation in the Philippines," *Irrigation Age* 15 (March 1901): 186–91; "Division of Forestry," ibid., 204–5; "A Traveler's Gift to the Farmers of America," ibid. (January 1901): 121–23; "Irrigation in Peru" (interview with Ramon Estacia), ibid. (January 1901): 125–26.
6. "News and Notes," *Forestry and Irrigation* 8 (November 1902): 440–44 at 441.
7. "The Delta Barrage," *Forestry and Irrigation* 9 (February 1903): 79–84 at 79 (quote); "News and Notes"; "The Nile Reservoir Dam at Assaun," ibid. 8 (December 1902): 494.
8. Donald C. Jackson, "Engineering in the Progressive Era: A New Look at Frederick Haynes Newell and the U.S. Reclamation Service," *Technology and Culture* 34 (July 1993): 539–74 at 574.

9. "Irrigation in Egypt," *Outlook* 90 (24 October 1908): 368; "The Date Palm in America," *Forestry and Irrigation* 9 (March 1903): 155–56; "News and Notes"; "Nile Reservoir Dam," 494; Means, "Delta Barrage"; Charles C. Colley, "The Desert Shall Blossom: North African Influence on the American Southwest," *Western Historical Quarterly* 14 (July 1983): 277–90; Charles C. Colley, *The Century of Robert H. Forbes: The Career of a Pioneer Agriculturist, Agronomist, Environmentalist, Conservationist and Water Specialist in Arizona and Abroad* (Tucson: Arizona Historical Society, 1977).
10. "Uncle Sam as Cement Manufacturer," *Forestry and Irrigation* 13 (March 1907): 142 (quote); "Begin Work on Big Dam," *Los Angeles Times*, 15 March 1908, III1; *Transactions of the American Society of Civil Engineers*, vol. 62 (New York: published by the Society, 1909), 30–32.
11. "Irrigationist Is Man of Many Virtues," *Daily Arizona Silver Belt*, 21 April 1910, 5.
12. Sydney Fisher to James Bryce, 17 January 1910, 23, book 30, James Bryce Papers, Bodleian Library, Oxford University.
13. Edward D. McQueen Gray, *Government Reclamation Work in Foreign Countries: Compiled from Consular Reports and Official Documents* (Washington, DC: Government Printing Office, 1909).
14. "Irrigationist Is Man of Many Virtues," 5 (quote); *Arizona Republican* (Phoenix), 2 October 1908, 1; Ralph Emerson Twitchell, ed., *Official Proceedings of the Sixteenth National Irrigation Congress: Held at Albuquerque, New Mexico, Sept. 29, 30, and Oct. 1–3, 1908* (Albuquerque, NM: Morning Journal, 1908); *Irrigation Age* 24 (July 1909): 334.
15. "Irrigationist Is Man of Many Virtues."
16. Tyrrell, *True Gardens*, chaps. 7–8; Governor Sir George I. Clarke to Newlands, 1 May 1902, fl. 43, 1902 May, box 4, Francis Griffith Newlands Papers, Sterling Memorial Library, Yale University.
17. Jessica B. Teisch, *Engineering Nature: Water, Development, and the Global Spread of American Environmental Expertise* (Chapel Hill: University of North Carolina Press, 2011); *Bisbee Daily Review*, 3 October 1908, 1; Dale C. Maluy, "Boer Colonization in the Southwest," *New Mexico Historical Review* 52 (April 1977): 98–100; Brian M. Du Toit, *Boer Settlers in the Southwest* (El Paso: Texas Western Press, 1995); Jeff Biggers, *In the Sierra Madre* (Urbana: University of Illinois Press, 2006), 158–59; John Hays Hammond, *The Autobiography of John Hays Hammond*, 2 vols. in one (1935; repr., New York: Arno Press, 1974), vol. 1, chaps. 13–18.
18. "Irrigation in Peru," 125–26.
19. *Daily Arizona Silver Belt*, 21 April 1910, 5; Sir George I. Clarke to Francis Newlands, 1 May 1902, fl. 43, 1902 May, box 4, Newlands Papers.
20. "Agua Liebre," *Los Angeles Times*, 16 September 1894, 20.
21. "Convention between the United States and the United States of Mexico Touching the International Boundary Line Where It Follows the Bed of the Rio Colorado, Washington, 1884, Concluded at Washington, November 12, 1884," www.ibwc.gov/Files/TREATY_OF_1884.pdf (accessed 1 May 2011). Despite the title, this treaty also included the Rio Grande.
22. Reprinted in 54th Cong., 1st Sess., S. Doc. No. 253, at 3 (quote) (11 May 1896). See also "The Irrigation Convention Adjourns," *New York Times*, 20 September 1895, 4 (a shorter version of the resolution); Chirakaikaran J. Chacko, *The International Joint Commission between the United States of America and the Dominion of Canada* (New York: Columbia University Press, 1932), 73.

23. Carl G. Winter, "The Boundary Waters Treaty," *Historian* 17 (Autumn 1954): 76; "Origins of the Boundary Waters Treaty," http://bwt.ijc.org/index.php?page=origins-of-the-boundaries-water-treaty&hl=eng (accessed 1 May 2011); *Report of the United States Deep Waterways Commission Prepared at Detroit, Michigan, December 18–22, 1896, by the Commissioners, James B. Angell, John E. Russell, Lyman E. Cooley . . .* (Washington, DC: Government Printing Office, 1897), 1–12; Chacko, *International Joint Commission*; Norris Hundley Jr., *Dividing the Waters: A Century of Controversy between the United States and Mexico* (Berkeley: University of California Press, 1966), 26–27.
24. For the Mexican outcome, see *Convention between the United States and Mexico: Equitable Distribution of the Waters of the Rio Grande, Signed at Washington, May 21, 1906 . . . Proclaimed, January 16, 1907* (Washington, DC: Government Printing Office, 1919); James A. Sandos, "International Water Control in the Lower Rio Grande Basin, 1900–1920," *Agricultural History* 54 (October 1980): 490–501; Hundley, *Dividing the Waters*, 30, 40.
25. Harold U. Faulkner, *The Decline of Laissez-Faire, 1897–1917* (New York: Holt, Reinhart and Winston, 1951), 315–17; Robert E. Lang, Deborah Epstein Popper, and Frank J. Popper, " 'Progress of the Nation': The Settlement History of the Enduring American Frontier," *Western Historical Quarterly* 26 (Autumn 1995): 289–307; William Willard Howard, "The Rush to Oklahoma," *Harper's Weekly* 33 (18 May 1889): 391–94 at 391.
26. Quoted in "The Irrigation Congress," *Los Angeles Times*, 11 October 1893, 4.
27. "National Irrigation Congress," *Irrigation Age* 15 (November 1900): 48 (quote); Donald J. Pisani, "Water Planning in the Progressive Era: The Inland Waterways Commission Reconsidered," *Journal of Policy History* 18, no. 4 (2006): 389–418 at 397–98. The Truckee River, Nevada, project was the first begun by the National Reclamation Service, in 1903.
28. William D. Rowley, *The Bureau of Reclamation: Origins and Growth to 1945*, vol. 1 (Denver, CO: Bureau of Reclamation, US Department of the Interior, 2006), 109; Shelly C. Dudley, "The First Five: A Brief Overview of the First Reclamation Projects Authorized by the Secretary of the Interior on March 14, 1903," in Brit Allan Storey, ed., *The Bureau of Reclamation: History Essays from the Centennial Symposium, Vols. I and II*, 292, http://www.usbr.gov/history/Symposium 2008/Historical_Essays.pdf.
29. *Theodore Roosevelt: An Autobiography*, new ed. (1913; New York: Charles Scribner's Sons, 1922), 393 (quote); *Arizona Republican* (Phoenix), 19 March 1911, reprinted in *Arizona Highways*, April 1961, 32.
30. *The Forester* 9 (December 1900): 289–90 has the full text. An abbreviated version is "Indorsed by Roosevelt," *San Francisco Chronicle*, 23 November 1900, 2.
31. George E. Mowry, *The Era of Theodore Roosevelt and the Birth of Modern America, 1900–1912* (New York: Harper and Row, 1954), 124–25; Rowley, *Reclaiming the Arid West*, 104.
32. Kathleen Dalton, *Theodore Roosevelt: A Strenuous Life* (New York: Knopf, 2002), 245.
33. Newlands to R. R. Crawford (Reno, NV), 23 January 1902, fl. 37, box 4, Newlands Papers. However, Roosevelt was not keen to acknowledge Newlands's influence, due to partisan rivalry. Roosevelt to Charles F. Loomis, 20 August 1902, in Elting E. Morison, ed., *The Letters of Theodore Roosevelt*, vol. 3 (Cambridge, MA: Harvard University Press, 1951), 317.
34. Pisani, *Water and American Government*.

35. Dalton, *Theodore Roosevelt*, 246.
36. Donald J. Pisani, "Frederick Haynes Newell," http://www.waterhistory.org/histories/newell/ (accessed 12 February 2012).
37. Pisani, *Water and American Government*, 293.
38. Donald W. Meinig, *On the Margins of the Good Earth: The South Australian Wheat Frontier, 1869–1884* (Chicago: Rand McNally, 1962); Tyrrell, *True Gardens*, 164, 166–67, 222. See also chapter 7, below, regarding dryland farming.
39. A. Hunter Dupree, *Science in the Federal Government: A History of Policies and Activities to 1940* (Cambridge, MA: Belknap Press of Harvard University Press, 1957), 249, 248 (first quote); Frederick Newell, "Discussion on Irrigation," in *Transactions of the American Society of Civil Engineers*, 10–15 at 11.
40. Julie A. Tuason, "The Ideology of Empire in *National Geographic Magazine*'s Coverage of the Philippines, 1898–1908," *Geographical Review* 89 (January 1999): 34–53 at 40–42; Biographies from American Society of Civil Engineers and Who's Who, fl. Frederick Haynes Newell, Letters Sent and Received, box 5, Frederick Haynes Newell Papers, LC.
41. Follett, comment in *Transactions of the American Society of Civil Engineers*, 33; Worster, *Rivers of Empire*, 143–50.
42. Lovett, *Conceiving the Future*, 113, for the intertwining of Roosevelt's domestic and foreign agendas regarding racial superiority; Laura Jensen, *Patriots, Settlers, and the Origins of American Social Policy* (New York: Cambridge University Press, 2002), 176n, 150 ("settler theory").
43. On Roosevelt's form of racism, see Lovett, *Conceiving the Future*, 112; Thomas Dyer, *Theodore Roosevelt and the Idea of Race* (Baton Rouge: Louisiana State University Press, 1980), 44–45; Dalton, *Theodore Roosevelt*, 126–27 (quote at 126). On Newlands, see Newlands to [John H.] Dennis [editor], *Nevada Journal*, 8 January 1902, fl. 37, 1902 Jan., box 4, Newlands Papers; Newlands to Joaquin Miller, 21 March 1902, fl. 40, 1902 March, box 4, Newlands Papers; Rowley, *Reclaiming the Arid West*, 105, 139; Act Appropriating the Receipts sec. 4. On Western labor and racism, see Alexander Saxton, *The Indispensable Enemy: Labor and the Anti-Chinese Movement in California* (Berkeley: University of California Press, 1975).
44. "Mr. Roosevelt Lauds the Irrigation Law," *New York Times*, 16 September 1903. See also *San Francisco Call*, 16 September 1903, 3, for the full text. The "international aspect" at the 1903 National Irrigation Congress was presented by French and Mexican representatives.
45. Lovett, *Conceiving the Future*, 113.
46. "Mr. Roosevelt Lauds the Irrigation Law"; Theodore Roosevelt to the National Irrigation Congress, 16 November 1900, in Morison, ed., *Letters*, 2:1421–42.
47. "Irrigation Exhibition in Paris," *Irrigation Age* 15 (January 1900): 109; W. J. McGee, "Water as a Resource," *Annals of the American Academy of Political and Social Science* 33 (May 1909): 37–50.
48. Newlands, "Irrigation in the West" (quote); Rowley, *Reclaiming the Arid West*, 83.
49. "To Colonize City Poor," *New York Times*, 17 September 1903.
50. Cf. Lovett, *Conceiving the Future*, 113.
51. Christian W. McMillen, "Rain, Ritual, and Reclamation: The Failure of Irrigation on the Zuni and Navajo Reservations, 1883–1914," *Western Historical Quarterly* 31 (Winter 2000): 434–56 at 443 (labor service); C. J. Blanchard, "The United States Reclamation Service," *Bulletin of the American Geographical Society* 37, no. 1 (1905): 1–14; Thomas R. McGuire, "Illusions of Choice in the Indian Irrigation Service: The Ak

Chin Project and an Epilogue," *Journal of the Southwest* 30 (Summer 1988): 200–222; Pisani, *Water and American Government*, 160–68, 186–87, 196–97; Donald J. Pisani, "Irrigation, Water Rights, and the Betrayal of Indian Allotment," *Environmental Review* 10 (Autumn 1986): 157–76, esp. 159–61.

52. Frederick Newell, *Water Resources: Present and Future Users* (1913; repr., New York: Arno Press, 1972), 35.
53. Ibid., 34.
54. Ibid., 35 (quote); McMillen, "Rain, Ritual, and Reclamation," 443, 451; McGuire, "Illusions of Choice," 200.
55. Cf. James C. Scott, *Seeing Like a State: How Certain Schemes to Improve the Human Condition Have Failed* (New Haven, CT: Yale University Press, 1998).
56. Pisani, "Indian Allotment," 161–66.
57. Frederick E. Hoxie, *A Final Promise: The Campaign to Assimilate the Indians, 1880–1920* (Lincoln: University of Nebraska Press, 1984).
58. Garfield diaries, entries for 6 and 7 July 1907 (p. 6), vol. 25, fl. 1907, box 8; Garfield to Theodore Roosevelt, 19 July 1907, fl. Theodore Roosevelt, 1906–10, box 119 (quotes); Roosevelt to Garfield, 6 July 1907, box 119; Roosevelt to Garfield, 23 May 1908, box 119; Roosevelt to Garfield, 16 February 1909 ("absolute honesty"), box 119; all in James R. Garfield Papers, LC; Theodore Roosevelt, *The Expansion of the White Races: Address at the Celebration of the African Diamond Jubilee of the Methodist Episcopal Church in Washington, D.C., January 18, 1909* (n.p., n.d.), http://www.theodore-roosevelt.com/images/research/txtspeeches/318.txt (accessed 14 August 2013); Walter L. Williams, "United States Indian Policy and the Debate over Philippine Annexation: Implications for the Origins of American Imperialism," *Journal of American History* 66 (March 1980): 810–31; Edward Charles Valandra, *Not without Our Consent: Lakota Resistance to Termination, 1950–59* (Urbana: University of Illinois Press, 2006), 107, 109.
59. Richard Pettigrew, *The Course of Empire* (New York: Boni and Liveright, 1920), 389.
60. *The Forester* 5 (August 1899): 172 (italics added).
61. *Irrigation Age* 15 (November 1900): 49; cf. Lovett, *Conceiving the Future*, 51.
62. Theodore Roosevelt, "First Annual Message," 3 December 1901, in Gerhard Peters and John T. Woolley, American Presidency Project, http://www.presidency.ucsb.edu/ws/?pid=29542 (accessed 19 July 2011).
63. For Republican "large policy," see Christopher Nichols, *Promise and Peril: America at the Dawn of a Global Age* (Cambridge, MA: Harvard University Press, 2011), 24–25.
64. Cf. Pisani, *Water and American Government*, 30–31.
65. Cf. Lovett, *Conceiving the Future*, 113.
66. Theodore Roosevelt, "Seventh Annual Message," 3 December 1907, in Gerhard Peters and John T. Woolley, American Presidency Project, http://www.presidency.ucsb.edu/ws/?pid=29548 (accessed 19 July 2011).
67. Ibid.
68. Cf. Pisani, "Water Planning in the Progressive Era," 399.
69. Ibid., 394–95. There was also the less developed question of the Atlantic Deep Waterway. William J. Roe, "The Atlantic Deep Waterway," *Arena* 41 (January 1909): 30–38.
70. W J McGee to Roosevelt, 4 February 1907, fl. Miscellaneous, October 22, 1906–April 10, 1907, Letterbook 3, box 26, W J McGee Papers, LC.
71. Hays, *Conservation and the Gospel of Efficiency*, 44, 102, 97–98. John Barrett, director of the International Bureau of American Republics (later the Pan American Union),

similarly argued that, with the canal, the Midwest could control trade with the west coast of South America. Barrett is cited in "Some Publications," *Iowa Journal of History and Politics* 6 (January 1908): 125–26. See also McGee to Officers and Members of the Latin American Club and Foreign Trade Association, [April?] 1906, box 26, McGee Papers.

72. W J McGee to Roosevelt, 4 February 1907, fl. Miscellaneous, October 22, 1906–April 10, 1907, Letterbook 3, box 26, McGee Papers; Roosevelt to McGee, 4 February 1907, ibid.
73. Pisani, "Water Planning in the Progressive Era," 392.
74. "Conserving Our Waterways as Natural Resources," *Outlook* 85 (20 April 1907): 868.
75. Quoted in "Inland Waterways Commission," *Independent* 62 (21 March 1907): 637.
76. James J. Hill, *Highways of Progress* (New York: Doubleday, Page, 1910), 207–9; "Conserving Our Waterways as Natural Resources," 868 (quote).
77. W. K. Kavanaugh, "Inland Waterways," *Independent* 64 (21 May 1908): 1144–45 at 1145 (quote).
78. Lee D. Baker, *From Savage to Negro: Anthropology and the Construction of Race, 1896–1954* (Berkeley: University of California Press, 1998), 65.
79. Donald Worster, *A River Running West: The Life of John Wesley Powell* (New York: Oxford University Press, 2001), 490.
80. Emma McGee, *Life of W J McGee: Distinguished Geologist, Ethnologist, Anthropologist, Hydrologist, Etc.* (Farley, IA: privately printed, 1915), 78.
81. Gifford Pinchot to Anita McGee, 22 March 1916, fl. Gen. Correspondence O–P, box 2, Anita McGee Papers, LC.
82. Newell, in "National Drainage Congress," *Forestry and Irrigation* 14 (February 1908): 102. "The effects of the operation of the Reclamation Act have been not only to reclaim public lands, but to break up large holdings, regulate the water supply to these and put upon the lands which otherwise are unproductive through *excess* [italics added] or deficiency of moisture a dense population of landowners and producers."
83. McGee, "Water as a Resource," 48 (quote); W J McGee to Roosevelt, 4 February 1907, fl. Miscellaneous, October 22, 1906–April 10, 1907, Letterbook 3, box 26, McGee Papers.
84. W J McGee, "The Conservation of Natural Resources," *Proceedings of the Mississippi Valley Historical Association*, in McGee, *Life of W J McGee*, 98–99; J. Leonard Bates, "Fulfilling American Democracy: The Conservation Movement, 1907 to 1921," *Mississippi Valley Historical Review* 44 (June 1957): 29–57 at 34, 39; Gifford Pinchot to Anita McGee, 22 March 1916, Anita McGee Papers (quote).
85. W J McGee to Roosevelt, 4 February 1907, fl. Miscellaneous, October 22, 1906–April 10, 1907, Letterbook 3, box 26, McGee Papers.
86. Gifford Pinchot, *The Fight for Conservation*, Americana Library, vol. 5 (1910; repr., Seattle: University of Washington Press, 1967), 44.
87. McGee, "Water as a Resource," 44, 48, 49 (quote).
88. Tuason, "Ideology of Empire," 40; Nichols, *Promise and Peril*, 24–25.
89. McGee, "The Growth of the United States," *National Geographic Magazine* 9 (September 1898): 377–386 at 386, quoted in Tuason, "Ideology of Empire," 41.
90. W J McGee, "National Growth and National Character," *National Geographic Magazine* 10 (June 1899): 185–206 at 204, quoted from the original (mistranscribed in Tuason, "Ideology of Empire," 41).
91. McGee, "National Growth," 205.
92. Ibid., 203.

93. Ibid., 206 ("Strong Man's burden"); Tuason, "Ideology of Empire," 42.
94. Bates, "Fulfilling American Democracy," 40; McGee, "The Five-fold Functions of Government," *Popular Science Monthly* 77 (September 1910), in McGee, *Life of W J McGee*, 194, 197.
95. W J McGee to Roosevelt, 4 February 1907, fl. Miscellaneous, October 22, 1906–April 10, 1907, Letterbook 3, box 26, McGee Papers.
96. Ian Tyrrell, "Public at the Creation: Place, Memory and Historical Practice in the Mississippi Valley Historical Association," *Journal of American History* 94 (June 2007): 39–40; McGee, "Conservation of Natural Resources," 88–100.
97. McGee to Roosevelt, 4 February 1907, fl. Miscellaneous, October 22, 1906–April 10, 1907, Letterbook 3, box 26, McGee Papers.
98. Frank Buffington Vrooman, *Theodore Roosevelt, Dynamic Geographer . . .* (London: Henry Frowde, Oxford University Press, 1909), 48–49 (quote).
99. W J McGee to Roosevelt, 4 February 1907, fl. Miscellaneous, October 22, 1906–April 10, 1907, Letterbook 3, box 26, McGee Papers.
100. McGee to Gifford Pinchot, 4 February 1907, ibid.
101. McGee to Robert Wakeman, 7 February 1907, ibid. (quote); Theodore Roosevelt to "Dear Sir," 14 March 1907, in *Preliminary Report of the Inland Waterways Commission* (Washington, DC: Government Printing Office, 1908), 16.
102. R. W. Douglas to McGee, 22 July 1909, fl. Gen. Corresp. D, 1900–1905, 1907–10, box 2, McGee Papers.
103. McGee, "Conservation of Natural Resources," 90.
104. McGee, "The Five-fold Functions of Government," 194, 197, 210–11; Roosevelt to McGee, 4 February 1907; McGee to Gifford Pinchot, 4 February 1907; McGee to Robert Wakeman, 7 February 1907; all in fl. Miscellaneous, October 22, 1906–April 10, 1907, Letterbook 3, box 26, McGee Papers; Pisani, "Water Planning in the Progressive Era," 394–95.
105. McGee, "Water as a Resource," 48.
106. Ibid., 38.
107. Stephen Ponder, *Managing the Press: Origins of the Media Presidency, 1897–1933* (New York: Palgrave Macmillan, 1998), 43; David Ryfe, *Presidents in Culture: The Meaning of Presidential Communication* (New York: Peter Lang, 2005).
108. Theodore Roosevelt, speech delivered at Vicksburg, Mississippi, 21 October 1907, http://www.theodore-roosevelt.com/images/research/txtspeeches/263.txt (accessed 10 October 2012).
109. Edmund Morris, *Theodore Rex* (New York: Random House, 2001), 496; Ponder, *Managing the Press*, 43–44 (quotes).
110. Morris, *Theodore Rex*, 497; Ryfe, *Presidents in Culture*, 39–40.
111. Donald J. Pisani, "A Conservation Myth: The Troubled Childhood of the Multiple-Use Idea," *Agricultural History* 76 (Spring 2002): 154–71 at 168; *Preliminary Report of the Inland Waterways Commission* (Washington, DC: Government Printing Office, 1908).
112. Hays, *Conservation and the Gospel of Efficiency*, 109; *Preliminary Report of the Inland Waterways Commission*, 30–31; John Howe Peyton, *The American Transportation Problem: A Study of American Transportation Conditions* (Louisville, KY: Courier-Journal Job Printing, 1909).
113. *Final Report of the National Waterways Commission* (Washington, DC: Government Printing Office, 1912).

114. Hays, *Conservation and the Gospel of Efficiency*, 139; *NCC Report*, vol. 2, esp. Herbert Knox Smith, "Transportation by Water," 13–66.
115. *Autobiography of John Hays Hammond*, 563.

NOTES TO CHAPTER SEVEN

1. Frederick Jackson Turner, *The Frontier in American History* (New York: Henry Holt, 1920), chap. 1; Robert E. Lang, Deborah Epstein Popper, and Frank J. Popper, "Progress of the Nation: The Settlement History of the Enduring American Frontier," *Western Historical Quarterly* 26 (Autumn 1995): 289–307 at 293.
2. David B. Danbom, *The Resisted Revolution: Urban America and the Industrialization of Agriculture, 1900–1930* (Ames: Iowa State University Press, 1979), 42.
3. Elizabeth Sanders, *Roots of Reform: Farmers, Workers, and the American State, 1877–1917* (Chicago: University of Chicago Press, 1999), esp. 151; Stuart W. Shulman, "The Origin of the Federal Farm Loan Act: Issue Emergence and Agenda-Setting in the Progressive Era Print Press," in Jane H. Adams, ed., *Fighting for the Farm: Rural America Transformed* (Philadelphia: University of Pennsylvania Press, 2003), 113–28.
4. Edmund Morris, *Theodore Rex* (New York: Random House, 2001), 250–51, 360. On Hill, see Albro Martin, *James J. Hill and the Opening of the Northwest* (New York: Oxford University Press, 1976); Michael Malone, *James J. Hill: Empire Builder of the Northwest* (Norman: University of Oklahoma Press, 1996).
5. "American Wastefulness," *Washington Post*, 21 January 1906.
6. *Proceedings of a Conference of Governors in the White House, Washington, D.C., May 13–15, 1908* (Washington, DC: Government Printing Office, 1908), 67.
7. "Hill Gives Farming Prizes," *New York Times*, 14 February 1906, 16.
8. Ibid.
9. James J. Hill, *Highways of Progress* (New York: Doubleday Page, 1910), 23 (quote); Malone, *James J. Hill*, 196.
10. *Conference of Governors*, 73, 74.
11. "Timely Suggestions: James J. Hill's Great Speech at the Minnesota State Fair on 'The Future of American Agriculture,' " *Southern Cultivator* 64 (15 October 1906): 4.
12. *New York Times*, 4 September 1906, 4.
13. "Agriculture the True Source of Our Wealth," *Scientific American* 95 (29 September 1906): 226.
14. Danbom, *Resisted Revolution*, 41 (second quote), 42 (first quote).
15. Cyril Hopkins, "The Conservation and Preservation of Soil Fertility," *Annals of the American Academy of Political and Social Science* 33 (May 1909): 147–62 at 148.
16. *Conference of Governors*, 25.
17. Ibid., 67.
18. Hopkins, "Conservation and Preservation of Soil Fertility," 149.
19. Dan O'Donnell, "The Pacific Guano Islands: The Stirring of American Empire in the Pacific Ocean," *Pacific Studies* 16 (March 1993): 43–66; Van Hise, in *NCC Report*, 1:180 (quote).
20. Hopkins, "Conservation and Preservation of Soil Fertility," 150.
21. F. B. Van Horn, "The Phosphate Deposits of the United States," *NCC Report*, 3:569.
22. Van Hise, in *NCC Report*, 1:180.
23. E. J. Russell, "Prof. Milton Whitney," *Nature* 121 (7 January 1928): 27; William R. Reed, "California Soil Mapping in the 1900s," http://www.pssac.org/SoilMapping History/1900smappinghistory.htm (accessed 18 December 2011); Douglas Helms,

Anne B. W. Effland, and Patricia J. Durana, eds., *Profiles in the History of the U.S. Soil Survey* (Ames: Iowa State University Press, 2002), 15, 20–22, 25–28, 32.

24. Milton Whitney, "Crop Yield and Soil Composition," *NCC Report*, 3:36, 29 (last quote).
25. Ibid., 36.
26. Ibid., 9. See also Milton Whitney, *Soil and Civilization: A Modern Concept of the Soil and the Historical Development of Agriculture* (New York: D. Van Nostrand, 1925).
27. Ellery Channing Chilcott, "The Agricultural Resources of the Eastern United States: Their Development and Conservation," *NCC Report*, 3:246–53 at 246, 252, 253.
28. Whitney, "Crop Yield," 28 (quote), 33–34; James C. Malin, *The Grassland of North America: Prolegomena to Its History* (Ann Arbor, MI: Edwards Brothers, 1947), 48, 224.
29. "Moisture in the Soil," *New York Times*, 18 November 1897; Reed, "California Soil Mapping in the 1900s"; Douglas Helms, "The Early Soil Survey: Engine for the Soil Conservation Movement," in D. E. Stott, R. H. Mohtar, and G. C. Steinhardt, eds., *Sustaining the Global Farm: Selected Papers from the 10th International Soil Conservation Organization Meeting Held May 24–29, 1999 at Purdue University and the USDA-ARS National Soil Erosion Research Laboratory* (n.p., 2001), 1031; Donald Worster, *Dust Bowl: The Southern Plains in the 1930s* (New York: Oxford University Press, 1979), 87–88.
30. Theodore Roosevelt, "Third Annual Message," 7 December 1903, in Gerhard Peters and John T. Woolley, American Presidency Project, http://www.presidency.ucsb.edu/ws/index.php?pid=29544#ixzz1hJSSNZ.
31. "To Save the Rural Towns," *Independent* 53 (16 May 1901): 1145, 1146. On Hamilton Holt, see http://peace.maripo.com/m_holt.htm (accessed 21 December 2011).
32. "To Save the Rural Towns" (quotes); Amy Kaplan, "Roman Fever: Imperial Melancholy in America" (paper presented at "Comparative Imperial Transformations" conference, University of Sydney, July 2008).
33. Roosevelt to Liberty Hyde Bailey, 10 August 1908, http://www.theodore-roosevelt.com/images/research/txtspeeches/299.txt (quote); *Theodore Roosevelt: An Autobiography*, new ed. (1913; New York: Charles Scribner's Sons, 1922), 413; *Report of the Commission on Country Life, with an Introduction by Theodore Roosevelt* (1911; repr., New York: Sturgis and Walton, 1917), 10.
34. *Special Message from the President of the United States Transmitting the Report of the Country Life Commission, February 9, 1909*, 60th Cong., 2nd Sess., S. Doc. No. 705, at 8 (1908–9).
35. Roosevelt to Pinchot, 10 August 1908, ser. 2, reel 350, Theodore Roosevelt Papers, LC.
36. *Theodore Roosevelt: An Autobiography*, 412.
37. *Special Message . . . Country Life*, 9; Roosevelt to Pinchot, 10 August 1908, series 2, reel 350, Roosevelt Papers ("farmer's wife").
38. *Special Message . . . Country Life*, 5. The gendered nature of these policies is well explored in Edith Ziegler, "'The Burdens and Narrow Life of Farm Women': Women, Gender, and Theodore Roosevelt's Commission on Country Life," *Agricultural History* 86 (Summer 2012): 77–103.
39. E.g., Roosevelt to Pinchot, 10 August 1908, 1 September 1908, and 9 November 1908, series 2, reels 350–52, Roosevelt Papers.
40. Scott J. Peters and Paul A. Morgan, "The Country Life Commission: Reconsidering a Milestone in American Agricultural History," *Agricultural History* 78 (Summer 2004): 289–316 at 292; Carl H. Moneyhon, "Environmental Crisis and American Politics, 1860–1920," in Lester J. Bilsky, ed., *Historical Ecology: Essays on Environment and Social Change* (Port Washington, NY: Kennikat Press, 1980), 140–55, esp. 152; Travis Koch,

"Two Sides of the Same Policy: Rural Improvement and Resource Conservation in the Progressive Era," http://www.stanford.edu/group/ruralwest/cgi-bin/drupal/content/rural-life-koch (accessed 22 November 2012); Timothy Collins, "Rural Community Development and Sustainability: The Lasting Legacy of Theodore Roosevelt's Commission on Country Life," Illinois Institute for Rural Affairs, http://www.iira.org/clc/clc_and_community.asp (accessed 22 January 2013); Clayton S. Ellsworth, "Theodore Roosevelt's Country Life Commission," *Agricultural History* 34 (Fall 1960): 155–72; Danbom, *Resisted Revolution*.

41. ASHPS, [Fourth] *Annual Report of the Society for the Preservation of Scenic and Historic Places and Objects, . . . 1899* (Albany, NY: State Printers, 1899), 11, 12 (quotes); Rebecca Conard, *Places of Quiet Beauty: Parks, Preserves, and Environmentalism* (Iowa City: University of Iowa Press, 1997), chap. 1.
42. *NCC Report*, 1:262.
43. Ibid., 155.
44. Statement by Liberty Bailey, National Conservation Congress, 1910, in ASHPS, *Sixteenth Annual Report 1911*, 210.
45. Ibid., 211.
46. Liberty Bailey, "A Reverie of Gardens," *Outlook* 68 (1 June 1901): 268.
47. L. H. Bailey, *The Outlook to Nature*, rev. ed. (1905; New York: Macmillan, 1911), esp. 79–81.
48. *Theodore Roosevelt: An Autobiography*, 413 (quote); Horace Plunkett, "Conservation and Rural Life: An Irish View of Two Roosevelt Policies," *Outlook* 94 (January 1910): 260–64; Margaret Digby, *Horace Plunkett: An Anglo-American Irishman* (Oxford: Basil Blackwell, 1949), 117–43.
49. Philip Bull, "Plunkett, Sir Horace Curzon (1854–1932)," in *Oxford Dictionary of National Biography* (Oxford University Press, online ed., January 2008), http://www.oxforddnb.com/view/article/35549 (accessed 11 October 2009); Lawrence M. Woods, *Horace Plunkett in America: An Irish Aristocrat on the Wyoming Range* (Norman, OK: Arthur H. Clark, 2010).
50. W. G. S. Adams, introduction to Digby, *Plunkett*, vii.
51. Digby, *Plunkett*, 118.
52. Woods, *Plunkett*, 151.
53. Digby, *Plunkett*, 121.
54. Roosevelt to Plunkett, 6 July 1906, Plunkett Papers, Plunkett Foundation, Oxford.
55. Memorandum of 29 November 1907, ibid.
56. *Theodore Roosevelt: An Autobiography*, 414.
57. Pinchot to Plunkett, 29 May 1908, Horace Plunkett Papers (microfilm ed.).
58. Roland A. White, Milo Reno: Farmers' Union Pioneer (Iowa City: Athens Press, 1941), 288.
59. Olaf F. Larson and Thomas B. Jones, "The Unpublished Data from Roosevelt's Commission on Country Life," *Agricultural History* 50 (October 1976): 583–99.
60. Plunkett to Pinchot, 6 February 1909, and Pinchot to Plunkett, 15 February 1909 (quote), Plunkett Papers (microfilm ed.).
61. *Special Message . . . Country Life*, 4–5 (quote at 5).
62. *Report of the Commission on Country Life*, 89.
63. Ibid., 90, 21 (first and last quotes).
64. Ibid., 72.
65. Ibid., 73. See, e.g., Joseph A. Holmes to Jane Sprunt Holmes, 4 March 1906, fl. 12, 1904–7, box 1, Joseph A. Holmes Papers, SHC.

66. *Report of the Commission on Country Life*, 23.
67. Ibid., 23, 69 (quote).
68. Ibid., 120.
69. Memorandum, 29 November 1907, Plunkett Papers.
70. *Report of the Commission on Country Life*, 129.
71. On the global diffusion of cooperative ideals, see Thomas Adam, *Intercultural Transfers and the Making of the Modern World, 1800–2000* (Basingstoke, Eng.: Palgrave Macmillan, 2012), chap. 1, esp. 34–36.
72. *Report of the Commission on Country Life*, 100.
73. Ibid., 125.
74. Larson and Jones, "Unpublished Data," 597.
75. Roosevelt to Pinchot, 9 November 1908, reel 352, Roosevelt Papers.
76. Liberty H. Bailey, unpaginated "Explanation" to *Report of the Commission on Country Life*, [p. 3].
77. Plunkett to Pinchot, 29 January 1910, Plunkett Papers.
78. Pinchot to Roosevelt, 23 April 1910, box 138, fl. Roosevelt, Theodore, Pinchot Papers.
79. Horace Plunkett, *The Rural Life Problem of the United States: Notes of an Irish Observer* (New York: Macmillan, 1910); Horace Plunkett, "Rural Regeneration," *North American Review* 214 (October 1921): 470–76.
80. Sanders, *Roots of Reform*, 175; Federal Farm Loan Act of 1916, Pub. L. No. 64-158, 39 Stat. 360 (1916).

NOTES TO CHAPTER EIGHT

1. Some preservationists complained that they had been sidelined. John Muir was not among the invitees to the conference. Donald Worster, *A Passion for Nature: The Life of John Muir* (New York: Oxford University Press, 2008), 430.
2. Douglas Brinkley, *The Wilderness Warrior: Theodore Roosevelt and the Crusade for America* (New York: HarperCollins, 2009), raises the issue; see esp. 789–90, 897 (quote). Kathleen Dalton, *Theodore Roosevelt: A Strenuous Life* (New York: Knopf, 2002), 239–48, tackles the problem more resolutely and analytically as a "pragmatic" and "political" compromise (240, 241). See also Robert W. Righter, *The Battle over Hetch Hetchy: America's Most Controversial Dam and the Birth of Modern Environmentalism* (New York: Oxford University Press, 2006), 65.
3. Brinkley, *Wilderness Warrior*, 828–30.
4. Ibid.; Theodore Roosevelt, *Wilderness Writings*, ed. Paul Schullery (Salt Lake City, UT: Gibbs M. Smith, 1986); Paul Russell Cutright, *Theodore Roosevelt: The Naturalist* (New York: Harper and Brothers, 1956); Paul Russell Cutright, *Theodore Roosevelt: The Making of a Conservationist* (Urbana: University of Illinois Press, 1985); W. Todd Benson, *President Theodore Roosevelt's Conservation Legacy* (Haverford, PA: Infinity Publishing, 2003), 101–6; Richard Slotkin, "Nostalgia and Progress: Theodore Roosevelt's Myth of the Frontier," *American Quarterly* 33, no. 5 (1981): 608–37; Roderick Nash, *Wilderness and the America Mind*, 4th ed. (New Haven, CT: Yale University Press, 2001), 149–52.
5. John F. Lacey, "Protection to Game and Birds in the United States," in L. H. Pammel, comp., *Major John F. Lacey: Memorial Volume* (Cedar Rapids: Iowa Park and Forestry Association, 1915), 169; ASHPS, *Sixteenth Annual Report 1911*, 206.
6. Melanie Hall, "Niagara Falls: Preservation and the Spectacle of Anglo-American Accord," in Melanie Hall, ed., *Towards World Heritage: International Origins of the Preservation Movement, 1870–1930* (Farnham, Eng.: Ashgate, 2009), 24.

7. David K. Prendergast and William M. Adams, "Colonial Wildlife Conservation and the Origins of the Society for the Preservation of the Wild Fauna of the Empire (1903–1914)," *Oryx* 37, no. 2 (2003): 251–60; Sherman Strong Hayden, *The International Protection of Wild Life* (New York: Columbia University Press, 1942), 36–37; Bernhard Gissibl, "German Colonialism and the Beginnings of International Wildlife Preservation in Africa," *GHI Bulletin Supplement* 3 (2006): 121–43 at 132; John MacKenzie, *The Empire of Nature: Hunting, Conservation and British Imperialism* (Manchester: Manchester University Press, 1989), 81, 202, 321; Richard White and Melissa Harper, "How National Were the First National Parks? Comparative Perspectives from the British Settler Societies," in Bernhard Gissibl, Sabine Höhler, and Patrick Kupper, eds., *Civilizing Nature: National Parks in Global Historical Perspective* (New York: Berghahn Books, 2012), 50–67; Tom Mels, *Wild Landscapes: The Cultural Nature of Swedish National Parks* (Lund: Lund University Press, 1999); Patrick Kupper, "Science and the National Parks: A Transatlantic Perspective on the Interwar Years," *Environmental History* 14, no. 1 (2009): 58–81.
8. Mark V. Barrow Jr., *Nature's Ghosts: Confronting Extinction from the Age of Jefferson to the Age of Ecology* (Chicago: University of Chicago Press, 2009), 106; E. Louise Peffer, *The Closing of the Public Domain: Disposal and Reservation Policies, 1900–50* (Stanford, CA: Stanford University Press, 1951), 15.
9. Brinkley, *Wilderness Warrior*, 13–14.
10. William T. Hornaday, "The Extermination of American Animals," *Chautauquan* 10 (December 1889): 304–7, in fl. Conservation Magazine Reprints, box 20, William T. Hornaday Papers, LC; William T. Hornaday, "The Destruction of Our Birds and Mammals," extracted from the *Second Annual Report of the New York Zoological Society* (New York: Office of the Society, 1898), box 23, Hornaday Papers; MacKenzie, *Empire of Nature*, chap. 8. Disappointing is Stephen Bechtel, *Mr. Hornaday's War: How a Peculiar Victorian Zookeeper Waged a Lonely Crusade for Wildlife That Changed the World* (Boston: Beacon Press, 2012), 46.
11. John F. Lacey, "On Forestry: Impromptu Address by Hon. John Lacey, Member of Congress from Iowa, at the American Forest Congress (403–09), Washington, DC, January 2–6, 1905," in Pammel, comp., *Lacey: Memorial Volume*, 96.
12. Carolyn Merchant, "Women of the Progressive Conservation Movement: 1900–1916," *Environmental Review* 8 (Spring 1984): 57–85, esp. 57–63; Carolyn Merchant, *American Environmental History: An Introduction* (New York: Columbia University Press, 2007), 142.
13. Theodore W. Cart, "The Lacey Act: America's First Nationwide Wildlife Statute," *Forest History* 17 (October 1973): 4–13 at 6.
14. Carolyn Merchant, "George Bird Grinnell's Audubon Society: Bridging the Gender Divide in Conservation," *Environmental History* 15 (January 2010): 3–30; Robin Doughty, *Feather Fashions and Bird Preservation: A Study in Nature Protection* (Berkeley: University of California Press, 1975), 97–115.
15. Barrow, *Nature's Ghosts*, 106; Executive Order, 13 March 1903, Pelican Island Reservation for Birds, http://www.theodore-roosevelt.com/trexecutiveorders.html.
16. Brinkley, *Wilderness Warrior*, 269; Mary Annette Gallagher, "John F. Lacey: A Study in Organizational Politics" (PhD diss., University of Arizona, 1970); Theodore W. Cart, "The Struggle for Wildlife Protection in the United States, 1870–1900: Attitudes and Events Leading to the Passage of the Lacey Act" (PhD diss., University of North Carolina, Chapel Hill, 1971); Cart, "Lacey Act," 4–13. See also Mary Gallagher, "Citizen of the Nation: John Fletcher Lacey, Conservationist," *Annals of Iowa* 46 (Summer

1981): 9–24; Rebecca Conard, "John F. Lacey: Conservation's Public Servant," in David Harmon, Francis McManamon, and Dwight T. Pitcaithley, eds., *The Antiquities Act: A Century of American Archaeology* (Tucson: University of Arizona Press, 2006), 48–63.

17. Alfred Runte, *National Parks: The American Experience*, 4th ed. (Lanham, MD: Taylor Trade, 2010), 191–92; Harlean James, *Romance of the National Parks* (1939; repr., New York: Arno Press, 1972), 68.
18. Lacey, "On Forestry," 93.
19. Rebecca Conard, *Places of Quiet Beauty: Parks, Preserves, and Environmentalism* (Iowa City: University of Iowa Press, 1997), chap. 1.
20. Gallagher, "John F. Lacey," 75–76.
21. Cart, "Struggle for Wildlife Protection," 114–15.
22. Gallagher, "John F. Lacey," viii, 340–44.
23. He praised the acquisition of the (still-unincorporated) territory of Alaska because of its resources, rejoicing that "time has vindicated the wisdom of Mr. Seward, and Alaska is no longer the least prized of our possessions" (Pammel, comp., *Lacey: Memorial Volume*, 326).
24. Ibid., 327.
25. John F. Lacey, "Forestry—the Tree Is the Mother of the Fountain—a Tree Is the Best Gift of Heaven to Man," in Pammel, comp., *Lacey: Memorial Volume*, 115.
26. Pammel, comp., *Lacey: Memorial Volume*, 194; Edmund J. James, *The Opportunity of Chicago to Become a Great City* (Chicago: World Review, [c. 1901]), 43, 46.
27. John F. Lacey, "Bird Protection" (1901), in Pammel, comp., *Lacey: Memorial Volume*, 172.
28. John F. Lacey, "Need of Forest Preservation," *Gunton's Magazine* 24 (January 1903): 85. Cf. George Perkins Marsh, *Man and Nature; or, Physical Geography as Modified by Man* (New York: Charles Scribner, 1864), 297.
29. Kurkpatrick Dorsey, *The Dawn of Conservation Diplomacy: U.S.-Canadian Wildlife Protection Treaties in the Progressive Era* (Seattle: University of Washington Press, 1998), 189, 191, 237.
30. L. H. Pammel, "Major John F. Lacey and the Conservation of Our Natural Resources," in Pammel, comp., *Lacey: Memorial Volume*, 39; Lacey, "Bird Protection," in ibid., 172. See also "Game Protectors Meet," *New York Times*, 14 January 1901.
31. Refuge Trespass Act of June 28, 1906, 18 U.S.C. §41.
32. George Gladden (secretary, Wild Life Protective Association), "Federal Protection for Migratory Birds: Genesis, Character and Purpose of the McLean Bill . . . ," *Outing* 162 (1913): 345–49; Doughty, *Feather Fashions and Bird Preservation*, 125–32; Cart, "Struggle for Wildlife Protection"; Cart, "Lacey Act," 4–13; Dorsey, *Dawn of Conservation Diplomacy*, chaps. 6–7.
33. An Act for the Preservation of American Antiquities, 16 U.S.C. §§431–33. The bill sailed through the Senate viva voce.
34. Brinkley, *Wilderness Warrior*, 755.
35. Robert W. Righter, "National Monuments to National Parks: The Use of the Antiquities Act of 1906, http://www.cr.nps.gov/history/hisnps/npshistory/righter.htm (accessed 29 February 2012).
36. Frank Norris, "The Antiquities Act and the Acreage Debates," *George Wright Forum* 23, no. 3 (2006): 6–16 at 11. Roosevelt's action was legally contested but affirmed by the Supreme Court in 1920.
37. John F. Lacey, "Forestry" (an address before the Iowa Federation of Women's Clubs, Waterloo, Iowa, 12 May 1905), in Pammel, comp., *Lacey: Memorial Volume*, 83;

Mrs. Lovell White (Calaveras Big Tree Committee, Outdoor Art League of California), "The Calaveras Grove of Big Trees: Reasons for Their Preservation by the Federal Government," *Forestry and Irrigation* 12 (February 1906): 102–3 at 103, regarding "the one million and a half of people who signed a petition in 1904 sent by the Outdoor Art League to President Roosevelt urging him to request Congress to pass the Big Tree bill." But the Calaveras Grove was not protected until 1931. Merchant, "Women of the Progressive Conservation Movement," 59–60.

38. Cf. Runte, *National Parks*, 49, 77–78 ("loss"), who stresses a tactical alliance rather than the application of aesthetic principles. See also Pammel, comp., *Lacey: Memorial Volume*, 83.
39. Edward Hagaman Hall, "Jamestown: Proposal That It Be Made a National Park," *New York Times*, 22 February 1902; James M. Lindgren, *Preserving the Old Dominion: Historic Preservation and Virginia Traditionalism* (Charlottesville: University of Virginia Press, 1993), 116. See also James M. Lindgren, *Preserving Historic New England: Preservation, Progressivism, and the Remaking of Memory* (New York: Oxford University Press, 1995).
40. Roy Rosenzweig and Elizabeth Blackmar, *The Park and the People: A History of Central Park* (Ithaca, NY: Cornell University Press, 1992), chap. 7; William R. Irwin, *New Niagara: Tourism, Technology, and the Landscape of Niagara Falls, 1776–1917* (University Park: Pennsylvania State University Press, 1996), 84.
41. Rosenzweig and Blackmar, *The Park and the People*, 348. For Green's role in Central Park, see ibid., chap. 7.
42. "George Frederick Kunz: Gem Expert," in *Twentieth Century Successful Americans* (n.p.: United Press Bureau, n.d.), no. 403C, http://www.archive.org/stream/twentieth century01unit/twentiethcentury01unit_djvu.txt (accessed 30 October 2010); *New York Times*, 4 July 1932, 10; Paul F. Kerr, "Memorial of George Frederick Kunz," *American Mineralogist* 18 (March 1933): 91–94, esp. 93.
43. On Howland, see "The Death of William B. Howland," *Outlook*, 7 March 1917, 399; obituary for Howland, *Independent*, 12 March 1917, 433–44, in fl. Howland Memorial, box 4, ACA General Correspondence, series 18, J. Horace McFarland Papers, PSA.
44. Cutright, *Roosevelt: Making of a Conservationist*, 238 (quote); Brinkley, *Wilderness Warrior*, 765–66.
45. Theodore Roosevelt, "First Annual Message," 3 December 1901, http://millercenter .org/scripps/archive/speeches/detail/3773; Theodore Roosevelt, "Eighth Annual Message," 9 December 1908, http://millercenter.org/scripps/archive/speeches/detail /3780 (accessed 11 December 2010).
46. Patrick Young, "A Tasteful Patrimony? Landscape Preservation and Tourism in the Sites and Monuments Campaign, 1900–1935," *French Historical Studies* 32 (Summer 2009): 447–77.
47. ASHPS, *Nineteenth Annual Report, 1914*, 316. On Belgian conservation, see Raf De Bont and Rajesh Heynickx, "Landscapes of Nostalgia: Life Scientists and Literary Intellectuals Protecting Belgium's 'Wilderness,' 1900–1940," *Environment and History* 18 (May 2012): 237–60 at 260.
48. ASHPS, [Fifth] *Annual Report, of the Society for the Preservation of Scenic and Historic Places and Objects, 1900* (Albany, NY: James B. Lyon, State Printer, 1900), 29. See also "Origin and Motives of the Society," in ASHPS, *Nineteenth Annual Report, 1914*, 13–14. By 1901 the ASHPS had a membership of "a few hundreds" and support from the *New York Times*, 20 April 1901, BR10. The name was now changed to American

Scenic and Historic Preservation Society. Cf. Michael Holleran, "America's Early Historic Preservation Movement (1850–1930) in a Transatlantic Context," in Hall, ed., *Towards World Heritage*, 197.

49. "Origin and Motives of the Society," 22; Thomas Lekan, *Imagining the Nation in Nature: Landscape Preservation and German Identity, 1885–1945* (Cambridge, MA: Harvard University Press, 2004), chap. 1.
50. Brinkley, *Wilderness Warrior*, 352; attachment to Minutes of Trustees, 9 December 1899, 78–79, book 1, box 6, ASHPS Records, NYPL.
51. Max Nicholson, *The New Environmental Age* (Cambridge: Cambridge University Press, 1987).
52. Brinkley, *Wilderness Warrior*, 766 (quote), 414.
53. "Origin and Motives of the Society"; [First] *Annual Report of the Society of the Trustees of Scenic and Historical* [*sic*] *Places and Objects* (Albany, NY: State Printers, 1896), 9 (quote); "Danger to Niagara Falls," *New York Times*, 18 December 1895.
54. See "Fight to Preserve Beauties of Niagara," *New York Times*, 23 December 1915, 9, where New York State assemblymen accused the Burton Act, and McFarland, of preserving a power monopoly and the hydroelectric concessionaires of being members of the ACA; Jean Strouse, *Morgan: American Financier* (New York: Random House, 1999), 395; *The American Scenic and Historic Preservation Society: A National Society for the Protection of Natural Scenery, the Preservation of Historic Landmarks and the Improvement of Cities* (New York: American Scenic and Historic Preservation Society, 1911), 3. ASHPS trustees included Edward Dean Adams and Morgan. Both had power company interests.
55. N. F. Dreisziger, "The Campaign to Save Niagara Falls and the Settlement of United States–Canadian Differences, 1906–1911," *New York History* 55 (October 1974): 437–58 at 446; George Kunz Address, Minutes of Trustees, 25 May 1908, 108, 109, book 1, box 10, ASHPS Records (quote).
56. "Statement of Mr. Charles M. Dow, President of the Commission of the New York State Reservation at Niagara," in *Preservation of Niagara Falls (H. R. 18024): Hearings before the Committee on Rivers and Harbors of the House of Representatives of the United States, Fifty-Ninth Congress, First Session* (Washington, DC: Government Printing Office, 1906), 165–67; ASHPS, *Nineteenth Annual Report, 1914*, 253; Irwin, *New Niagara*, 85, 211; Gail E. H. Evans, "Storm over Niagara: A Catalyst in Reshaping Government in the United States and Canada during the Progressive Era," *Natural Resources Journal* 32 (Winter 1992): 27–54.
57. "Beautiful America: Conducted by J. Horace McFarland, President of the American League for Civic Improvement," *Ladies' Home Journal* 29 (March 1904): 15; Julian Chambliss, "Perfecting Space: J. Horace McFarland and the American Civic Association," *Pennsylvania History: A Journal of Mid-Atlantic Studies* 77, no. 4 (2010): 486–97; Ernest Morrison, *J. Horace McFarland: A Thorn for Beauty* (Harrisburg: Pennsylvania Historical and Museum Commission, 1995); William H. Wilson, "J. Horace McFarland and the City Beautiful Movement," *Journal of Urban History* 7 (May 1981): 315–34.
58. J. Horace McFarland, "The Last Call on Billboards," *Ladies' Home Journal* 22 (August 1905): 28; J. Horace McFarland, "A Direct Talk about Billboards," ibid. 22 (July 1905): 18; J. Horace McFarland, "Why Billboard Advertising as at Present Conducted Is Doomed," *Chautauquan* 51 (June 1908): 19 (pointing to the existence of "sane regulation" of billboards in "many European cities").
59. ASHPS, *Nineteenth Annual Report, 1914*, 253; ASHPS, [First] *Annual Report, 1896*, 9 (quote); Irwin, *New Niagara*, 85, 211; Evans, "Storm over Niagara: A Catalyst";

Richard Watrous to William Frederick Badè, 29 August 1910, http://cdn.calisphere.org/data/13030/7s/kt6s20367s/files/kt6s20367s-d3e256258.jpg (accessed 30 July 2012).

60. J. Horace McFarland, "Shall We Make a Coal-Pile of Niagara?," *Ladies' Home Journal* 22 (September 1905): 19.
61. McFarland to John Muir, 22 December 1908, fl. Mc–Mu, box 16, series 23, Hetch Hetchy General Correspondence, 1908–15, McFarland Papers.
62. McFarland to the *Independent,* 2 March 1917, 3–4, fl. Howland Memorial, box 4, series 18, ACA General Correspondence, McFarland Papers.
63. J. Horace McFarland, "A Brief Summary of the Niagara Campaign," *Chautauquan* 47 (August 1907): 279–86 at 280.
64. J. Horace McFarland, "The Desecration of Niagara," *Ladies' Home Journal* 23 (January 1906): 27.
65. J. Horace McFarland, "How the Power Companies Beautify Niagara," *Ladies' Home Journal* 23 (October 1906): 39 (first quote); Evans, "Storm over Niagara: A Catalyst," 42 (second quote).
66. McFarland, "How the Power Companies Beautify Niagara," 39 (first three quotes); McFarland, "Desecration of Niagara," 27; Theodore Roosevelt, "Fifth Annual Message," 5 December 1905, in Gerhard Peters and John T. Woolley, American Presidency Project, http://www.presidency.ucsb.edu/ws/index.php?pid=29546#ixzz1njpwPplW (accessed 28 January 2012), (last quote).
67. Minutes of the Queen Esther Circle of Grace Methodist Episcopal Church, Harrisburg, PA, 5–29, box 8, series 15, Miss Helen McFarland, 1903–56, McFarland Papers.
68. Morrison, *McFarland: A Thorn for Beauty,* 4–5, 269; J. Horace McFarland, *The Awakening of Harrisburg* (Harrisburg: n.p., 1907), 5 (quote), 13.
69. McFarland, *Awakening of Harrisburg,* 13.
70. Ian Tyrrell, *Reforming the World: The Creation of America's Moral Empire* (Princeton, NJ: Princeton University Press, 2010), chaps. 6–7.
71. Cf. Samuel P. Hays, *Conservation and the Gospel of Efficiency: The Progressive Conservation Movement, 1890–1920* (Cambridge, MA: Harvard University Press, 1959), 141–42.
72. *Christian Science Monitor,* 7 March 1910, 7.
73. "Taft Stirs Laymen," *Washington Post,* 12 November 1909, 1.
74. John Wood (Domestic and Foreign Missionary Society of the Protestant Episcopal Church in the United States of America) to Gifford Pinchot, 12 October 1909, fl. Conservation/GC/1909, Oct.–Nov., box 465, Pinchot Papers, LC.
75. *Chautauquan* 40 (June–August 1907): 293; "News of the Arroyo," http://www.arroyoseco.org/newsfull.php?artic=917 (accessed 21 March 2012).
76. McFarland, "Brief Summary of the Niagara Campaign," 279 (quote); McFarland, "How the Power Companies Beautify Niagara," 39.
77. McFarland, "Shall We Make a Coal-Pile of Niagara?," 19.
78. McFarland to the *Independent,* 2 March 1917, fl. Howland Memorial, box 4, series 18, ACA General Correspondence, McFarland Papers.
79. McFarland, "Shall We Make a Coal-Pile of Niagara?," 19 (quotes); Evans, "Storm over Niagara: A Catalyst," 43.
80. McFarland, "Shall We Make a Coal-Pile of Niagara?," 19.
81. McFarland, "Brief Summary of the Niagara Campaign," 279. Though this campaign may have suited J. P. Morgan's existing interests, he did not take an active part in any of the campaigning on this issue, and McFarland and the ACA made it clear that the choice was a pragmatic one, not to benefit Morgan: "The provision to prohibit

importation seems a check on Canadian diversion, as you can readily understand, and it is the only efficient check that is possible at this time." McFarland to Charles Dow, 16 April 1906, fl. Dow, 1906–7, box 24, series 28, ACA Niagara Falls General Correspondence A–F, McFarland Papers.

82. J. Horace McFarland, "Shall We Have Ugly Conservation?," *Outlook* 91 (13 March 1909): 594–99 at 596.
83. *Proceedings of a Conference of Governors in the White House, Washington, D.C., May 13–15, 1908* (Washington, DC: Government Printing Office, 1908), 153–56.
84. McFarland, "Shall We Have Ugly Conservation?," 597.
85. Chambliss, "Perfecting Space"; William H. Wilson, *The City Beautiful Movement* (Baltimore, MD: John Hopkins University Press, 1989); Stanley K. Schultz, *Constructing Urban Culture: American Cities and City Planning, 1800–1920* (Philadelphia: Temple University Press, 1989).
86. Worster, *Passion for Nature*, 430; George Kunz, "The Preservation of Scenic Beauty," in *Conference of Governors*, 408–19; ASHPS submission by Kunz in *NCC Report*, 1:262; Charles Evans Hughes, in *Conference of Governors*, 322; James, *Romance of the National Parks*, 69. Bailey's most forceful statement came at the National Conservation Congress meeting in Saint Paul, Minnesota, in 1910, representing the ASHPS. Liberty Hyde Bailey, reported in ASHPS, *Sixteenth Annual Report, 1911*, 205.
87. McFarland, "Shall We Have Ugly Conservation?," 594. This formulation is similar to the idea of "prudence" suggested by Ethan Fishman, but the latter connotes a balancing act, whereas the scenic-beauty critique attempted a resolution of the problem. Fishman, "The Quality of Roosevelt's Environmentalism," in Serge Ricard, ed., *A Companion to Theodore Roosevelt* (Malden, MA: Wiley-Blackwell, 2012), 173–85.
88. ASHPS, *Sixteenth Annual Report, 1911*, 205.
89. Thomas C. Chamberlain, "Soil Wastage," in *Conference of Governors*, 75, 76.
90. *Conference of Governors*, 193 (quote); McFarland, draft letter to the governors, 15 July 1908, and "Some Reminiscences of the White House Conference in the Conservation of Natural Resources, May 12–15, 1908," both in fl. White House conference, 1908, box 21, series 25, ACA, National Parks, General Corresp. Smo-Yosemite, McFarland Papers; Minutes of Trustees, 25 October 1909, 24, book 2, box 10, ASHPS Records.
91. McFarland, "Shall We Have Ugly Conservation?," 596.
92. Liberty Hyde Bailey, in ASHPS, *Sixteenth Annual Report, 1911*, 205–11, esp. 206 ("latent beauty"); Declaration of Principles, reprinted in Treadwell Cleveland Jr., "The North American Conservation Conference," *Conservation* 15 (March 1909): 165.
93. *NCC Report*, 1:141; Pinchot to McFarland, 22 April 1909, and 24 November 1909, fl. Pinchot, box 16, series 23, Hetch Hetchy General Correspondence, 1908–15, McFarland Papers; *Conference of Governors*, 155; Minutes of Trustees, 25 October 1909, 24, book 2, box 10, ASHPS Records; Circular Letters to NCC members and chairmen of the state conservation commissions, 6 and 10 March 1909, fl. 1, Corresp. 1908–15, Harry A. Slattery Papers, David M. Rubenstein Rare Book and Manuscript Library, Duke University, Durham, NC.
94. Evans, "Storm over Niagara: A Catalyst," 21, 53 (quote). See also Nandor Dreisziger, "The Great Lakes in United States–Canadian Relations: The First Stock-Taking," *Inland Seas: The Quarterly Journal of the Great Lakes Historical Society* 28 (Winter 1972): 259–71; Dreisziger, "Campaign to Save Niagara Falls"; Gail E. H. Evans, "Storm over Niagara: A Study of the Interplay of Cultural Values, Resource Politics, and Environmental Policy in an International Setting, 1670s–1950" (PhD diss.,

University of California, Santa Barbara, 1991); McFarland, "Desecration of Niagara," 27.

95. Irwin, *New Niagara*, 114.
96. Righter, *Hetch Hetchy*, 57–61, 166.
97. Worster, *Passion for Nature*, esp. 413–14; Morrison, *McFarland: A Thorn for Beauty*, 168.
98. McFarland to Charles Ingersoll, 3 July 1908, fl. Gen. Corresp. F–Y, box 15, series 19, ACA Billboards, McFarland Papers.
99. Pinchot to McFarland, 22 April 1909, and 24 November 1909 (quote); McFarland to Pinchot, 26 November 1909, fl. Pinchot, box 16, series 23, Hetch Hetchy General Correspondence, 1908–15, McFarland Papers; McFarland to Agnes Laut, 8 December 1909, fl. Johnson-1910-L, ibid.
100. William B. Howland to McFarland, 22 February 1916, fl. Howland, Wm. B. 1915–16 Mar., box 4, ACA General Correspondence, series 18, McFarland Papers.
101. Stephen Fox, *John Muir and His Legacy: The American Conservation Movement* (Boston: Little, Brown, 1981), 144; ASHPS, *Nineteenth Annual Report, 1914*, 25; *Hearing before the Committee on the Public Lands, House of Representatives, Sixty-Fourth Congress, First Session, on H.R. 434 and H.R. 8668, Bills to Establish a National Park Service and for Other Purposes, April 5 and 6, 1916* (Washington, DC: Government Printing Office, 1916), 89–90 (hereafter *Congressional Hearing, 1916*).
102. ASHPS, *Sixteenth Annual Report, 1911*, 19.
103. ASHPS, *Eighteenth Annual Report, 1913*, 269; *New York Times* editorial, 5 October 1913, in ASHPS, *Nineteenth Annual Report, 1914*, 252–54.
104. *Congressional Hearing, 1916*, 89. On Bush-Brown, see http://gettysburgsculptures.com/gen_sedgwick_equestrian_monument/hk_bush_brown_sculptor_sedgwick_equestrian_monument (accessed 11 November 2011).
105. *Congressional Hearing, 1916*, 90. See Kupper, "Science and the National Parks," on the Swiss National Park.
106. Marguerite S. Shaffer, *See America First: Tourism and National Identity, 1880–1940* (Washington, DC: Smithsonian Institution Press, 2001), 26, 40, 101–2, 111.
107. *Proceedings of the National Parks Conference Held at Berkeley, California, March 11, 12, and 13, 1915* (Washington, DC: Government Printing Office, 1915), 16.
108. Ibid., 15.
109. ASHPS, *Seventeenth Annual Report, 1912*, 273; E. J. Hart, *J. B. Harkin: Father of Canada's National Parks* (Calgary: University of Alberta Press, 2009), 33.
110. Howard Douglas, chief superintendent of parks, Dominion of Canada, quoted in ASHPS, *Seventeenth Annual Report, 1912*, 273; superintendent of Yellowstone Park, reported in the statement of the commissioner of parks, Dominion of Canada, ibid., 278 (last quote).
111. Preamble to an Act to Establish a National Park Service, and for Other Purposes, Approved August 25, 1916, 39 Stat. 535.
112. These treaty negotiations are already well covered in Dorsey, *Dawn of Conservation Diplomacy*, 102.
113. Chirakaikaran J. Chacko, *The International Joint Commission between the United States of America and the Dominion of Canada* (New York: Columbia University Press, 1932), 73; "Origins of the Boundary Waters Treaty," http://bwt.ijc.org/index.php?page=origins-of-the-boundaries-water-treaty&hl=eng. (accessed 6 July 2012); Carl G. Winter, "The Boundary Waters Treaty," *Historian* 17 (Autumn 1954): 76–96 at 76; James A. Sandos, "International Water Control in the Lower Rio Grande Basin,

1900–1920," *Agricultural History* 54 (October 1980): 490–501 at 490–91; Convention between the United States and Mexico Providing for the Equitable Distribution of the Waters of the Rio Grande for Irrigation Purposes, May 21, 1906, U.S.–Mex., 34 Stat. 2953.

114. Peter Neary, "Grey, Bryce, and the Settlement of Canadian-American Differences, 1905–1911," *Canadian Historical Review* 49, no. 4 (1968): 357–80 at 357–58; Dreisziger, "Campaign to Save Niagara Falls," 449.
115. *Report of the United States Deep Waterways Commission: Prepared at Detroit, Michigan, December 18–22, 1896, by the Commissioners, James B. Angell, John E. Russell, Lyman E. Cooley* (Washington, DC: Government Printing Office, 1897); Minutes of Board of Trustees, 22 April 1901, 39–44, book 2, box 6, ASHPS Records; Brinkley, *Wilderness Warrior*, 414; Irwin, *New Niagara*, 210; *The American Scenic and Historic Preservation Society*, 16–17; *The American Scenic and Historic Preservation Society: A National Society for the Protection of Natural Scenery, the Preservation of Historic Landmarks and the Improvement of Cities* (n.p., 1907), 16–17, 26–27; Winter, "Boundary Waters," 76–84.
116. H. V. Nelles, *The Politics of Development: Forests, Mines and Hydro-electric Power in Ontario, 1849–1941*, 2nd ed. (Montreal: McGill-Queen's University Press, 2005), 314.
117. Forrest Crissey, *Theodore E. Burton, American Statesman* (Cleveland: World, [1956]), 118; Roosevelt to James R. Garfield, 5 September 1906, box 119, James R. Garfield Papers, LC; "New Job for Burton," *New York Times*, 13 June 1908, 6.
118. "To Outlook Dec. 7, 1912," fl. Addresses and articles by J. H. McFarland, A–Ch, box 14, series 19, ACA Minutes, Programmes, etc., 1903–21, 1930–33, McFarland Papers.
119. Peter Neary, "Gibbons, Sir George Christie," in *Dictionary of Canadian Biography Online*, http://www.biographi.ca/009004-119.01-e.php?BioId=41513&query= (quote) (accessed 23 May 2011); Neary, "Grey, Bryce," 372–76.
120. "Origins of the Boundary Waters Treaty."
121. Neary, "Grey, Bryce," 372–76; Dreisziger, "Campaign to Save Niagara Falls," 448; Dreisziger, "The Great Lakes in United States–Canadian Relations," 262 (quote); Tony McCulloch, "Theodore Roosevelt and Canada: Alaska, the 'Big Stick' and the North Atlantic Triangle, 1901–1909," in Ricard, ed., *Companion*, 294–95.
122. Winter, "Boundary Waters," 82.
123. Ibid., 82.
124. Nelles, *Politics of Development*, 316–17.
125. Quoted ibid., 314; Dorsey, *Dawn of Conservation Diplomacy*, 139, notes Root's interest in conservation.
126. Neary, "Grey, Bryce," 372–76; Winter, "Boundary Waters," 83 (quote). Canadians wanted water drawn from Lake Michigan for the Chicago River diversion to be counted and linked to the Niagara diversion allocations. Root to McFarland, 24 October 1906, file 597, Numerical and Minor Files of the Department of State, RG 59, NARA; George Gibbons to Chandler Anderson, 10 December 1908, 185–86, book 28, James Bryce Papers, Bodleian Library, Oxford University (hereafter BP).
127. Gibbons to Anderson, 10 December 1908, 185–86, book 28, BP; Winter, "Boundary Waters," 85–86, 94–95.
128. Winter, "Boundary Waters," 94.
129. "Treaty between the United States and Great Britain Relating to Boundary Waters, and Questions Arising between the United States and Canada," http://ijc.org/en_/BWT#text (quote) (accessed 4 April 2014); David Lemarquand, "The International Joint Commission and Changing Canada–United States Boundary Relations,"

Natural Resources Journal 33 (Winter 1993): 59–91 at 67; L. M. Bloomfield and Gerald F. Fitzgerald, *Boundary Waters Problems of Canada and the United States (The International Joint Commission, 1912–1958)* (Toronto: Carswell, 1958), 172–73.

130. Josiah Bailey to McFarland, 15 January 1906; McFarland to Bailey, 1 March 1909; George Perkins to William Frederic Badè, 22 June 1909; all in fl. B.–Ban, box 24, series 28, ACA Niagara Falls General Correspondence A–F, McFarland Papers.
131. *Preservation of Niagara Falls: Hearings before the Committee on Foreign Affairs, House of Representatives, January 16, 18, 19, 20, 23, 26, and 27, 1912, on H.R. 6746 . . . and H.R. 7694 . . .* (Washington, DC: Government Printing Office, 1912), 11.
132. Cashbook, 374, box 28A, ASHPS Records.
133. Evans, "Storm over Niagara: A Catalyst," 52–54 (quote at 54).
134. Dreisziger, "The Great Lakes in United States–Canadian Relations"; Dreisziger, "Campaign to Save Niagara Falls."
135. J. Horace McFarland, "Are National Parks Worth While?," in *The American Civic Association's Movement for a National Parks Bureau*, American Civic Association, National Parks, pamphlet series 11, no. 6 (Washington, DC: American Civic Association, 1912), 30.
136. "Statement of Mr. Charles M. Dow," 171.
137. Bloomfield and Fitzgerald, *Boundary Waters Problems*, 245; Chacko, *International Joint Commission*, 12.

NOTES TO CHAPTER NINE

1. *NCC Report*, 3:620–751; Irving Fisher, *Bulletin 30 of the Committee of One Hundred on National Health, Being a Report on National Vitality: Its Wastes and Conservation* (Washington, DC: Government Printing Office, 1909), 68 (hereafter Fisher, *National Vitality*).
2. William Jay Schieffelin, *Work of the Committee of One Hundred on National Health*, American Academy of Political and Social Science, Publication no. 628 (n.p., [1911]), 80.
3. Irving Fisher and Eugene Lyman Fisk, *How to Live: Rules for Healthful Living Based on Modern Science* (New York: Funk and Wagnalls, 1915).
4. Report of the Inter-departmental Committee on Physical Deterioration (1904 Report, Cd. 2175; Witnesses and Minutes of Evidence, Cd. 2210; Appendix and General Index, Cd. 2186); Richard Soloway, "Counting the Degenerates: The Statistics of Race Deterioration in Edwardian England," *Journal of Contemporary History* 17 (January 1982): 137–64; Debórah Dwork, *War Is Good for Babies and Other Young Children: A History of the Infant and Child Welfare Movement in England, 1898–1918* (London: Tavistock, 1987), 56.
5. F. M. L. Thompson, *The Rise of Respectable Society: A Social History of Victorian Britain, 1830–1900* (Cambridge, MA: Harvard University Press, 1988), 132–33.
6. Rowena Hammal, "How Long before the Sunset? British Attitudes to War, 1871–1914," *History Review*, 2010, http://www.historytoday.com/rowena-hammal/how-long-sunset-british-attitudes-war-1871-1914 (quotes) (accessed 30 November 2011); Ruth Brewster Sherman, "The Discussion on Tuberculosis: Second Paper (Continued)," *American Journal of Nursing* 5 (April 1905): 427–32 at 427.
7. Fisher, *National Vitality*, 54; Anna Davin, "Imperialism and Motherhood," *History Workshop*, no. 5 (Spring 1978): 9–65.
8. Allan M. Brandt, *No Magic Bullet: A Social History of Venereal Disease in the United States since 1880* (New York: Oxford University Press, 1985), 97–98 (quote at 98).

9. Fisher, *National Vitality*, 36: "Among the troops stationed in the Philippines, the venereal morbidity during the year 1904 was 297 per 1,000, largely exceeding the morbidity from malarial fevers and diarrhea."
10. Peter Curson and Kevin McCracken, *Plague in Sydney: The Anatomy of an Epidemic* (Sydney: University of New South Wales Press, 1989); John McNeill, *Mosquito Empires: Ecology and War in the Greater Caribbean, 1620–1914* (New York: Cambridge University Press, 2010); Marilyn Chase, *The Barbary Plague: The Black Death in Victorian San Francisco* (New York: Random House, 2004).
11. Michael Willrich, *Pox: An American History* (New York: Penguin Press, 2011), 165.
12. Ibid., 10. "We in this country should profit by the experience of older countries in respect to school hygiene. Switzerland has led the nations in its concern for the physical welfare of its children" (Fisher, *National Vitality*, 76).
13. Franklin Giddings, *Democracy and Empire; with Studies of Their Psychological, Economic, and Moral Foundations* (New York: Macmillan, 1900), chap. 17.
14. Schieffelin, *Work of the Committee of One Hundred*; Arthur Ekirch Jr., "The Reform Mentality, War, Peace, and the National State: From the Progressives to Vietnam," *Journal of Libertarian Studies* 3, no. 1 (1979): 55–72 at 58–59.
15. Fisher, *National Vitality*, 68.
16. Irving Fisher, *The Modern Crusade against Consumption: Being, in Part, an Address Delivered at United Church, New Haven, Conn., April 26, 1903 . . .*, fl. 365, box 19, Irving Fisher Papers, Sterling Memorial Library, Yale University; *Address Delivered before the Association of Life Insurance Presidents, New York, February 5, 1909*, fl. 366, ibid.
17. "Address of Professor Irving Fisher at Meeting of American Academy of Political and Social Science, January 19, 1911," 1, fl. 369 Eugenics and Hygiene 1911 Jan. 19, box 19, Fisher Papers (quote); Pinchot to Charles W. Eliot, 18 December 1915, fl. Dr. Charles W. Eliot, box 489, Pinchot Papers, LC.
18. "Urge Reforms in White Plague Fight," *New York Times*, 2 October 1908, 6.
19. Fisher, *National Vitality*, 62 ("administration"), 126 (all other quotes).
20. Theodore Roosevelt, "Seventh Annual Message," 3 December 1907, in Gerhard Peters and John T. Woolley, American Presidency Project, http://www.presidency.ucsb.edu/ws/index.php?pid=29548#ixzz1gfa8pWEL (accessed 28 June 2010).
21. Theodore Roosevelt, "Eighth Annual Message," 8 December 1908, in ibid.
22. Fisher, *National Vitality*, 5.
23. Daniel T. Rodgers, *Atlantic Crossings: Social Politics in a Progressive Age* (Cambridge, MA: Belknap Press of Harvard University Press, 1998), 245–56, esp. 247; Fisher, *National Vitality*, 61 (quote).
24. Fisher, *National Vitality*, 5 (quotes), 47 (industrial accidents). This statement was reproduced in other publications. See, e.g., M. G. Overlock, *The Working People: Their Health and How to Protect It* (Boston: Massachusetts Health Book, 1911), 107.
25. Fisher, *National Vitality*, 5, 46.
26. Ibid., 12.
27. Cf. Rodgers, *Atlantic Crossings*; David Gutzke, ed., *Britain and Transnational Progressivism* (Basingstoke, Eng.: Palgrave Macmillan, 2008).
28. Fisher, *National Vitality*, 31, 12.
29. Ibid., 63.
30. Theodore Roosevelt, "Sixth Annual Message," 3 December 1906, in Peters and Woolley, American Presidency Project, http://www.presidency.ucsb.edu/ws/index.php?pid=29547#ixzz1hDTirrzA (accessed 3 December 2011).
31. Fisher, *National Vitality*, 126.

32. Amy Kaplan, "Roman Fever: Imperial Melancholy in America" (paper presented at "Comparative Imperial Transformations" conference, University of Sydney, July 2008).
33. Fisher, *National Vitality*, 126.
34. Ibid.
35. Ibid., 125 (quotes); Charles S. Howe, "The Function of the Engineer in the Conservation of the Natural Resources of the Country," *Science* 28 (23 October 1908): 537–48.
36. Charles Wohlforth, "Conservation and Eugenics: The Environmental Movement's Dirty Secret," *Orion*, July/August 2010, http://www.orionmagazine.org/index.php /articles/article/5614 (accessed 12 November 2012); Gregg Mittman, "In Search of Health: Landscape and Disease in American Environmental History," *Environmental History* 10 (April 2005): 184–209 at 200–201.
37. Irving Fisher, *Eugenics* (Battle Creek, MI: Good Health Publishing, 1913), esp. 3, 17. However, in *National Vitality*, 50, Fisher described the hereditary principle as being only among "some of the topics" requiring closer examination.
38. On eugenics in the United States, see Daniel J. Kevles, *In the Name of Eugenics: Genetics and the Uses of Human Heredity* (New York: Alfred A. Knopf, 1985), 64; Richmond Hobson, *Alcohol and the Human Race* (New York: Fleming H. Revell, 1919), 105, 118 (quote). Fisher (*National Vitality*, 50) cited Weismann and Darwin as well as Mendel, arguing that ideas of heredity needed further study "in relation to their practical application." Prohibitionists in the 1920s used the racial poison concept. Bartlett C. Jones, "Prohibition and Eugenics, 1920–1933," *Journal of the History of Medicine and Allied Sciences* 18, no. 2 (1963): 158–73 at 160; S. J. Mennell, "Prohibition: A Sociological View," *Journal of American Studies* 3 (December 1969): 159–175 at 166.
39. *Acts 1907, Laws of the State of Indiana, Passed at the Sixty-Fifth Regular Session of the General Assembly* (Indianapolis: William B. Burford, 1907), 377. Similar laws in the South proscribed interracial sex or sought to limit the size of poor families.
40. Aristide R. Zolberg, *A Nation by Design: Immigration Policy in the Fashioning of America* (Cambridge, MA: Harvard University Press, 2006), 248–51.
41. The chief polemic has been Anna Bramwell, *Ecology in the Twentieth Century: A History* (New Haven, CT: Yale University Press, 1989). For more on the controversy over connections between conservation and Nazis, see Franz-Josef Brüggemeier, Mark Cioc, and Thomas Zeller, eds., *How Green Were the Nazis? Nature, Environment, and Nation in the Third Reich* (Athens: Ohio University Press, 2005); Gary Brechin, "Conserving the Race: Natural Aristocracies, Eugenics, and the U.S. Conservation Movement," http://graybrechin.net/articles/1990s/conserving.html (accessed 1 April 2014).
42. Kevles, *In the Name of Eugenics*, 64.
43. Alexandra Stern, *Eugenic Nation: Faults and Frontiers of Better Breeding in Modern America* (Berkeley: University of California Press, 2005), 53–54.
44. *Proceedings of the First National Conference on Race Betterment, January 8, 9, 10, 11, 12, 1914* (n.p.: Race Betterment Foundation, n.d. [1914]), 1, 56; Harold K. Steen, ed., *The Conservation Diaries of Gifford Pinchot* (Washington, DC: Island Press, 2001), 10; Fisher, *Modern Crusade against Consumption*, 1, 14.
45. Stern, *Eugenic Nation*, 53; Lawrence M. Woods, *Horace Plunkett in America: An Irish Aristocrat on the Wyoming Range* (Norman, OK: Arthur H. Clark, 2010), 234 (quote).
46. Kevles, *In the Name of Eugenics*, 64.
47. Madison Grant, *The Passing of the Great Race; or, The Racial Basis of European History* (New York: Charles Scribner's Sons, 1916), viii.

48. Jonathan Peter Spiro, *Defending the Master Race: Conservation, Eugenics, and the Legacy of Madison Grant* (Burlington: University of Vermont Press, 2009), xii, 149.
49. William T. Hornaday, *Our Vanishing Wildlife: Its Extermination and Preservation* (New York: Charles Scribner's Sons, 1913), 102.
50. Fisher, *National Vitality*, 129.
51. Theodore Roosevelt, "A Premium on Race Suicide," *Outlook* 105 (27 September 1913): 163; Roosevelt to Charles Davenport, 3 January 1913, quoted in Edwin Black, *War against the Weak: Eugenics and America's Campaign to Create a Master Race* (New York: Four Walls Eight Windows, 2003), 99. The text can be found at http://www.dnalc.org/view/11219-T-Roosevelt-letter-to-C-Davenport-about-degenerates-reproducing-.html (quote) (accessed 3 December 2012).
52. Fisher, *Eugenics*, 16, 17–18.
53. Kellogg lacks an adequate biography. One, by Professor Ronald Numbers, is stated to be in progress. See also Richard W. Schwarz, *John Harvey Kellogg, M.D.: Pioneering Health Reformer* (n.p.: Review and Herald Publishing Association, 2006).
54. Fisher, *National Vitality*, 50.
55. Ibid., 129 ("marriage choices"), 51 (Japan).
56. Roosevelt, "Sixth Annual Message"; James R. Holmes, *Theodore Roosevelt and World Order: Police Power in International Relations* (Washington, DC: Potomac Books, 2007), 257; Theodore Roosevelt to Cecil Spring Rice, 16 June 1905, series 1, reel 55, Theodore Roosevelt Papers, LC.
57. David S. Patterson, "Japanese-American Relations: The 1906 California Crisis, the Gentlemen's Agreement, and the World Cruise," in Serge Ricard, ed., *A Companion to Theodore Roosevelt* (Malden, MA: Wiley-Blackwell, 2011), 391–416, esp. 398–402; Edmund Morris, *Theodore Rex* (New York: Random House, 2001), 483.
58. Roosevelt, "Twisted Eugenics," *Outlook* 106 (3 January 1914): 30–34.
59. Kathleen Dalton, *Theodore Roosevelt: A Strenuous Life* (New York: Knopf, 2002), 312.
60. Theodore Roosevelt, "Race Decadence," *Outlook* 97 (8 April 1911): 763–68 at 768.
61. Thomas G. Dyer, *Theodore Roosevelt and the Idea of Race* (Baton Rouge: Louisiana State University Press, 1980), 163.
62. Marilyn Lake and Henry Reynolds, *Drawing the Global Colour Line: White Men's Countries and the International Challenge of Racial Equality* (Cambridge: Cambridge University Press, 2008), 98–101.
63. Fisher, *National Vitality*, 54.
64. Roosevelt, "Sixth Annual Message"; Annie L. Cot, " 'Breed out the Unfit and Breed in the Fit': Irving Fisher, Economics, and the Science of Heredity," *American Journal of Economics and Sociology* 64 (July 2005): 793–826; Theodore Roosevelt, "Citizenship in a Republic," speech delivered at the Sorbonne, Paris, France, on 23 April 1910, 7, http://www.theodore-roosevelt.com/trsorbonnespeech.html (accessed 10 January 2011).
65. Roosevelt, "Twisted Eugenics," 32.
66. Fisher, *National Vitality*, 129.
67. Fisher, *Eugenics*, 16, 17–18; Roosevelt, "Twisted Eugenics."
68. Theodore Roosevelt, "The New Nationalism," speech delivered at Osawatomie, KS, on 31 August 1910, http://www.theodore-roosevelt.com/images/research/speeches/trnationalismspeech.pdf (author's italics) (accessed 20 January 2011).
69. Platform of the Progressive Party, http://www.pbs.org/wgbh/americanexperience/features/primary-resources/tr-progressive/12 (accessed 3 December 2011).

70. Ronald Hamowy, *Government and Public Health in America* (Northampton, MA: Edward Elgar, 2007), 28.
71. Charles W. Eliot to Pinchot, 6 December 1915, fl. Dr. Charles W. Eliot, box 489, Pinchot Papers.
72. Pinchot to Eliot, 18 December 1915, ibid. See also Pinchot to Eliot, 12 October 1915, ibid.

NOTES TO CHAPTER TEN

1. Theodore Roosevelt, *African Game Trails: An Account of the African Wanderings of an American Hunter-Naturalist* (London: John Murray, 1910), 457–59; "The African Hunt," *New York Times*, 7 February 1913, 6. Edmund Morris, *Colonel Roosevelt* (New York: Random House, 2010), 593, lists nearly 11,400 animal species, about 10,000 plants, "and a small collection of ethnological objects."
2. Morris, *Colonel Roosevelt*, 22 (4,000 mammals). Many have written on the African hunt. See J. Lee Thompson, *Theodore Roosevelt Abroad: Nature, Empire, and the Journey of an American President* (New York: Palgrave Macmillan, 2010), chaps. 2–5; Morris, *Colonel Roosevelt*, 3–39; Kathleen Dalton, *Theodore Roosevelt: A Strenuous Life* (New York: Alfred A. Knopf, 2002), 347–58; Patricia O'Toole, *When Trumpets Call: Theodore Roosevelt after the White House* (New York: Simon and Schuster, 2005), chaps. 2–3; Sarah Watts, *Rough Rider in the White House: Theodore Roosevelt and the Politics of Desire* (Chicago: University of Chicago Press, 2003), 183–91; Monica Rico, *Nature's Noblemen: Transatlantic Masculinities and the Nineteenth-Century American West* (New Haven, CT: Yale University Press, 2013), 192–211. See also the colorful account in Mark Sullivan, *Our Times: The United States, 1900–1925*, vol. 4, *The War Begins, 1909–1914* (New York: Charles Scribner's Sons, 1932), chap. 22.
3. Theodore Roosevelt to Edward North Buxton, 8 December 1902, Theodore Roosevelt Papers, LC, Theodore Roosevelt Digital Library, Dickinson State University, http://www.theodorerooseveltcenter.org/Research/Digital-Library/Record.aspx?libID=0183699; Roosevelt, *African Game Trails*, 393 (quote), 404–5, 11 (African game reserves).
4. Dalton, *Theodore Roosevelt*, 348; Morris, *Colonel Roosevelt*, 11.
5. Roosevelt, *African Game Trails*, 502.
6. Dalton, *Theodore Roosevelt*, 349; *Bay of Plenty Times* (N.Z.), 12 July 1909, 2; *Iowa City Citizen*, 31 March 1909, 7; Gary Rice, "Trailing a Celebrity: Press Coverage of Theodore Roosevelt's African Safari, 1909–1910," *Theodore Roosevelt Association Journal* 22 (Fall 1996): 4–16 at 6, 11; Robert Wooster, *Nelson A. Miles and the Twilight of the Frontier Army* (Lincoln: University of Nebraska Press, 1993), 250 ("driven"); *New York Sun*, 1 April 1909, 1 (quotes). Miles did not mention Roosevelt by name. He also endorsed Taft as making "an exceptionally good president." Roosevelt called Miles a "brave peacock." Spencer C. Tucker, James Arnold, and Roberta Wiener, eds., *The Encyclopedia of North American Indian Wars, 1607–1890: A Political, Social, and Military History* (Santa Barbara, CA: ABC-CLIO, 2011), 495.
7. For Burroughs, see "California Has Won Eminent Naturalist," *San Francisco Call*, 27 April 1909, editorial; Dalton, *Theodore Roosevelt*, 242.
8. "The African Hunt," *New York Times*, 17 February 1913, 6. These justifications were common to an elite Anglo-American hunting ethic. John MacKenzie, *The Empire of Nature: Hunting, Conservation and British Imperialism* (Manchester: Manchester University Press, 1989), 298.
9. "The African Hunt," 6.

10. Roosevelt, *African Game Trails*, 404–5; cf. W. Robert Foran, *Kill: Or Be Killed; The Rambling Reminiscences of an Amateur Hunter* (1933; repr., New York: St. Martin's Press, 1988), 209.
11. "The African Hunt," 6; Roosevelt, *African Game Trails*, 12–13, 393, 404–5.
12. Roosevelt, *African Game Trails*, 409.
13. Roosevelt to Warrington Dawson, 9 August 1909, fl. Letters 1909, box 7, Warrington Dawson Papers, David M. Rubenstein Rare Book and Manuscript Library, Duke University, Durham, NC; http://www.dailymail.co.uk/news/article-2066763/Teddy-Roosevelts-curious-collectables-revealed-6-2m-renovation.html#ixzz1gTXPLJdu (accessed 30 January 2013).
14. Gail Bederman, *Manliness and Civilization: A Cultural History of Gender and Race in the United States, 1880–1917* (Chicago: University of Chicago Press, 1995), 207–13; O'Toole, *When Trumpets Call*, 42 (quote).
15. "Meeting Roosevelt in Mombasa," *San Francisco Call*, 22 August 1909, 6.
16. Bederman, *Manliness*, 213.
17. Dalton, *Theodore Roosevelt*, 242; *Theodore Roosevelt: An Autobiography*, new ed. (1913; New York: Charles Scribner's Sons, 1922), 401.
18. The African visit differed in its multiple agendas from Roosevelt's adventure of 1914 in Brazil, commonly called the River of Doubt (Rio da Dúvida) trip, which indeed focused on daring exploration. Theodore Roosevelt, *Through the Brazilian Wilderness* (New York: Charles Scribner's Sons, 1914); Candice Millard, *River of Doubt: Theodore Roosevelt's Darkest Journey* (New York: Doubleday, 2005); David M. Wrobel, *Global West: Travel, Empire, and Exceptionalism from Manifest Destiny to the Great Depression* (Albuquerque: University of New Mexico Press, 2013), 103–11.
19. Reported in Warrington Dawson to Theodore Roosevelt, 14 October 1909, fl. Letters 1909, box 7, Dawson Papers.
20. Rice, "Trailing a Celebrity," 13.
21. Roosevelt to Warrington Dawson, 1 July 1909, and Dawson to Roosevelt, 14 October 1909, fl. Letters 1909, box 7, Dawson Papers; see also "Addendum," "Written from memory the next day," in "Atmosphere and Theodore Roosevelt," fl. Theodore Roosevelt Materials, box 35, Dawson Papers ("care of animals"); Rice, "Trailing a Celebrity," 12; "Editor's Note," in Foran, *Kill: Or Be Killed*, [3–4, unpaginated]; Foran, *Kill: Or Be Killed*, 57–59, 94–95, 137–39.
22. "Meeting Roosevelt in Mombasa," 6.
23. Axel Lunderberg and Frederick Seymour, *The Great Roosevelt African Hunt and the Wild Animals of Africa . . . with Thrilling, Exciting, Daring and Dangerous Exploits of Hunters of Big Game in Wildest Africa* (n.p.: D. B. McCurdy, 1909); Marshall Everett, *Roosevelt's Thrilling Experiences in the Wilds of Africa Hunting Big Game . . .* ([Chicago?]: A. Hamming, 1910); Bederman, *Manliness*, 215.
24. Shiori Hasegawa, "Sensational Africa: Roosevelt's Cultural Politics and Expeditionary Filmmaking of 1909–1910," *Inter Faculty* (University of Tsukuba) 1 (2010), https://journal.hass.tsukuba.ac.jp/interfaculty/article/view/8/10 (accessed 14 March 2012).
25. Ibid.; Jan Olsson, "Trading Places: Griffith, Patten and Agricultural Modernity," *Film History: An International Journal* 17, no. 1 (2005): 39–65 at 54.
26. *Hunting Big Game in Africa*, http://www.wildfilmhistory.org/film/27/27.html?filmid=27 (accessed 14 March 2012).
27. MacKenzie, *Empire of Nature*, 162 (quote), 218–19; Morris, *Colonel Roosevelt*, 12–13.
28. Warrington Dawson, *Pittsburgh Press*, 5 May 1909, 1 (first quote); Roosevelt, *African Game Trails*, 99 (last quote); Sullivan, *Our Times*, 422. On the contest for the Repub-

lican Party nomination in 1912, the *Pittsburgh Press* (10 April 1912, 1) stated that "Bwana Tumbo makes mincemeat of opposition in sucker state" and "sinks Taft's Ship in Illinois."

29. Cf. Roosevelt, *African Game Trails*, 409.
30. O'Toole, *When Trumpets Call*, 47.
31. Dalton, *Theodore Roosevelt*, 349.
32. MacKenzie, *Empire of Nature*, 247, 248; Edward I. Steinhart, *Black Poachers, White Hunters: A Social History of Hunting in Colonial Kenya* (Oxford: James Currey, 2006), 116 (celebrity).
33. "Atmosphere and Theodore Roosevelt," fl. Theodore Roosevelt Materials, box 35, Dawson Papers.
34. Roosevelt, *African Game Trails*, 30, 34 (quote), 40; Theodore Roosevelt to Corinne Robinson, 19 May 1909, in Correspondence and Compositions, 1772–1946, MS Am 1540 (150–194), Theodore Roosevelt Collection, Houghton Library, Harvard University, http://pds.lib.harvard.edu/pds/view/21513848?n=29&imagesize=1200&jp2Res=.25&printThumbnails=no; Rico, *Nature's Noblemen*, 199. Cf. Jeremy M. Johnston, "Theodore Roosevelt's Quest for Wilderness: A Comparison of Roosevelt's Visits to Yellowstone and Africa," in Alice Wondrak Biel, ed., *Beyond the Arch: Community and Conservation in Greater Yellowstone and East Africa* (Yellowstone National Park, WY: Yellowstone Center for Resources, 2003), 149–67 at 164.
35. Cherry Kearton, *With Roosevelt in Africa* (1910). For extant parts of this film, see http://memory.loc.gov/cgi-bin/query/r?ammem/papr:@filreq%28@field%28NUMBER+@band%28trmp+4102s1%29%29+@field%28COLLID+roosevelt%29%29 (accessed 31 March 2013).
36. MacKenzie, *Empire of Nature*, 50–51.
37. Sherman Strong Hayden, *The International Protection of Wild Life* (New York: Columbia University Press, 1942), 36–37; Bernhard Gissibl, "German Colonialism and the Beginnings of International Wildlife Preservation in Africa," *GHI Bulletin Supplement* 3 (2006): 131–43 at 132; MacKenzie, *Empire of Nature*, 81, 202, 321; David K. Prendergast and William M. Adams, "Colonial Wildlife Conservation and the Origins of the Society for the Preservation of the Wild Fauna of the Empire (1903–1914)," *Oryx* 37, no. 2 (2003): 251–60; Theodore Roosevelt, "Extract from Message from the Hon. Theodore Roosevelt, President of the United States," *Journal of the Society for the Preservation of the Wild Fauna of the Empire* 4 (1908): 8.
38. MacKenzie, *Empire of Nature*, 50.
39. *Journal of the Society for the Preservation of the Wild Fauna of the Empire* 3 (1907): 12–13.
40. Theodore Roosevelt, *African and European Addresses, with an Introduction Presenting a Description of the Conditions under Which the Addresses Were Given during Mr. Roosevelt's Journey in 1910 from Khartoum through Europe to New York, by Lawrence F. Abbott* (New York: G. P. Putnam's Sons, 1910), 157–72 at 159 ("waste"), 160 ("wild man"); "Roosevelt and the Ptolemies," *New York Times*, 6 June 1910, editorial.
41. *The Truth about British East Africa: Being a Speech Delivered by the Hon. Col. Theodore Roosevelt at Nairobi, August, 3rd, 1909* (n.p.: Leader of British East Africa, n.d.), fl. Roosevelt Papers: Nairobi Speeches and Printed Materials, box 36, Dawson Papers; Michael Twaddle, "The Settlement of South Asians in East Africa," in Robin Cohen, ed., *The Cambridge Survey of World Migration* (Cambridge: Cambridge University Press, 1995), 74–76; Roosevelt, *African Game Trails*, 37–41. Roosevelt did advise the audience: "Don't get mixed up in differences of creed or national origin."

After all, Roosevelt was of Dutch, not British or technically "Anglo-Saxon," ancestry. But his statement was not more specific, and it did not appear in the original text. The *Nairobi Leader* editor added Roosevelt's extemporaneous comments about the English-speaking peoples, but Roosevelt "did not at all mind the publication." Dawson manuscript (p. 6), fl. Miscellaneous Writings about Africa, box 35, Dawson Papers.

42. Roosevelt, *African Game Trails*, 12–13.
43. Ibid. On African hunting, see MacKenzie, *Empire of Nature*; William Beinart, "Empire, Hunting and Ecological Change in Southern and Central Africa," *Past and Present*, no. 128 (August 1990): 162–86.
44. Roosevelt, *African Game Trails*, 11–12.
45. "First Photos of Roosevelt in Africa," *San Francisco Call*, 10 October 1909, 5.
46. Ruth A. Tucker, *From Jerusalem to Irian Jaya: A Biographical History of Christian Missions*, rev. ed. (Grand Rapids, MI: Zondervan, 2004), 348; Roosevelt, *African Game Trails*, 144.
47. Roosevelt, *African Game Trails*, 367–72. In Kampala, he visited "Mother Paul," an American, in the Church of England mission (370) and proclaimed any missionary work better than "Stygian darkness" (368).
48. "Address of Colonel Roosevelt Made at the Laying of Corner Stone of School Building, Kijabe, B.E. Africa, August 4th 1909," http://www.wheaton.edu/bgc/archives/exhibits/ohistory/ora118a.htm (accessed 12 December 2012).
49. Roosevelt, *African Game Trails*, 454 (quote); Thompson, *Theodore Roosevelt Abroad*, 39.
50. *San Francisco Call*, 10 October 1909, 5.
51. "Address of Colonel Roosevelt . . . Corner Stone."
52. Roosevelt, *African Game Trails*, 31.
53. Theodore Roosevelt, foreword to Roosevelt, *African and European Addresses*, iv–v; Theodore Roosevelt, "Peace and Justice in the Sudan: An Address at the American Mission in Khartum, March 16, 1910," in ibid., 3–11 at 3–4.
54. Roosevelt, foreword, in ibid., v ("noisy" radicals); Theodore Roosevelt, "Law and Order in Egypt: An Address before the National University in Cairo, March 28, 1910," in ibid., 24 (last quote); Morris, *Colonel Roosevelt*, 38.
55. *Los Angeles Herald*, 18 March 1910, 16.
56. *Daily Mail*, reprinted in "A Man of Action: Mr. Roosevelt's Record," *Times of India*, 26 March 1909, 7 (quote); "Man of Action," *Star* (Christchurch, N.Z.), 3 May 1909, 2; Serge Ricard, "A Hero's Welcome: Theodore Roosevelt's Triumphant Tour of Europe in 1910," in Hans Krabbendam and John M. Thompson, eds., *America's Transatlantic Turn: Theodore Roosevelt and the "Discovery" of Europe* (New York: Palgrave Macmillan, 2012), 143–58.
57. Ricard, "Hero's Welcome," 156.
58. Thompson, *Theodore Roosevelt Abroad*, 144.
59. Theodore Roosevelt, "The World Movement," in Roosevelt, *African and European Addresses*, 131.
60. Ibid., 122.
61. Ibid.
62. See chapter 11, below.
63. Ricard, "Hero's Welcome," 154.
64. See, e.g., *Welcome Home, a Dinner to Theodore Roosevelt, June Twenty Second, Nineteen Hundred and Ten, at Sherry's New York* (n.p.: privately printed, 1910); Morris, *Colonel Roosevelt*, 78–79.

65. Everett, *Roosevelt's Thrilling Experiences*; cf. Bederman, *Manliness*, 215.
66. Peter MacQueen, in Everett, *Roosevelt's Thrilling Experiences*, 367. Lawton died in battle.
67. *Peter MacQueen: Traveler, Explorer, War Correspondent, Lecturer* (1910?), http://digital.lib.uiowa.edu/cdm/compoundobject/collection/tc/id/22734/rec/4 (accessed 1 December 2012); Peter MacQueen, *Raconteur, Author, Lecturer* (1909?), http://digital.lib.uiowa.edu/cdm/compoundobject/collection/tc/id/36969/rec/2 (quotes) (accessed 13 March 2013).
68. MacQueen, in Everett, *Roosevelt's Thrilling Experiences*, 374–75.
69. Cf. Bederman, *Manliness*, 211–15.

NOTES TO CHAPTER ELEVEN

1. Harry Slattery, "From Roosevelt to Roosevelt, 1900–1946, Part 1," 13–14, 25, fl. Autobiography, box 28, Harry A. Slattery Papers, David M. Rubenstein Rare Book and Manuscript Library, Duke University, Durham, NC; George Britt, *Forty Years, Forty Millions: The Career of Frank A. Munsey* (New York: Farrar and Rinehart, [1935]), 168 ("seldom"), 172–74.
2. *Washington Times*, 20 February 1909, 2 (quotes). See also "Their Task Broadens," *Washington Post*, 21 February 1909, 13; *Chicago Tribune*, 20 February 1909, 7; *Christian Science Monitor*, 20 February 1909, 1; *Ogden Morning Examiner*, 28 February 1909, 6; *Raleigh (NC) Herald*, 11 March 1909, 2.
3. For press publicity, see Kathleen Dalton, *Theodore Roosevelt: A Strenuous Life* (New York: Knopf, 2002), 242; Stephen Ponder, "News Management in the Progressive Era, 1898–1909: Gifford Pinchot, Theodore Roosevelt and the Conservation Crusade" (PhD diss., University of Washington, 1985), 4.
4. Treadwell Cleveland Jr, "The North American Conservation Conference," *Conservation* 15 (March 1909): 159–68 at 165, 168; "Conservators End Work," *Washington Post*, 25 February 1909, 14.
5. Roosevelt to William T. Hornaday, 29 December 1908, series 2, reel 353, Roosevelt Papers, LC; *Washington Herald*, 20 February 1909, 9 (quotes).
6. "Saving of America," *Washington Post*, 19 February 1909, 1; Cleveland, "North American Conservation Conference," 167, 168 (quotes).
7. Memorandum of Elihu Root, 6 January 1909, 17321/4–5, reel 989, Numerical and Minor Files of the Department of State, RG 59, NARA; copy also in book 28, BP. This was followed up with a circular (written by Secretary of State Robert Bacon), Alvey Adee to Certain Diplomatic Officers Abroad, 19 February 1909, 1–2, http://digicoll.library.wisc.edu/cgi-bin/FRUS/FRUS-idx?type=turn&entity=FRUS.FRUS1909.p0073&id=FRUS.FRUS1909&isize=M&q1=conservation.
8. "Proceedings of the Joint Conservation Conference," in *NCC Report*, 1:123–267; *Theodore Roosevelt: An Autobiography*, new ed. (1913; New York: Charles Scribner's Sons, 1922), 409–10.
9. "Sister Lands Invited: Mexico and Canada Asked to Confer on Resources," *Washington Post*, 28 December 1908, 2; *Forestry and Irrigation* 12 (January 1906): 35. Canada had sent delegations to US forestry conferences before. "News and Notes," *Forestry and Irrigation* 10 (January 1904): 23. "A matter of satisfaction and encouragement . . . is the attitude of the Canadians." Canadians joined the American forestry association "in substantial numbers and [had] given it both financial and personal support"; "News and Notes," *Forestry and Irrigation* 12 (October 1906): 444. At the Vancouver meeting of 25–27 September 1906, Overton W. Price, associate forester of

the United States, was "the representative of President Roosevelt, and addressed the convention on the subject of forest work in the United States"; *Forestry and Irrigation* 12 (November 1906): 442.

10. *Washington Times*, 10 December 1908, 1.
11. W. C. Edwards to Wilfred Laurier, 31 December 1908, frames 15048–55, reel C-872, Wilfred Laurier Papers, Archives Canada, Ottawa; Laurier to Roosevelt, 24 December 1908, frame 149353, reel C-870, ibid.
12. Samuel P. Hays, *Conservation and the Gospel of Efficiency: The Progressive Conservation Movement, 1890–1920* (Cambridge, MA: Harvard University Press, 1959), 148–49. Hays traces the break with Taft to 22 December 1909; Roosevelt to Governor-General Earl Grey, 24 December 1908, inclosure 3 in no. 78, FO 414/214, *Part VI: Further Correspondence Respecting Proceedings for the Settlement of Questions between the United States and Canada, January to June 1909*, 67; James Bryce to Sir Edward Grey, 25 September 1908, no. 84, FO 414/207, *Part V: Further Correspondence Respecting Proceedings for the Settlement of Questions between the United States and Canada, July to December 1908*, 79a, National Archives, UK, Kew (all references that follow are from the digitized correspondence).
13. *The Citizen* (Berea, KY), 17 December 1908, 1 ("drive"). The paper explained Roosevelt's reasoning: "Just now Mr Roosevelt has not succeeded in stirring the people up enough to force Congress to pass his measures. But he has done a great deal toward arousing such feelings throughout the country and by the time Taft comes in most of the desires of the President will be forced upon Congress by the public opinion of the country at large."
14. James Bryce to Sir Edward Grey, 8 January 1909, no. 53, FO 414–214, *Part VI: Further Correspondence*, 33; and inclosure 2, no. 53, Memorandum, 6 January 1909, with Elihu Root to Bryce, 7 January 1909, ibid., 34 (quote). See also James Brown Scott, *Robert Bacon, Life and Letters* (Garden City, NY: Doubleday, Page, 1923), 134.
15. Roosevelt to Grey, 24 December 1908, inclosure 3 in no. 78, FO 414/214, *Part VI: Further Correspondence*, 67; "Our American Letter," *Sydney Morning Herald*, 27 March 1909, 11; "All World Asked to Save Wealth: Roosevelt Calls Meeting of Great Nations to Conserve," *Chicago Daily Tribune*, 20 February 1909, 7.
16. "All World Asked to Save Wealth," 7. The Netherlands was, however, irritated, and objected to the circulation of the memorandum soliciting interest in the conference because although the memorandum proposed the Netherlands as the host country, the Dutch had not been given sufficient briefing. Philander C. Knox to diplomatic officials, 21 April 1909, file 19632, encl. Aide-Memoire, 15 April 1909, US Department of State, *Papers Relating to the Foreign Relations of the United States with the Annual Message of the President Transmitted to Congress December 7, 1909*, http://digicoll.library.wisc.edu/cgi-bin/FRUS/FRUS-idx?type=turn&entity=FRUS.FRUS1909.p0075&id=FRUS.FRUS1909&isize=M&q1=conservation (accessed 7 November 2012).
17. *Washington Times*, 20 February 1909, 2. See also *Ogden Morning Examiner*, 28 February 1909, 6.
18. "The World Movement: An Address Delivered at the University of Berlin, May 12, 1910," in Theodore Roosevelt, *History as Literature and Other Essays* (New York: Charles Scribner's Sons, 1913), 125–26.
19. Theodore Roosevelt, "Eighth Annual Message," 8 December 1908, in Gerhard Peters and John T. Woolley, American Presidency Project, http://www.presidency.ucsb.edu/ws/index.php?pid=29549#ixzz1ci97brDC (accessed 11 August 2012).

20. James R. Holmes, *Theodore Roosevelt and World Order* (Dulles, VA: Potomac Books, 2007), views Roosevelt as restrained rather than warmongering and imperialist, but not rooted in a specifically internationalist stance.
21. http://www.nobelprize.org/nobel_prizes/peace/laureates/1906/roosevelt-bio.html, (accessed 30 January 2013).
22. Memorandum, 6 January 1909, with Elihu Root to Bryce, 7 January 1909, inclosure 2, no. 53, FO 414-214, *Part VI: Further Correspondence*, 34 (quote).
23. "Address to the North American Conference," fl. North American Conservation Conference, 2, box 533, Pinchot Papers, LC; Gifford Pinchot, Robert Bacon, James Rudolph Garfield, Sydney Fisher, et al., "North American Conservation Conference," *Chautauquan* 55 (June 1909): 107–11.
24. Proceedings of the North American Conservation Conference, 36, fl. North American Conservation Conference, box 533, Pinchot Papers.
25. "Cherry Trees and Conservation," *Outlook* 91 (9 January 1909): 50 (quote); "Continental Conservation," ibid. (27 February 1909): 412.
26. *Washington Times*, 20 February 1909, 2. Cf. Benjamin Kidd, *Control of the Tropics* (London: Macmillan, 1898), 46; and William Elliot Griffis, "America in the Far East," *Outlook* 60 (10 December 1898): 902–6.
27. Robert Bacon to Dear Sir, 19 February 1909, with no. 363, the Earl of Crewe to Sir Edward Grey, 28 May 1909, fl. 119, box 1094, RG 25, Archives Canada, Ottawa (quote; italics added).
28. Robert Bacon to Dear Sir, 19 February 1909; "Robert Bacon, Athlete, Financier, Man of the World," *New York Times*, 10 September 1905 ("old friend"). For modern American reformulations of economics to incorporate sustainability, see Herman Daly, *Steady-State Economics* (San Francisco: W. H. Freeman, 1977); Herman Daly and J. B. Cobb Jr., *For the Common Good: Redirecting the Economy toward Community, the Environment, and a Sustainable Future* (Boston: Beacon Press, 1994).
29. Robert Bacon to Dear Sir, 19 February 1909 (first quote); *Washington Times*, 20 February 1909, 2 (last quote).
30. "Conservation, a World Movement," *Forestry and Irrigation* 14 (September 1908): 496–98; William Beinart and Lotte Hughes, *Environment and Empire*, Oxford History of the British Empire: Companion Series (Oxford: Oxford University Press, 2007); Tom Griffiths and Libby Robin, eds., *Ecology and Empire: Environmental History of Settler Societies* (Melbourne: Melbourne University Press, 1997); Thomas Dunlap, *Nature and the English Diaspora: Environment and History in the United States, Canada, Australia, and New Zealand* (New York: Cambridge University Press, 1999), 118–23.
31. Caroline Ford, "Nature's Fortunes: New Directions in the Writing of European Environmental History," *Journal of Modern History* 79 (March 2007): 112–33; Caroline Ford, "Nature, Culture and Conservation in France," *Past and Present*, no. 183 (May 2004): 173–98.
32. Huntington Wilson, Memorandum, 550.8 Al/161, with Wilson to Philander Knox, 30 March 1910, Minor and Numerical Files of the Department of State, box 5486, RG 59, NARA.
33. William Beinart, *The Rise of Conservation in South Africa* (Oxford: Oxford University Press, 2003), 19.
34. *Official Report of the International Irrigation Congress Held at Los Angeles, California, October, 1893* (Los Angeles: Chamber of Commerce, 1893); *Los Angeles Herald*, 10 October 1893, International Irrigation Congress, Special Edition; "The Eleventh National Irrigation Congress," *Forestry and Irrigation* 9 (October 1903): 481; *Official*

Proceedings of the Third National Irrigation Congress, Held at Denver, Colorado, Sept. 3rd to 8th, 1894 (Denver: n.p., n.d.), 47–49; Ian Tyrrell, *True Gardens of the Gods: Californian-Australian Environmental Reform, 1860–1930* (Berkeley: University of California Press, 1999), 108.

35. Amalia Ribi Forclaza, "New Target for International Social Reform: The International Labour Organization and Working and Living Conditions in Agriculture in the Inter-war Years," special issue, *Contemporary European History* 20 (2003): 307–29 at 313; Asher Hobson, *The International Institute of Agriculture: An Historical and Critical Analysis of Its Organization, Activities and Policies of Administration* (Berkeley: University of California Press, 1931).
36. "A Great Movement," *Chicago Daily Tribune*, 7 March 1909, B4.
37. Mary Wilma M. Hargreaves, *Dry Farming in the Northern Great Plains, 1900–1925* (Cambridge, MA: Harvard University Press, 1957); Mary W. M. Hargreaves, "Hardy Webster Campbell (1850–1937)," *Agricultural History* 32 (January 1958): 62–65; Hardy Webster Campbell, *Campbell's 1907 Soil Culture Manual: A Complete Guide to Scientific Agriculture as Adapted to the Semi-arid Regions . . .* (Lincoln, NE: H. W. Campbell, 1907); James C. Malin, *The Grassland of North America: Prolegomena to Its History* (Ann Arbor, MI: Edwards Brothers, 1947), 237–42; John A. Widtsoe, *Dry-Farming: A System of Agriculture for Countries under Low Rainfall* (New York: Macmillan, 1920), 98.
38. Pierre-Yves Saunier, "Learning by Doing: Notes about the Making of the Palgrave Dictionary of Transnational History," *Journal of Modern European History* 6, no. 2 (2008): 159–80 at 162; Sarah T. Phillips, "Lessons from the Dust Bowl: Dryland Agriculture and Soil Erosion in the United States and South Africa, 1900–1950," *Environmental History* 4 (April 1999): 245–66 at 250; *Agriculture and Irrigation: Information Collected by Hon. J. H. McColl, MP. during His Recent Trip through America* (Bendigo: T. Cambridge, 1906), 3–5; Hargreaves, "Hardy Webster Campbell"; Parliament of the Commonwealth, no. 39, *Votes and Proceedings of the House of Representatives*, Tuesday, 31st August, 1909, Papers, Mr. Deakin presented, by command of His Excellency the Governor-General, *Dry Farming: Report by Senator J. H. McColl of Proceedings at Third Trans-Missouri Dry Farming Congress, Cheyenne, Wyoming, February, 1909; and Further Investigations in America* (Melbourne: printed for the Commonwealth of Australia by J. Kemp, 1909); William MacDonald, *Dry-Farming in America: Being a Report Presented to the Transvaal Government* (Pretoria: Government Printing and Stationery Office, 1909); *Brisbane Courier* (Qld.), 14 June 1909, 4; *Sydney Morning Herald*, 16 June 1909, 9.
39. Bryce (along with William T. Hornaday) was present at the North American Conservation Conference when the delegates were presented at the White House on 18 February 1909. "Nations of the World Invited to Confer," *Washington Herald*, 20 February 1909, 9; "Saving of America," 1.
40. James Bryce, *The American Commonwealth* (London: Macmillan, 1888).
41. Quoted in Bradford Perkins, *The Great Rapprochement: England and the United States, 1895–1914* (New York: Atheneum, 1968), 277; Edmund Ions, *James Bryce and American Democracy* (London: Macmillan, 1967), 200; Peter Neary, "Grey, Bryce, and the Settlement of Canadian-American Differences, 1905–1911," *Canadian Historical Review* 49 (December 1968): 357–80.
42. James Bryce, "National Parks—the Need of the Future" (20 November 1912), American Civic Association, National Parks, Pamphlet Series 11, no. 6 (Washington, DC: n.p., 1912), 6–13.

43. William F. Badè to Bryce, 29 November 1912, 160, book 11, BP. See also Badè to Viscount Bryce, 1 February 1915, 164, ibid.; James Bryce to Sierra Club, 15 February 1915, http://www.sierraclub.org/john_muir_exhibit/life/bryce_tribute_scb_1916.aspx (accessed 30 January 2013).
44. James Bryce to Sir Edward Grey, 26 May 1908, no. 11, FO 414/202, *Part III: Further Correspondence Respecting the Affairs of North America, 1908*, 31.
45. "Canada to Remain Loyal, Says Bryce," *New York Times*, 15 May 1908.
46. Bryce to Sir Edward Grey, 22 December 1908, no. 7, FO 414/214, *Part VI: Further Correspondence*, 33.
47. Ibid., 5.
48. Sydney Fisher [paraphrasing Bryce] to Bryce, 17 January 1910, 23, book 30, BP.
49. "Canada to Remain Loyal, Says Bryce."
50. Governor-General Earl Grey to the Earl of Crewe, 12 January 1909, inclosure 2 in no. 78, FO 414/214, *Part VI: Further Correspondence*, 67; Roosevelt to Grey, 24 December 1908, inclosure 3 in ibid. (quote).
51. Sydney Fisher to Bryce, 26 February 1909, 285, book 28, BP (quote); W J McGee, "Current Progress in Conservation Work," *Science*, n.s., 29 (26 March 1909): 490–96 at 491.
52. Sydney Fisher to Bryce, 26 February 1909, 285, book 28, BP.
53. R. Peter Gillis and Thomas R. Roach, "American Influence on Conservation in Canada: 1899–1911," *Journal of Forest History* 30 (October 1986): 160–74; David J. Hall, *Clifford Sifton: A Lonely Eminence, 1901–1929* (Vancouver: University of British Columbia Press, 1981), 50; Gifford Pinchot, *Breaking New Ground* (New York: Harcourt Brace, 1947), 366–72.
54. Char Miller, *Gifford Pinchot and the Making of Modern Environmentalism* (Washington, DC: Shearwater Press, 2001), 91. Fernow was dean of the Faculty of Forestry in the University of Toronto from 1907 to 1919. James P. Hull, "Fernow, Bernhard Eduard," in *Dictionary of Canadian Biography Online*, http://www.biographi.ca/009004-119.01-e.php?&id_nbr=8131&terms=death (quote) (accessed 9 December 2012).
55. R. Peter Gillis and Thomas R. Roach, *Lost Initiatives: Canada's Forest Industries, Forest Policy, and Forest Conservation* (Westport, CT: Greenwood, 1986), 75.
56. Ibid., 77.
57. Gillis and Roach, "American Influence on Conservation in Canada"; Sydney Fisher to Bryce, 17 January 1910, 23, book 30, BP.
58. D. J. Hall, "Sifton, Sir Clifford," http://www.biographi.ca/en/bio/sifton_clifford_15E.html (quote) (accessed 30 September 2011); Hall, *Sifton*, 237. See also Michel F. Girard, *L'écologisme retrouvé: Essor et declin de la Commission de la conservation au Canada* (Ottawa: University of Ottawa Press, 1994); Michel F. Girard, "The Commission of Conservation as a Forerunner to the National Research Council, 1909–1921," *Scientia Canadensis: Canadian Journal of the History of Science, Technology and Medicine* 15, no. 2 (1991): 19–40; "Conservation Means Development," *Conservation* 9 (July–August 1920): 26.
59. Hall, *Sifton*, 244.
60. Ibid., 245.
61. Ibid., 237.
62. Ibid., 246.
63. W. C. Edwards to Laurier, 31 December 1908, 6–7, frames 15053–54, Laurier Papers; "Sister Lands Invited," 2.

64. Hall, *Sifton*, 244; Walter Scott (Liberal premier of Saskatchewan) to Bryce, 21 November 1912, 204, book 33, BP.
65. Edward Porritt, "Canada and the Payne Bill," *North American Review* 189 (May 1909): 688–94.
66. James Bryce to Sir Edward Grey, 8 January 1909, no. 53, FO 414/214, *Further Correspondence*, 33; Sydney Fisher to Bryce, 17 January 1910, 23, book 30, BP.
67. Girard, "Commission of Conservation as a Forerunner," 38–39.
68. James A. Sandos, "International Water Control in the Lower Rio Grande Basin, 1900–1920," *Agricultural History* 54 (October 1980): 490–501; "Their Task Broadens," 13 (first quote); statements by Senator Francis Newlands (30–32) and Rómulo Escobar Zerman (21), in Proceedings of the North American Conservation Conference; Norris Hundley, *Dividing the Waters: A Century of Controversy between the United States and Mexico* (Berkeley: University of California Press, 1966), 38–39; "Synopsis of an Interview between the President of Mexico and Mr. Gifford Pinchot, at Mexico City, on Thursday, January 20, 1909," 6, fl. Mexico PS, box 625, Pinchot Papers (second quote).
69. Gifford Pinchot, quoted in José G. Godoy, *Porfirio Díaz, President of Mexico* (New York: G. P. Putnam's Sons, 1910), 178.
70. Thomas B. Davis, "Porfirio Díaz in the Opinion of His North American Contemporaries," *Revista de historia de América*, no. 63/64 (January–December 1967): 79–116 at 94.
71. "Synopsis of an Interview," 10, 11.
72. Ibid., 12.
73. Ibid., 12–13.
74. Ibid., 12 ("race"), 14 (first two quotes).
75. "Conservation, a World Movement," 496.
76. Lane Simonian, *Defending the Land of the Jaguar: A History of Conservation in Mexico* (Austin: University of Texas Press, 1995), 75, 77.
77. Ibid., 75.
78. "Wants Nations to Aid," *Washington Post*, 20 February 1909, 3 (quote); Sandos, "International Water Control in the Lower Rio Grande Basin"; "Their Task Broadens," 13; statements by Senator Francis Newlands (30–32) and Rómulo Escobar Zerman (21), in Proceedings of the North American Conservation Conference; "Synopsis of an Interview," 6.
79. "Synopsis of an Interview," 11.
80. Simonian, *Defending the Land of the Jaguar*, 78. See fl. Mexico PS, box 625, Pinchot Papers, for newspaper clippings on the Mexican Revolution; e.g., "76 Americans Killed by Mexicans in 3 Years," *New York Sun*, 18 February 1916; "Exploiters Once Worked Hand in Glove with Diaz," *Rochester Times*, 12 December 1919 (letter to editor by "Common Good").
81. Sandos, "International Water Control in the Lower Rio Grande Basin," 493; Hundley, *Dividing the Waters*, 29–30; Emily Wakild, "Border Chasm: International Boundary Parks and Mexican Conservation, 1935–1945," *Environmental History* 14 (July 2009): 453–75.
82. James Bryce, conversation with Theodore Roosevelt, reported in Bryce to Sir Edward Grey, 8 January 1909, no. 53, FO 414/214, *Part VI: Further Correspondence*, 34.
83. Ibid., 33, 34; "Our American Letter," *Sydney Morning Herald*, 27 March 1909, 11.
84. Evan Charlton, "Reed, Sir (Herbert) Stanley (1872–1969)," revised version, in *Oxford Dictionary of National Biography* (Oxford University Press, online ed., 2004), http://www.oxforddnb.com/view/article/35709 (accessed 15 September 2011).

85. "An Epic of Waste," *Times of India*, 19 April 1909, 6 (quotes); Rudolf Cronau, "A Continent Despoiled," *McClure's Magazine* 32 (April 1909): 639–48.
86. "An Epic of Waste," 6.
87. "The Conference at Pusa," in "Indian Agriculture," *Times of India*, 23 November 1911, 8 (quote). In *Times of India*, see also editorials, 2 September 1910, 6; "Forestry in America," 30 August 1907, 9.
88. "National [*sic*] Resources: Mr. Taft on Their Conservation," ibid., 7 September 1910, 7; India Office to Secretary of State [Sir Edward Grey], 17 June 1909, no. 291, FO 414/214, *Part VI: Further Correspondence*, xxvii.
89. Minute by Third Assistant Secretary of State William Phillips, 11 May 1909, file 17231/84, reel 989, RG 59, Numerical and Minor Files of the Department of State, NARA; Bryce to Sir Edward Grey, 8 January 1909, no. 53, FO 414/214, *Part VI: Further Correspondence*, 34; Sir Joseph Pope to Sydney Fisher, 21 December 1909, fl. 119, box 1094, Archives Canada, Ottawa.
90. "The World's Resources," *Poverty Bay Herald* (Gisborne, N.Z.), 21 October 1909. For Australia, see fl. A6661 Governor-General's Office, Correspondence 1172, Proposed International Conference for the Preservation of Natural Resources (at the Hague), Australian Archives, Canberra, esp. PM 09/2727, Alfred Deakin to Governor-General (Earl of Dudley), 18 August 1909; no. 148, Marquis of Crewe (Secretary of State for the Colonies) to Governor-General of Australia, 21 April 1910; no. 376, Crewe to Dudley, 29 October 1909.
91. William Phillips memorandum, 1 October 1909, file 17231/142, reel 989, Minor and Numerical Files of the Department of State, RG 59, NARA. "It would appear, therefore, that this government is pretty well committed to pursue the conference which it inaugurated."
92. Ibid.; Bryce to Grey, 16 February 1909, no. 101, FO 414/214, *Part VI: Further Correspondence*, 88 (for France); Siegfried von Ciriacy-Wantrup, *Resource Conservation Economics and Policies* (Berkeley: University of California Press, 1952), 315 (for Germany). On France see also the statement by J. J. Jusserand, "The Canals: A Glory of France," *Forestry and Irrigation* 14 (January 1908): 47 (address at the National Rivers and Harbors Congress, 3–5 December 1907). For Britain's colonies and dominions, see James Bryce to Sir Edward Grey, 31 March 1910, no. 9, FO 414/218, *Part V: Further Correspondence Respecting the Affairs of North America, 1910*, 11.
93. Sydney Fisher to Bryce, 1 April 1909, 28–30, book 29, BP (first quote); editorial, *Conservation* 14 (September 1908): 496–98 at 498. "The fact of the matter is, the rest of the world is doing exactly the same thing, except that it isn't calling upon all creation to take notice."
94. "Ignore Roosevelt Plan," *New York Times*, 10 August 1910, 16; Pinchot, *Breaking New Ground*, 366; Ciriacy-Wantrup, *Resource Conservation Economics and Policies*, 315. Pinchot's agitation for a World Congress was not a whim later forgotten. He persistently returned to it. For later attempts to revive the idea of international action, see "Experts to Take World-Wide View of Conservation," *Christian Science Monitor*, 4 September 1912, 8; "Hidden Wealth of U.S.," *Washington Post*, 5 January 1913, 11; Pinchot, *Breaking New Ground*, 367–68; and the Epilogue, below.
95. James Bryce to Sir Edward Grey, 31 March 1910, no. 9, FO 414/218, *Part V: Further Correspondence Respecting the Affairs of North America, 1910*, 11; Alfred Mitchell Innes to Sir Edward Grey, 12 September 1910, ibid., 47; Hays, *Conservation and the Gospel of Efficiency*, 148–49.
96. "Notes on a meeting with 'Pres.' " [Taft], 31 March 1910, 58a, book 30, BP.

97. Jonkheer J. Loudon to Philander Knox, 13 January 1911, and Knox to Loudon, 21 January 1911, file 550.8 Al/166, box 5486, RG 59, NARA (first quote); "Ignore Roosevelt Plan" (quotes). Ciriacy-Wantrup, *Resource Conservation Economics and Policies*, 315, pointed out from State Department records that ultimately twenty-nine agreed. Pinchot, *Breaking New Ground*, 366, claimed thirty countries had accepted.
98. Bryce's views, reported by Sydney Fisher, 17 January 1910, 23, book 30, BP.
99. Cf. James Belich, *Replenishing the Earth: The Settler Revolution and the Rise of the Anglo-World, 1783–1939* (Oxford: Oxford University Press, 2009).
100. Susana Bandieri, "Pensar una Patagonia con dos océanos: El proyecto de desarrollo de Ezequiel Ramos Mexía," *Quinto Sol*, no. 13 (2009): 47–71 at 49; Pedro Navarro Floria, "La Comisión del Paralelo 41° y los límites del "progreso" liberal en los Territorios Nacionales del Sur argentino (1911–1914)," *Revista Electrónica de Geografía y Ciencias Sociales* 12, no. 264, http://www.ub.edu/geocrit/sn/sn-264.htm (accessed 18 August 2012); *Northern Patagonia: Character and Resources*, text and maps by the Comisión de estudios hidrológicos, Bailey Willis, director, 1911–14 (New York: Scribner Press, 1914), vi (hereafter Willis, *Northern Patagonia*); Bailey Willis, *A Yanqui in Patagonia* (Stanford, CA: Stanford University Press, [1947]); Silvana A. Palermo, "The Nation Building Mission: The State-Owned Railways in Modern Argentina, 1870–1930" (PhD diss., State University of New York at Stony Brook, 2001), esp. 101–5, 135, 139–40, 149–52; Bailey Willis, "Recent Surveys in Northern Patagonia," *Geographical Journal* 40 (December 1912): 607–15; Roy Hora, *The Landowners of the Argentine Pampas: A Social and Political History, 1860–1945* (New York: Oxford University Press, 2001), esp. 79, 118–19.
101. Argentina had passed its own Homestead Act in 1884, which was ineffectual in breaking up estates. John Weaver, *The Great Land Rush and the Making of the Modern World* (Montreal: McGill-Queen's University Press, 2005), 16, 326. Patagonia provided an opportunity to start national park reserves carte blanche when a 1902 agreement with Chile settled the disputed territorial division of the region.
102. Theodore Catton, "The Campaign to Establish Mount Rainier National Park, 1893–1899," *Pacific Northwest Quarterly* 88 (Spring 1997): 70–81; Jerry DeSanto, "Foundation for a Park: Explorer and Geologist Bailey Willis in the Area of Glacier National Park," *Forest and Conservation History* 39 (July 1995): 130–137 at 136.
103. Willis, *Northern Patagonia*, vi–vii, 290; "Willis Off to Argentina," *Washington Post*, 5 February 1911, 11 (quote); Willis, *Yanqui in Patagonia*; Bandieri, "Pensar una Patagonia," 59–63.
104. Willis, *Yanqui in Patagonia*, viii, ix, x (quote).
105. Willis, *Northern Patagonia*, x (copy in UCLA's Charles E. Young Research Library). The online version at Google Books is missing part of p. x. However, the full text mysteriously appears under a wrong cataloging. See https://archive.org/details/geografadepanam00valdgoog (accessed 16 April 2014).
106. Willis, *Northern Patagonia*, 9.
107. Ibid., 11, 13 ("anxious"), 372 ("judicious").
108. Willis, *Yanqui in Patagonia*, 105.
109. Willis, *Northern Patagonia*, 393 (quote); [director general of agriculture, Dr. Julio López Mañan], "The National Park of the South," in ibid., 410–13.
110. Ibid., 393.
111. Ibid., 286.
112. Ibid., 13 (quote), 401; Willis to Ezequiel Ramos Mexia, 6 January 1913, fl. 62, box 30, Bailey W. Willis Papers, Huntington Library, San Marino, CA.

113. Willis, *Northern Patagonia*, 286.
114. Palermo, "The Nation Building Mission," 149 ("consolidate"); Bailey Willis, "The Physical Basis of the Argentine Nation," *Journal of Race Development* 4 (April 1914): 342 (second quote); Willis to Ramos Mexia, 6 January 1913, fl. 62, box 30, Willis Papers; Floria, "La Comisión del Paralelo 41°," at fig. 5 and nn. 17–18 ("El plan de colonización y los sujetos del progreso").
115. Ramos Mexia to Willis, 14 July 1914, fl. 4, box 20, Willis Papers.
116. Floria, "La Comisión del Paralelo 41°"; Palermo, "The Nation Building Mission," esp. 139–40, 151; Ezequiel Ramos Mexia, *Programme of Public Works and Finance for the Argentine Republic . . .* (n.p., 1913), 1–7, 50–51, 56, 89–100, 144.
117. Hora, *Landowners of the Argentine Pampas*, 157–58.
118. Lourenço Baeta-Neves, *Third Dry Farming Congress Cheyenne, Wyo., U.S.A., Feb. 23, 24, and 25, 1909: Addresses by the Special and Official Brazilian Delegate, L. Baeta-Neves* (n.p., 1909), 18.
119. *Conférence internationale pour la protection de la nature: Recueil de procès-verbaux* (Berne: K. J. Wyss, 1914), 6; "Exposé introductif de M. Paul Sarasin," in ibid., 24–56; Fiona Paisley, "Mock Justice: World Conservation and Australian Aborigines in Interwar Switzerland?," *Transforming Cultures eJournal* 3, no. 1 (2008): 1–31; *Conference de Berne 1913 pour la protection de la nature internationale: Imprimes* (Proceedings of the Berne Conference for the International Protection of Nature), 19 November 1913, Swiss Archives, Berne; Paul Sarasin, "Über Weltnaturschutz," in Rudolf Ritter von Stumer-Träunfels, ed., *Verhandlungen des VIII. Internationalen Zoologen-Kongresses zu Graz* (Jena: Verlag von Gustave Fischer, 1910), 240–55 (I thank Nadine Kavanagh for the translation); Andreas Kley, "Die Weltnaturschutzkonferenz 1913 in Bern," *Umweltrecht in der Praxis*, Sonderheft zu Grundsatzfragen des Umweltrechts, 7 (2007): 685–705, esp. 692–99.

NOTES TO CHAPTER TWELVE

1. See http://www.csmonitor.com/About/The-Monitor-difference (accessed 12 December 2012).
2. "The Net Profit of Conservation," *Christian Science Monitor*, 30 April 1910, D12. See also "Waterpower Sites and Their Uses," *Christian Science Monitor*, 21 January 1910, 12.
3. For the last meeting, where a split occurred, see *Washington Post*, 7 May 1916, 6.
4. Lewis L. Gould, *The William Howard Taft Presidency* (Lawrence: University Press of Kansas, 2009), 65–66, 160. See also *Theodore Roosevelt: An Autobiography*, new ed. (1913; New York: Charles Scribner's Sons, 1922), 361, regarding the original withdrawal of these lands. On the Ballinger affair more generally, see James Penick, *Progressive Politics and Conservation: The Ballinger-Pinchot Affair* (Chicago: University of Chicago Press, 1968).
5. Louis R. Glavis, "The Whitewashing of Ballinger," *Collier's Magazine* 44 (13 November 1909): 15–17.
6. *Washington Post*, 7 January 1910, 1, 4; *New York Sun*, 11 January 1910, 2.
7. *Investigation of the Department of the Interior and of the Bureau of Forestry*, vol. 1, *Report of the Committee*, 61st Cong., 3rd Sess., S. Doc. No. 719 (1910–11).
8. "The New Nationalism," 31 August 1910, http://www.theodore-roosevelt.com/trspeechescomplete.html (accessed 9 January 2012); Yanek Mieczkowski, *Gerald Ford and the Challenges of the 1970s* (Lexington: University Press of Kentucky, 2005), 304 ("1912"). On the 1912 campaign, see Sidney M. Milkis, *Theodore Roosevelt, the Progressive Party, and the Transformation of American Democracy* (Lawrence: University

Press of Kansas, 2009); Lewis L. Gould, *Four Hats in the Ring: The 1912 Election and the Birth of Modern American Politics* (Lawrence: University Press of Kansas, 2012).

9. Mr. [E. H.] Madison [R-Kansas], "Views of Mr. Madison," in *Investigation of the Department of the Interior*, vol. 1, *Report of the Committee*, 188.
10. Robert W. Righter, *The Battle over Hetch Hetchy: America's Most Controversial Dam and the Birth of Modern Environmentalism* (New York: Oxford University Press, 2006), 167–70.
11. Frederic C. Howe, "The White Coal of Switzerland," *Outlook* 94 (22 January 1910): 158. Howe praised the Swiss for "planning to subjugate nature to the service of the people" (ibid., 151). See also "The Nation's White Coal," *Outlook* 103 (15 February 1913): 338; and Pinchot's remarks in "Synopsis of an Interview between the President of Mexico and Mr. Gifford Pinchot, at Mexico City, on Thursday, January 20, 1909," 14, fl. Mexico PS, box 625, Pinchot Papers, LC.
12. *Investigation of the Department of the Interior*, vol. 1, *Report of the Committee*, 180.
13. Penick, *Progressive Politics and Conservation*, 181, xiii. Gould, *Taft*, 65–66, sees differences from Roosevelt over conservation as crucial.
14. Donald F. Anderson, *William Howard Taft: A Conservative's Conception of the Presidency* (Ithaca, NY: Cornell University Press, 1968), 230–31 (quote); Jonathan C. Lurie, *William Howard Taft: The Travails of a "Progressive Conservative"* (New York: Cambridge University Press, 2012), 112–16 at 114.
15. See his own account in Gifford Pinchot, *Breaking New Ground* (New York: Harcourt Brace, 1947), 391–420. Support in the western states for conservation was, as Penick, *Progressive Politics and Conservation*, 34, points out, "diverse and complex and not simplistic." Older studies more readily accepted an anticonservationist stance for Taft. Rose Mildred Stahl, "The Ballinger-Pinchot Controversy," *Smith College Studies in History* 11 (January 1926): 134.
16. Christopher Johnson, *This Grand and Magnificent Place: The Wilderness Heritage of the White Mountains* (Lebanon: University of New Hampshire Press, 2006), 172, 196–97; Harold K. Steen, *The U.S. Forest Service: A History* (Seattle: University of Washington Press, 1977), 124–28; Kurkpatrick Dorsey, *The Dawn of Conservation Diplomacy: U.S.-Canadian Wildlife Protection Treaties in the Progressive Era* (Seattle: University of Washington Press, 1998), 78–79, 138–39, 159–64.
17. Brian Balogh, "Scientific Forestry and the Roots of the Modern American State: Gifford Pinchot's Path to Progressive Reform," *Environmental History* 7 (April 2002): 198–225; Bruce J. Schulman, "Governing Nature, Nurturing Government: Resource Management and the Development of the American State, 1900–1912," *Journal of Policy History* 17, no. 4 (2005): 375–403 at 375–76.
18. Text of Taft's speech, Second National Conservation Congress Convention, Saint Paul, MN, in *New York Times*, 6 September 1910, 7.
19. Gifford Pinchot, *The Fight for Conservation*, Americana Library, vol. 5 (1910; repr., Seattle: University of Washington Press, 1967), 43.
20. Samuel P. Hays, *Conservation and the Gospel of Efficiency: The Progressive Conservation Movement, 1890–1920* (Cambridge, MA: Harvard University Press, 1959), 148–49.
21. Daniel T. Rodgers, "In Search of Progressivism," *Reviews in American History* 10 (December 1982): 113–32; Hays, *Conservation and the Gospel of Efficiency*, 170–74; Charles B. Forcey, *The Crossroads of Liberalism: Croly, Weyl, Lippmann, and the Progressive Era, 1900–1925* (New York: Oxford University Press, 1961), xvi, 133.
22. File Memos, Reports, etc., Conservation Congress, box 1; and file Correspondence re Distribution of Proceedings, Conservation Congress, 1909, also Lists, etc., box 2,

Records Relating to the National Conservation Congress, 1909–14, entry 25, RG 95, NARA.

23. Martin L. Fausold, *Gifford Pinchot: Bull Moose Progressive* (Syracuse, NY: Syracuse University Press, 1961), 38, depicted Pinchot's *Fight for Conservation* as the clarion call that "demonstrated the propriety and necessity for fighting for [Pinchot's] conservation principles in the political arena."
24. Pinchot to C. A. Grasselli, 3 July 1912, fl. Gra–Gre, box 473, Pinchot Papers. C. A. Grasselli was a Cleveland, Ohio, industrialist, philanthropist, and chemist. See *Encyclopedia of Cleveland History*, http://ech.case.edu/ech-cgi/article.pl?id=GCA.
25. Pinchot to Harry A. Slattery, 5 January 1916, and Slattery to Pinchot, 4 January 1916, fl. Gifford Pinchot, box 489, Pinchot Papers.
26. Harry A. Slattery to Pinchot, 13 November 1915 (quotes), Slattery to Pinchot, 14 April 1916, and Pinchot to Slattery, 24 April 1916, ibid.
27. "Shields Bill Upheld," *Washington Post*, 7 May 1916, 6; Hays, *Conservation and the Gospel of Efficiency*, 188.
28. Douglas Brinkley, *The Wilderness Warrior: Theodore Roosevelt and the Crusade for America* (New York: HarperCollins, 2009), 636; Righter, *Hetch Hetchy*, 167–70.
29. Righter, *Hetch Hetchy*, 57–61, 166–67.
30. Slattery to Pinchot, 12 January 1916, fl. Gifford Pinchot, box 489, Pinchot Papers.
31. "Oil Bulletin Just Issued," *Los Angeles Times*, 21 January 1912, VI13 (quote); "Outlook for Oil Industry Bright," *San Francisco Chronicle*, 19 January 1914, 15; Diana Davids Olien and Roger M. Olien, "Running Out of Oil: Discourse and Public Policy, 1909–1929," *Business and Economic History* 22 (Winter 1999): 36–66.
32. Price Fishback, *Soft Coal, Hard Choices: The Economic Welfare of Bituminous Miners* (New York: Oxford University Press, 1992), table 3-1, 20–21; David Stradling, *Smokestacks and Progressives: Environmentalists, Engineers, and Air Quality in America, 1881–1951* (Baltimore, MD: Johns Hopkins University Press, 1999), 187; Sam H. Schurr and Bruce C. Netschert, *Energy in the American Economy, 1850–1975* (Baltimore, MD: Johns Hopkins University Press, 1960), 508–9, 40 (quote); *Historical Statistics of the United States* (1976 ed.), 585, 588.
33. Stradling, *Smokestacks*, 97.
34. "Doctor Holmes, the Pinchot of the Mines," *Chicago Tribune*, 9 October 1910, fl. 3, box 1, Joseph A. Holmes Papers, SHC; see generally fls. 21–22, 26 (memorials, tributes), box 2, ibid.; J. A. Holmes to W J McGee, 18 November 1901, and Holmes to McGee, 22 August 1909, fl. General Corresp. H, box 4, W J McGee Papers, LC.
35. *Proceedings of a Conference of Governors in the White House, Washington, D.C., May 13–15, 1908* (Washington, DC: Government Printing Office, 1908), 439 (quote), 443–44.
36. Ibid.
37. Chap. 240, §1, 36 Stat. 369; "J. A. Holmes Dies: Martyr to Miners," *New York Times*, 14 July 1915.
38. Holmes to McGee, 22 August 1909, fl. General Corresp. H, box 4, McGee Papers; Holmes to Jane Sprunt Holmes, 27 July 1910, fl. 13, box 2, Holmes Papers.
39. Holmes to Jane Sprunt Holmes, 27 July 1910, fl. 13, box 2, Holmes Papers; Daniel T. Rodgers, *Atlantic Crossings: Social Politics in a Progressive Age* (Cambridge, MA: Belknap Press of Harvard University Press, 1998), 245.
40. The Act of February 25, 1913, chap. 72, sec. 3, 37 Stat. 681; *The Bureau of Mines: Its History, Activities and Organization*, Institute for Government Research, Service Monographs, US Government, no. 3 (New York: D. Appleton, 1922), 5. On mining accidents and the Progressive response, see Rodgers, *Atlantic Crossings*, 245–47.

41. Van H. Manning, *Yearbook of the Bureau of Mines, 1916* (Washington, DC: Government Printing Office, 1917), 66–69, 83, http://digital.library.unt.edu/ark:/67531/metadc12346/ (accessed 17 November 2011).
42. J. A. Holmes, "Production and Waste of Mineral Resources and Their Bearing on Conservation," *Annals of the American Academy of Political and Social Science* 33 (May 1909): 202–14, esp. 205–6.
43. Stradling, *Smokestacks*, 184.
44. Harold F. Williamson, "Prophecies of Scarcity or Exhaustion of Natural Resources in the United States," *American Economic Review* 35 (May 1945): 97–109 at 108.
45. Imperial Conference, 1911, *Minutes of Proceedings of the Imperial Conference, 1911*, Dominions no. 7 (London: HMSO, 1911), 340; Dominions Royal Commission, *Final Report* (London: HMSO, 1918), 413 (quote).
46. C. B. Schedvin, *Shaping Science and Industry: A History of Australia's Council for Scientific and Industrial Research, 1926–49* (Sydney: Allen and Unwin, 1987), 18.
47. "Fifth British Empire Forestry Conference," *Nature* 164 (6 August 1949): 222–23; J. M. Powell, "'Dominion over Palm and Pine': The British Empire Forestry Conferences, 1920–1947," *Journal of Historical Geography* 33, no. 4 (2007): 852–77; *Official Yearbook of the Commonwealth of Australia*, no. 22, *1929* (Melbourne: Commonwealth Government Printer, n.d.), 742.
48. *Addresses and Proceedings of the Second National Conservation Congress Held at Saint Paul, Minnesota, September 5–8, 1910* (Washington, DC: National Conservation Congress, 1911), 89.
49. "Address before the Second National Conservation Congress at St. Paul, September 5–9, 1910, by John Barrett, Director General of the Pan American Union, Washington, D.C.," 1, fl. Address, Second National Conservation Congress, St. Paul, Minn., Sept. 5–9, 1910, box 102, John Barrett Papers, LC.
50. *Addresses and Proceedings of the Second National Conservation Congress*, 80 (quote), 240.
51. "Extracts from the Address of John Barrett, Director General of the Pan American Union, Washington, D.C., Formerly United States Minister to Siam, in Asia," box 112, Barrett Papers; "[1908] Annual Report of the Director of the International Bureau of the American Republics," 3, fl. Dec. 1908, box 27, ibid.; untitled proceedings of a dinner meeting [16 April 1909], fl. General Correspondence, April 1909, box 28, ibid.; Barrett to William Howard Taft, 4 December 1908, fl. Dec. 1908, box 27, ibid.
52. "Pan-American Scientific Congress to Bind Western Republics Closer," *Washington Post*, 13 June 1915, 7. See also *Second Pan American Scientific Congress, Held in the City of Washington in the United States of America December 27, 1915–January 8, 1916: The Final Act and Interpretative Commentary Thereon Prepared by James Brown Scott, . . . Reporter General of the Congress* (Washington, DC: Government Printing Office, 1916), 19–20, 198–99.
53. Raphael Zon, "South American Timber Resources and Their Relation to the World's Timber Supply," *Geographical Review* 2 (October 1916): 256–66 at 263, 266 (quote).
54. *Proceedings of the Second Pan-American Scientific Congress, Section IV, Part I* (Washington, DC: Government Printing Office, 1917), 232, 240, 283.
55. Ibid., 280; James Brown Scott, *Robert Bacon: Life and Letters* (Garden City, NY: Doubleday, Page, 1923).
56. The chair of the conservation section was George Rommel, chief of the Bureau of Animal Husbandry in the Department of Agriculture and a member of the Cosmos Club. "Resigns Government Post," *New York Times*, 22 October 1921.

57. "Pan-American Scientific Congress to Bind Western Republics Closer," 7; *Second Pan American Scientific Congress Held in the City of Washington in the United States of America December 27, 1915–January 8, 1916: The Report of the Secretary General Prepared by John Barrett, Secretary General and Glen Levin Swiggett, Assistant Secretary General* (Washington, DC: Government Printing Office, 1917), 3, 268.
58. *Report of the Secretary General*, 4.
59. Ibid. See also "Barrett Fears Mexico," *Washington Post*, 3 May 1916, 2.
60. *Report of the Secretary General*, 67 (quotes), 71, 130.
61. Lawrence Rakestraw, "George Patrick Ahern and the Philippine Bureau of Forestry, 1900–1914," *Pacific Northwest Quarterly* 58 (July 1967): 149 (quote); Harry N. Whitford, *The Forests of the Philippines*, pt. 1, *Forest Types and Products*, Bureau of Forestry Bulletin no. 10 (Manila: Bureau of Printing, 1903).
62. "Notes," *Journal of Forestry* 17 (January 1919): 106 (quote); Tom Gill, "Parana Pine—a Source of Wood for Reconstruction in Europe," *Unasylva* 1 (September–October 1947), http://www.fao.org/docrep/x5340e/x5340e0a.htm (accessed 11 December 2011).
63. "Forest Resources of the World," *Unasylva* 2 (July–August 1948), http://www.fao.org/docrep/x5345e/x5345e03.htm (accessed 28 September 2013); Gill, "Parana Pine"; "A Summary of the Career of Tom Gill, International Forester: An Interview Conducted by Amelia R. Fry" (Berkeley, CA: Regional Oral History Office, 1969), 32 (quote). The FAO Products Division undertook a comprehensive global forestry inventory after 1947, drawing upon the efforts of Zon, Pinchot, and others in the US Forest Service. See Raphael Zon and William A. Sparhawk, with a foreword by Gifford Pinchot, *Forest Resources of the World*, 2 vols. (New York: McGraw-Hill, 1923). See also Stuart McCook, "'The World Was My Garden': Tropical Botany and Cosmopolitanism in American Science, 1898–1935," in Alfred McCoy and Francisco Scarano, eds., *Colonial Crucible: Empire in the Making of the Modern American State* (Madison: University of Wisconsin Press, 2009), 499–507.
64. Olivia Agresti, *David Lubin: A Study in Practical Idealism* (Boston: Little, Brown, 1922), 86, 188–91, 206–15, 229–31, 268–69.
65. Philip A. Grant, "Senator Hoke Smith, Southern Congressmen, and Agricultural Education, 1914–1917," *Agricultural History* 60 (Spring 1986): 111–22.
66. David Lubin, *The International Institute of Agriculture: Its Organization—Its Work—Its Results* (Rome: Printing Office of the Institute, 1914), 30–31 (quote); *An Outline of the European Co-operative Credit Systems*, 2nd ed. (Rome: Printing Office of the Institute, 1913); *Systems of Rural Cooperative Credit: An Outline of the European Cooperative Credit Systems from Bulletins of Economic and Social Intelligence Published by the International Institute of Agriculture* (Washington, DC: Government Printing Office, 1912).
67. Horace Plunkett, "American Agricultural Commission," *Economic Journal* 23 (June 1913): 291–93.
68. Lubin, *International Institute*, 30, 31 (quote).
69. The US Congress's delegates were tasked with focusing on the credit issue. The group appointed by the Southern Commercial Congress was the "American Commission" but the two groups traveled together and did joint research. *Agricultural Cooperation and Rural Credit in Europe: Report of the American Commission, Consisting of Delegates from Different States in the United States and Different Provinces of Canada, Assembled for the Purpose of Investigating in European Countries Cooperative Agricultural Finance, Production, Distribution, and Rural Life*, pt. 1, *Observations* (Washington, DC: Government Printing Office, 1914), 9–12; *Agricultural Credit . . . Report of the United States Commission to Investigate and Study in European Countries Cooperative Land-Mortgage*

Banks, Cooperative Rural Credit Unions, and Similar Organizations and Institutions Devoting Their Attention to the Promotion of Agriculture and the Betterment of Rural Conditions, 63rd Cong., 2nd Sess., S. Doc. No. 380 (1914).

70. Editorial, *Times* (London), reported in "Blames Our Farmers," *New York Times*, 9 July 1913.
71. "Germans Resent American Inquiry," *New York Times*, 20 June 1913.
72. *Agricultural Cooperation and Rural Credit in Europe*, 11, 15 (quote), 16; Lubin, *International Institute*, 31; Alexander Nüztzenadel, "A Green International? Food Markets and Transnational Politics (c. 1850–1914)," in Alexander Nüztzenadel and Frank Trentmann, eds., *Food and Globalisation: Consumption, Markets and Politics in the Modern World* (Oxford: Berg, 2008), 153–72.
73. Rodgers, *Atlantic Crossings*, 339 (quote), 340.
74. Martin J. Sklar, *The Corporate Reconstruction of American Capitalism, 1890–1916: The Market, the Law, and Politics* (New York: Cambridge University Press, 1988), 153.
75. Schulman, "Governing Nature."
76. Benedict J. Colombi, "Salmon Migrations, Nez Perce Nationalism, and the Global Economy," in Erika Marie Bsumek, David Kinkela, and Mark Atwood Lawrence, eds., *Nation-States and the Global Environment: New Approaches to International Environmental History* (New York: Oxford University Press, 2013), 214–19; James C. Scott, *Seeing Like a State: How Certain Schemes to Improve the Human Condition Have Failed* (New Haven, CT: Yale University Press, 1998); Donald J. Pisani, *Water and American Government: The Reclamation Bureau, National Water Policy, and the West, 1902–1935* (Berkeley: University of California Press, 2002); Mark David Spence, *Dispossessing the Wilderness: Indian Removal and the Making of the National Parks* (New York: Oxford University Press, 1999); Robert H. Keller and Michael F. Turek, *American Indians and National Parks* (Tucson: University of Arizona Press, 1998), 189, 239.
77. See chapters 4 and 6, above.
78. Christopher McKnight Nichols, *Promise and Peril: America at the Dawn of a Global Age* (Cambridge, MA: Harvard University Press, 2011), puts Rooseveltian "large policy" in its place, giving a sensible discussion of anti-imperial and isolationist thought. For polemics, in descending order of reliability, see Thomas E. Woods Jr., "The Anti-imperialist League and the Battle against Empire," http://mises.org/daily/2408 (accessed 14 August 2012); Andrew Napolitano, *Theodore and Woodrow: How Two American Presidents Destroyed Constitutional Freedoms* (Nashville, TN: Thomas Nelson, 2012); James Bradley, *The Imperial Cruise: A Secret History of Empire and War* (Boston: Little, Brown, 2009).

NOTES TO EPILOGUE

1. Gifford Pinchot, *The Fight for Conservation*, Americana Library, vol. 5 (1910; repr., Seattle: University of Washington Press, 1967), 43.
2. *Addresses and Proceedings of the Second National Conservation Congress Held at Saint Paul, Minnesota, September 5–8, 1910* (Washington, DC: National Conservation Congress, 1911), 92.
3. Harold Hotelling, "The Economics of Exhaustible Resources," *Journal of Political Economy* 39 (April 1931): 137–75.
4. Lewis Cecil Gray, "The Economic Possibilities of Conservation," *Quarterly Journal of Economics* 26 (May 1913): 497–519, esp. 517–18. See also Lewis Cecil Gray, "Rent under the Assumption of Exhaustibility," ibid. 28 (May 1914): 466–89.

5. Gray, "Economic Possibilities," 514. For rare exceptions to the neglect, see José Luis Ramos Gorostiza, "Ethics and Economics: Lewis Gray and the Conservation Question," *Documentos de trabajo de la Facultad de ciencias económicas y empresariales*, no. 6 (2002): 1; Joan Martinez-Alier, *Ecological Economics: Energy, Environment and Society* (Oxford: Basil Blackwell, 1990), 163.
6. Hotelling, "Economics of Exhaustible Resources."
7. Cf. Brian Balogh, "Scientific Forestry and the Roots of the Modern American State: Gifford Pinchot's Path to Progressive Reform," *Environmental History* 7 (April 2002): 199.
8. Richard T. Ely, "Conservation and Economic Theory," in Richard T. Ely et al., *The Foundations of National Prosperity: Studies in the Conservation of Permanent National Resources* (New York: Macmillan, 1917), 34.
9. Quoted in ibid.
10. Ibid. The other sections of this book are pt. 2, "Conservation and Economic Evolution," by Ralph H. Hess; pt. 3, "Conservation of Certain Mineral Resources," by Charles K. Leith; and pt. 4, "Conservation of Human Resources," by Thomas Nixon Carver.
11. Harry A. Slattery to Pinchot, 3 March 1916, fl. Gifford Pinchot, box 489, Pinchot Papers, LC. See also Pinchot to Maurice Deutch, 28 November 1916, box 484, regarding Deutch, "Preparedness and Water Power Development in North and South America"; *Washington Post*, 3 May 1916, 2.
12. *Washington Post*, 3 May 1916, 2 (quotes); Thomas Hughes, *Networks of Power: Electrification in Western Society, 1880–1950* (Baltimore, MD: Johns Hopkins University Press, 1983), 293–95.
13. "Proposed interview regarding conservation situation, to be given to R. B. Smith of the International News Service" [1916], fl. Gifford Pinchot, box 489, Pinchot Papers.
14. Harry A. Slattery to Pinchot, 26 April 1916, fl. Gifford Pinchot, box 489, Pinchot Papers.
15. John T. Baker to "Gentlemen" [NCA], 16 November 1917, untitled folder, box 490, Pinchot Papers.
16. "Industrial Conservation in the First World War," *Monthly Labor Review* 54 (January 1942): 16–27.
17. Char Miller, *Gifford Pinchot and the Making of Modern Environmentalism* (Washington DC: Shearwater Press, 2001), 246.
18. James P. Johnson, "The Wilsonians as War Managers: Coal and the 1917–18 Winter Crisis," *Prologue* 9 (Winter 1977): 193–208.
19. Julian E. Zelizer, *The American Congress: The Building of Democracy* (Princeton, NJ: Princeton University Press, 2004), 437.
20. Nuno Luis Madureira, "The Anxiety of Abundance: William Stanley Jevons and Coal Scarcity in the Nineteenth Century," *Environment and History* 18 (August 2012): 395–420.
21. Henry J. Limperich to Harry Garfield, 20 December 1919, box 128, and George A. Thompson to Garfield, n.d. [January 1920], box 129, Harry Augustus Garfield Papers, LC.
22. Pinchot to Clarence W. Barron, 7 December 1917, and Barron to Pinchot, 17 November 1917, untitled folder, box 490, Pinchot Papers; Pinchot to Mary A. Burnham, 22 May 1917, fl. 294, ibid.
23. J. Leonard Bates, "Fulfilling American Democracy: The Conservation Movement, 1907 to 1921," *Mississippi Valley Historical Review* 44 (June 1957): 29–57 at 55.

24. "Conservation for Utilization," *San Francisco Chronicle*, 3 September 1920, 20 (italics added). For the Teapot Dome scandal, see J. Leonard Bates, *The Origins of Teapot Dome* (Urbana: University of Illinois Press, 1963); Burl Noggle, *Teapot Dome: Oil and Politics in the 1920s* (Baton Rouge: Louisiana State University Press, 1962), 4, 225; Burl Noggle, "The Origins of the Teapot Dome Investigation," *Mississippi Valley Historical Review* 44 (September 1957): 237–66.
25. Fisher, "National Vitality," *NCC Report*, 3:620–751.
26. *New York Times*, 3 January 1915, 12.
27. Miller, *Gifford Pinchot*, 252.
28. Ely et al., *Foundations of National Prosperity*, vi (Ely quote), 329 (Carver quote). On Carver, see Lawrence T. Nichols, "The Establishment of Sociology at Harvard: A Case of Organizational Ambivalence and Scientific Vulnerability," in Clark A. Elliott and Margaret W. Rossiter, eds., *Science at Harvard University: Historical Perspectives* (Cranbury, NJ: Associated Universities Press, 1992), 197–99.
29. On the 1920s rise of Fordism and consumerism, see Victoria de Grazia, *Irresistible Empire: America's Advance through 20th-Century Europe* (Cambridge, MA: Harvard University Press, 2005).
30. Arthur S. Link, "What Happened to the Progressive Movement in the 1920's?," *American Historical Review* 64 (July 1959): 833–51.
31. Quoted in Kendrick A. Clements, *Hoover, Conservation, and Consumerism: Engineering the Good Life* (Lawrence: University Press of Kansas, 2000), 78, from William Hard, "Giant Negotiations for Giant Power: An Interview with Herbert Hoover," *Survey Graphic* 51 (1 March 1924): 577–80 at 577; Kendrick A. Clements, *The Life of Herbert Hoover: Imperfect Visionary, 1918–1928* (New York: Palgrave Macmillan, 2010), esp. 41, 43–45.
32. "Engineers Endorse Hoover Work Plan," *New York Times*, 11 September 1921, 30; Committee on Elimination of Waste in Industry of the Federal American Engineering Societies, *Waste in Industry* (New York: McGraw Hill, 1921), 16; Herbert Hoover, "Industrial Waste, Address before Executive Board of the American Engineering Council, February 14, 1921," *Bulletin of the Taylor Society* 6 (April 1921): 71, 78–79.
33. Clements, *Hoover, Conservation, and Consumerism*, 44; Harold F. Williamson, "Prophecies of Scarcity or Exhaustion of Natural Resources in the United States," *American Economic Review* 35 (May 1945): 97–109 at 107–8; Evan B. Metcalf, "Secretary Hoover and the Emergence of Macroeconomic Management," *Business History Review* 49 (Spring 1975): 60–80 at 65; Samuel Haber, *Efficiency and Uplift: Scientific Management and the Progressive Era, 1890–1920* (Chicago: University of Chicago Press, 1964), 162. Stuart Chase, *The Tragedy of Waste* (New York: Macmillan, 1925), esp. 36–37, was an exception in dissenting from this business hegemony.
34. Richard S. Kirkendall, "L. C. Gray and the Supply of Agricultural Land Reviewed," *Agricultural History* 37 (October 1963): 206–14; Sarah T. Phillips, *This Land, This Nation: Conservation, Rural America, and the New Deal* (New York: Cambridge University Press, 2007), 138. Gray's influence was manifest in the National Land Utilization Conference of 1931. Albert Z. Guttenberg, *The Language of Planning: Essays on the Origins and Ends of American Planning Thought* (Urbana: University of Illinois Press, 1993), 120–21; Lewis C. Gray et al., "The Utilization of Our Land for Crops, Pasture and Forest," in US Department of Agriculture, *Yearbook of Agriculture* (Washington, DC: Government Printing Office, 1923), 415–506.
35. Stuart Chase, *Rich Land, Poor Land: A Study in Waste of the Natural Resources of America* (New York: Whittlesey House, 1936), 231.

36. Chase, *Tragedy of Waste*, 233; cf. Clements, *Hoover, Conservation, and Consumerism*, 78.
37. "Barrett Fears Mexico," *Washington Post*, 3 May 1916, 2; see also Clarence W. Barron, *The Mexican Problem* (Boston: Houghton Mifflin, 1917), 23.
38. Alfred Eckes, *The United States and the Global Struggle for Minerals* (Austin: University of Texas Press, 1979), 24–25; John Mason Hart, *Empire and Revolution: The Americans in Mexico since the Civil War* (Berkeley: University of California Press, 2002); Jeffry A. Frieden, "The Economics of Intervention: American Overseas Investments and Relations with Underdeveloped Areas, 1890–1950," *Comparative Studies in Society and History* 31 (January 1989): 55–80 at 65; Diana Davids Olien and Roger M. Olien, "Running Out of Oil: Discourse and Public Policy, 1909–1929," *Business and Economic History* 22 (Winter 1999): 36–66.
39. *Conditions in the Philippine Islands: Message from the President of the United States Transmitting a Report by Colonel Carmi A. Thompson on the Conditions in the Philippine Islands Together with Suggestions with Reference to the Administration and Economic Development of the Islands* (1926), printed in *New York Times*, 23 December 1926, 7; "Thompson Predicts Success in Rubber," *New York Times*, 29 August 1926, 13; Harry N. Whitford, "The Crude Rubber Supply: An International Problem," *Foreign Affairs* 2 (15 June 1924): 613–21; Greg Grandin, *Fordlandia: The Rise and Fall of Henry Ford's Forgotten Jungle City* (New York: Metropolitan Books, 2009); Richard Tucker, *Insatiable Appetite: The United States and the Ecological Degradation of the Tropical World* (Berkeley: University of California Press, 2000), 146, 193, and chap. 5.
40. Eckes, *Global Struggle for Minerals*, 49–50 (iron ore); Neal Potter and Francis T. Christy Jr., *Trends in Natural Resource Commodities: Statistics of Prices, Output, Consumption, Foreign Trade, and Employment in the United States, 1870–1957* (Baltimore, MD: published for Resources for the Future by Johns Hopkins University Press, 1962), 38, 30 (quote); Robert B. Pettengill, "The United States Foreign Trade in Copper: 1790–1932," *American Economic Review* 25 (September 1935): 426–41 at 434.
41. Eckes, *Global Struggle for Minerals*, 23 (quote); Gifford Pinchot, "Follow Roosevelt Ideals, Plea to League Congress," *North American* (Philadelphia), 6 February 1919, fl. League of Nations Clippings, box 613, Pinchot Papers; Ely et al., *Foundations of National Prosperity*, pt. 3.
42. Eckes, *Global Struggle for Minerals*, 46–48; Anna-Katharina Wöbse, "Oil on Troubled Waters? Environmental Diplomacy in the League of Nations," *Diplomatic History* 32 (September 2008): 519–37 at 536 ("minimal"); see generally fl. League of Nations Clippings, box 613, Pinchot Papers; Gifford Pinchot, *Breaking New Ground* (New York: Harcourt Brace, 1947), 367–68; "Hoover Asked to Call Conservation Parley," *New York Times*, 26 March 1929, 12.
43. Edward B. Barbier, *Scarcity and Frontiers: How Economies Have Developed through Natural Resource Exploitation* (New York: Cambridge University Press, 2011), 394–402; Gavin Wright, "The Origins of American Industrial Success, 1879–1940," *American Economic Review* 80 (September 1990): 651–81.
44. Robert Bacon to Dear Sir, 19 February 1909, with no. 363, the Earl of Crewe to Sir Edward Grey, 28 May 1909, fl. 119, box 1094, RG 25, Archives Canada, Ottawa. For energy economics, see Herman Daly, *Steady-State Economics* (San Francisco: W. H. Freeman, 1977); Herman Daly and J. B. Cobb Jr., *For the Common Good: Redirecting the Economy toward Community, the Environment, and a Sustainable Future* (Boston: Beacon Press, 1994); Martinez-Alier, *Ecological Economics*.
45. Rhys Isaac, "History Made from Stories Found: Seeking a Microhistory That Matters," http://www.common-place.org/vol-06/no-01/author/ (accessed 12 October 2010).

INDEX

Italicized page numbers refer to illustrations.